Advances in Intelligent Systems and Computing

Volume 1427

AF899892

The series "Advances in Intelligent Systems and Computing" contains publications on theory, applications, and design methods of Intelligent Systems and Intelligent Computing. Virtually all disciplines such as engineering, natural sciences, computer and information science, ICT, economics, business, e-commerce, environment, healthcare, life science are covered. The list of topics spans all the areas of modern intelligent systems and computing such as: computational intelligence, soft computing including neural networks, fuzzy systems, evolutionary computing and the fusion of these paradigms, social intelligence, ambient intelligence, computational neuroscience, artificial life, virtual worlds and society, cognitive science and systems, Perception and Vision, DNA and immune based systems, self-organizing and adaptive systems, e-Learning and teaching, human-centered and human-centric computing, recommender systems, intelligent control, robotics and mechatronics including human-machine teaming, knowledge-based paradigms, learning paradigms, machine ethics, intelligent data analysis, knowledge management, intelligent agents, intelligent decision making and support, intelligent network security, trust management, interactive entertainment, Web intelligence and multimedia.

The publications within "Advances in Intelligent Systems and Computing" are primarily proceedings of important conferences, symposia and congresses. They cover significant recent developments in the field, both of a foundational and applicable character. An important characteristic feature of the series is the short publication time and world-wide distribution. This permits a rapid and broad dissemination of research results.

Indexed by DBLP, INSPEC, WTI Frankfurt eG, zbMATH, Japanese Science and Technology Agency (JST).

All books published in the series are submitted for consideration in Web of Science.

For proposals from Asia please contact Aninda Bose (aninda.bose@springer.com).

More information about this series at https://link.springer.com/bookseries/11156

Roman Szewczyk · Cezary Zieliński ·
Małgorzata Kaliczyńska
Editors

Automation 2022: New Solutions and Technologies for Automation, Robotics and Measurement Techniques

Springer

Editors
Roman Szewczyk
Łukasiewicz Research Network – Industrial Research Institute for Automation and Measurements PIAP
Warsaw, Poland

Cezary Zieliński
Łukasiewicz Research Network – Industrial Research Institute for Automation and Measurements PIAP
Warsaw, Poland

Małgorzata Kaliczyńska
Łukasiewicz Research Network – Industrial Research Institute for Automation and Measurements PIAP
Warsaw, Poland

ISSN 2194-5357 ISSN 2194-5365 (electronic)
Advances in Intelligent Systems and Computing
ISBN 978-3-031-03501-2 ISBN 978-3-031-03502-9 (eBook)
https://doi.org/10.1007/978-3-031-03502-9

This Springer imprint is published by the registered company Springer Nature Switzerland AG
The registered company address is: Gewerbestrasse 11, 6330 Cham, Switzerland

Foreword

In 1992, American political scientist Francis Fukuyama published the book "The End of History and the Last Man." He presented the idea of the final universalization of Western liberal democracy worldwide as the final and stable form of economic and political evolution. Despite the fact that Fukuyama's ideas were commonly accepted for decades, we now observe a breakdown of stable global political environment. During the last years, due to the COVID-19 pandemic and the increase of political tensions, we have faced unexpected and devastating military actions in Europe.

As a result, the global economy is devastated and significantly disrupted in its key areas, causing the risk of poverty and social unrest for a large part of the world population. Now, our economy requires urgent reorganization and further automatization of production, delivery chains, and services to respond to those problems. This reorganization has to lead to a radical increase in efficiency of resource use and the robustness of global production in an unstable environment. In addition, security and defense technologies seem to become the focal point of recent technological development.

This volume presents the result of discussions among interdisciplinary specialists tackling recent industrial and economic challenges. It contains 38 papers authored by both scientists and practitioners focused on an interdisciplinary approach to developing measuring techniques, robotic and mechatronic systems, industrial automation, numerical modeling and simulation, as well as application of artificial intelligence techniques required by the transformation leading to Industry 4.0. We strongly believe that the solutions and guidelines presented in this volume will be useful to both researchers and engineers facing problems associated with developing cyber-physical systems for global development. As a result, we expect to give input to reconstructing the world economy, which is still suffering from the pandemic and military upheaval, and lead it to further stable development and prosperity.

February 2022

Roman Szewczyk
Cezary Zieliński
Małgorzata Kaliczyńska

AUTOMATION 2022 Conference

Organizing Committee

Chairman

Stanisław Kaczanowski

Secretaries

Jadwiga Konopa
Przemysław Dąbek

Editor

Małgorzata Kaliczyńska

Program Committee

Chairman

Janusz Kacprzyk

Members

Zbigniew Banaszak
Andrzej Bartoszewicz
Grzegorz Bocewicz
Adam Borkowski
Maciej Cader
Andrzej Dzieliński
Stanisław Kaczanowski
Tadeusz Kaczorek
Andrzej Kasiński
Józef Korbicz
Jan Maciej Kościelny
Zdzisław Kowalczuk
Andrzej Masłowski
Zbigniew Nahorski
Mariusz Olszewski
Krzysztof Oprzędkiewicz
Jacek Salach
Piotr Skrzypczyński
Roman Szewczyk
Piotr Szynkarczyk
Mirosław Świercz
Piotr Tatjewski

Contents

Robotics

Measuring Techniques and Systems

About the Editors

Prof. Roman Szewczyk received both his PhD and DSc in the field of mechatronics. He specializes in modeling of properties of magnetic materials as well as in sensors and sensor interfacing, in particular magnetic sensors for security applications. He leads the development of: a sensing unit for a mobile robot developed for the Polish Police Central Forensic Laboratory, and methods of nondestructive testing based on magnetoelastic effect. Professor Szewczyk has been involved in over 10 European Union-funded research projects within the FP6 and FP7 as well as projects financed by the European Defence Organization. Moreover, he has led two regional and national scale technological foresight projects and was active in the organization and implementation of technological transfer between companies and research institutes. Roman Szewczyk is Secretary for Scientific Affairs in the Industrial Research Institute for Automation and Measurements PIAP. He is also Associate Professor at the Faculty of Mechatronics, Warsaw University of Technology, and Vice-chairman of the Academy of Young Researchers of the Polish Academy of Sciences.

Prof. Cezary Zieliński received his M.Sc./Eng., Ph.D. and habilitation degrees in control and robotics from the Faculty of Electronics and Information Technology, Warsaw University of Technology (WUT), Warsaw, Poland. He is Full Professor of WUT, Head of the Robotics Research Group, Director of the Institute of Control and Computation Engineering and Professor of Industrial Research Institute for Automation and Measurement PIAP. His research concentrates on robot control and programming methods. His research interests focus on robotics, in general, and in particular include: robot programming methods, formal approach to the specification of architectures of multi-effector and multi-receptor systems, robot kinematics, robot position–force control, visual servo control and design of digital circuits. He is Author/Co-author of over 200 conference and journal papers as well as books concerned with the above-mentioned research subjects.

Dr. Małgorzata Kaliczyńska received her M.Sc./Eng. degree in cybernetics from the Faculty of Electronics, Wrocław University of Technology, and her Ph.D. degree in the field of fluid mechanics from the Faculty of Mechanical and Power Engineering in the same university. Now, she is Assistant Professor in the Industrial Research Institute for Automation and Measurement PIAP and Editor of the scientific and technological magazine "Measurements Automation Robotics." Her areas of research interest include distributed control systems, Internet of Things, Industry 4.0, information retrieval and webometrics.

Control and Automation

Invariant Subspaces of Fractional Linear Continuous-Time Systems

Tadeusz Kaczorek(✉)

Białystok University of Technology, Wiejska 45D, 15-351 Białystok, Poland
t.kaczorek@pb.edu.pl

Abstract. The A-invariant and (A,B)-invariant linear subspaces are introduced for fractional linear continuous-time linear systems. The basic properties of these invariant linear subspaces are extended to the fractional linear systems and illustrated on simple examples. An extension to fractional descriptor linear systems is also given.

Keywords: A-invariant and (A,B)-invariant linear subspace · Fractional · Linear · Continuous-time · Descriptor · System

1 Introduction

The geometrical approach has been an interesting and powerful method of analysis and synthesis of continuous-time and discrete-time linear systems [5, 15, 16]. An interesting introduction to the methods of geometrical approach is given in the monograph [15]. In recent years the known analysis and synthesis methods have been extended to the fractional orders linear continuous-time and discrete-time systems [7, 9–11, 14, 15]. An extension to different fractional linear systems has been given in [6–9, 12]. Descriptor (singular) linear systems have been analyzed in many books and papers [1–4, 9, 13, 14, 17].

In this paper the geometrical approach will be extended to the fractional continuous-time linear systems. The paper is organized as follows. In Sect. 2 a new version of the necessary and sufficient conditions for the controllability and observability for fractional systems are given. The A-invariant subspaces of the fractional linear continuous-time systems are introduced in Sect. 3 and the (A,B)-invariant linear subspaces in Sect. 4. The properties of (A,E)-invariant linear subspaces of descriptor fractional linear continuous-time systems are analyzed in Sect. 5. Concluding remarks are given in Sect. 6.

The following notations will be used: $\Re$ – the set of real numbers, $\Re^{n\times m}$ – the set of $n\times m$ real matrices and $\Re_+^n=\Re_+^{n\times 1}$, I_n – the $n\times n$ identity matrix.

2 Controllability and Observability of Fractional Linear Systems

Consider the fractional linear continuous-time system

$$\frac{d^\alpha x(t)}{dt^\alpha}=Ax(t)+Bu(t),\quad 0<\alpha<1 \tag{1a}$$

R. Szewczyk et al. (Eds.): AUTOMATION 2022, AISC 1427, pp. 3–12, 2022.
https://doi.org/10.1007/978-3-031-03502-9_1

$$y(t) = Cx(t), \tag{1b}$$

where

$$ {}_0D_t^\alpha f(t) = \frac{d^\alpha f(t)}{dt^\alpha} = \frac{1}{\Gamma(1-\alpha)} \int_0^t \frac{\dot{f}(\tau)}{(t-\tau)^\alpha} d\tau \tag{1c}$$

is Caputo definition of the fractional derivative of α order, $\dot{f}(\tau) = \frac{df(\tau)}{d\tau}$, $\Gamma(x) = \int_0^\infty t^{x-1}e^{-t}dt$, $\mathrm{Re}(x) > 0$ is the Euler gamma function, $x(t) \in \Re^n$, $u(t) \in \Re^m$, $y(t) \in \Re^p$ are the state, input and output vectors and $A \in \Re^{n\times n}$, $B \in \Re^{n\times m}$, $C \in \Re^{p\times n}$.

To simplify the notation it is assumed that $0 < \alpha < 1$, the consideration can be easily extended to the case $0 < \alpha < 2$ [7, 9].

Theorem 1. [7, 9] The solution to the Eq. (1a) has the form

$$x(t) = \Phi_0(t)x_0(0) + \int_0^t \Phi(t-\tau)Bu(\tau)d\tau \tag{2a}$$

where

$$\Phi_0(t) = \sum_{k=0}^\infty \frac{A^k t^{k\alpha}}{\Gamma(k\alpha+1)}, \tag{2b}$$

$$\Phi(t) = \sum_{k=0}^\infty \frac{A^k t^{(k+1)\alpha-1}}{\Gamma[(k+1)\alpha]}. \tag{2c}$$

Definition 1. The fractional system (1a) is called controllable if for any given initial state $x_0 \in \Re^n$ and final state $x_f \in \Re^n$ there exists an input $u(t)$, $t \in [0, t_f]$ which steers the state of the system from x_0 to $x_f = x(t_f)$.

Without loss of generality we may assume that $x_0 = 0$.

Theorem 2. The fractional system (1a) is controllable if and only if

$$\mathrm{rank}\begin{bmatrix} B\ AB \ldots A^{n-1}B \end{bmatrix} = n, \tag{3}$$

Proof. Let

$$\det[I_n\lambda - A] = \lambda^n + a_{n-1}\lambda^{n-1} + \ldots + a_1\lambda^1 + a_0. \tag{4}$$

be the characteristic polynomial of the matrix $A \in \Re^{n\times n}$. Then by the Cayley-Hamilton theorem we have

$$A^n = -a_{n-1}A^{n-1} - \ldots - a_1A - a_0I_n. \tag{5}$$

Using (5) we may reduce (2c) to the form

$$\Phi(t) = \sum_{k=0}^{n-1} A^k c_k(t), \qquad c_k(t) = [t, a_i, \Gamma(i, a)], \qquad k = 0, 1, ..., n-1 \tag{6}$$

and

$$x(t_f) = \int_0^{t_f} \Phi(t_f - \tau)Bu(\tau)d\tau = \sum_{k=0}^{n-1} A^k B\overline{u}_k(t_f)$$
$$= \left[B\ AB \ldots A^{n-1}B\right]\begin{bmatrix} \overline{u}_0(t_f) \\ \overline{u}_1(t_f) \\ \vdots \\ \overline{u}_{n-1}(t_f) \end{bmatrix} \tag{7}$$

By Kronecke-Cappelly theorem there exist $\overline{u}_k(t_f)$, $k = 0, 1, \ldots, n-1$ for any given $x_f = x(t_f)$ if and only if the condition (3) is satisfied.

It is well-known [7] that the observability is the dual notion to the controllability.

Definition 2. The fractional system (1a) is called observable in time $\{0, t_f]$ if knowing its input $u(t)$ and output $y(t)$ for $t \in [0, t_f]$, it is possible to find its unique initial state x_0.

Theorem 3. The fractional system (1a) is observable if and only if

$$\text{rank}\begin{bmatrix} C \\ CA \\ \vdots \\ CA^{n-1} \end{bmatrix} = n, \tag{8}$$

The proof is dual to the proof of Theorem 2.

3 A-invariant Subspaces of Fractional Linear Systems

Definition 3. Let $A : X \to X$ be a map. A linear subspace P of X is called A-invariant if

$$Ax \in P \ \text{ for all } x \in P. \tag{9}$$

Example 1. Let $x_1, x_2, \ldots, x_n$ be the eigenvectors of the matrix $A \in \Re^{n \times n}$ corresponding to different its eigenvalues $s_1, s_2, \ldots, s_n$

$$Ax_i = s_i x_i, \ i = 1, 2, \ldots, n. \tag{10}$$

It is well-known that the set of all linear combinations of $x_1, x_2, \ldots, x_n$ of the form

$$x = \sum_{i=1}^{n} a_i x_i, \ a_i F \ \ (F\text{—the field of complex numbers}) \tag{11}$$

is the n-dimensional A-invariant subspace of F^n.

From (10) and (11) we have

$$Ax = \sum_{i=1}^{n} Aa_i x_i = \sum_{i=1}^{n} a_i s_i x_i = \sum_{i=1}^{n} b_i x_i \in P \tag{12}$$

where $b_i = a_i s_i$.

From (12) it follows that the set (11) satisfies the condition (10) and it is A-invariant linear subspace of F^n.

Lemma 1. If P is A-invariant linear subspace then P is also:

1) A-invariant linear subspace for $k = 2, 3, \ldots$
2) e^A-invariant linear subspace.

Proof. Using Definition 3 by induction we obtain

$$A^k P = A^{k-1}(AP) \subset A^{k-1}P \subset A^{k-2}P \subset \ldots \subset AP = P. \tag{13}$$

Therefore, P is A^k-invariant linear subspace.

From (5) we have

$$e^A = \sum_{k=1}^{n} d_k A^{k-1} \tag{14}$$

where $d_k \in F$ for $k = 1, 2, \ldots, n$.

Using (14) and (13) we obtain

$$e^A P = \left(\sum_{k=1}^{n} d_k A^{k-1} \right) P = \sum_{k=1}^{n} d_k A^{k-1} P \subset P \tag{15}$$

Therefore, P is the e^A-invariant linear subspace.

Theorem 4. If P is A-invariant linear subspace then P is also

$$\Phi_0(t) = \sum_{k=0}^{\infty} \frac{A^k A^{k\alpha}}{\Gamma(k\alpha + 1)} \ \text{– invariant linear subspace} \tag{16}$$

$$\Phi(A) = \sum_{k=0}^{\infty} \frac{A^k A^{(k+1)\alpha - 1}}{\Gamma[(k+1)\alpha]} \ \text{– invariant linear subspace.} \tag{17}$$

Proof. Using (5) to (16) we obtain

$$\Phi_0(A) = \sum_{k=0}^{n-1} h_0(t)A^{k-1}, h_0(t) = h_0(t, \alpha, \Gamma) \tag{18}$$

From (18) and (9) we have

$$\Phi_0(A)P = \left(\sum_{k=0}^{n-1} h_0(t)A^{k-1}\right)P = \sum_{k=0}^{n-1} h_0(t)A^{k-1}P \subset P. \tag{19}$$

The proof for $\Phi(A)$ is similar.

4 (A,B)-invariant Linear Subspaces of Fractional Linear Systems

Consider the fractional linear system (1a–1c).

Definition 4. Let $A : X \to X$ and $B : U \to X$ be linear spaces of the system (1a–1c). Linear subspace V of the linear space is called (A,B)-invariant if

$$AX \subset V + \operatorname{Im} B \tag{20}$$

where Im B is the image defined by

$$\operatorname{Im} B = \{x : x \in X \exists u \quad U, x = Bu\} \qquad (\exists - \textit{there exists}). \tag{21}$$

From Definition 4 it follows that any linear subspace A-invariant is always (A,B)-invariant for every B.

Note that for given maps A and B there exist many different (A,B)-invariant subspaces of X [5, 15].

Example 2. Consider the fractional system (1a–1c) with the matrices

$$A = \begin{bmatrix} 1 & -1 & 0 \\ 0 & 1 & 1 \\ 0 & 0 & 1 \end{bmatrix}, \qquad B = \begin{bmatrix} 0 & 0 \\ 1 & 1 \\ 0 & 1 \end{bmatrix}. \tag{22}$$

For (22) we have the following (A,B)-invariant linear subspaces of R:

$$\begin{aligned}
V_1 &= \left\{ x : x = \begin{bmatrix} 1 \\ 0 \\ 0 \end{bmatrix} a \quad \text{for} \quad a \in \Re \right\} \\
V_2 &= \left\{ x : x = \begin{bmatrix} 1 \\ 0 \\ 0 \end{bmatrix} a + \begin{bmatrix} 0 \\ 1 \\ 0 \end{bmatrix} b \quad \text{for} \quad a, b \in \Re \right\} \\
V_3 &= \left\{ x : x = \begin{bmatrix} 1 \\ 0 \\ 0 \end{bmatrix} a + \begin{bmatrix} 0 \\ 1 \\ 0 \end{bmatrix} b + \begin{bmatrix} 0 \\ 0 \\ 1 \end{bmatrix} c \quad \text{for} \quad a, b, c \in \Re \right\}
\end{aligned} \tag{23}$$

Theorem 5. V is the (A,B)-invariant linear subspace of X of the system (1a–1c) if and only if there exists a state feedback $K : X \to U$ such that V is the $(A + BK)$-invariant linear subspace, i.e.

$$(A + BK)V \subset V. \tag{24}$$

Proof. Sufficiency: Let V be the (A,B)-invariant linear subspace of X and the vectors $v_1, v_2, \ldots, v_p$ be the basis of V, $p \le n = \dim X$. From (4.4) it follows that for every v_i $(i = 1, \ldots, p)$ there exists $w_i \in V$ and $u_i \in U$ such that

$$Av_i = w_i - Bu_i \text{ for } i = 1, \ldots, p. \tag{25}$$

We choose K such that

$$Kv_i = u_i \text{ for } i = 1, \ldots, p. \tag{26}$$

Substituting (26) into (25) we obtain

$$(A + BK)v_i = w_i \text{ for } i = 1, \ldots, p. \tag{27}$$

which is equivalent to (4.4).

Therefore, V is $(A + BK)$-invariant linear subspace.

Necessity: Assume that there exists a feedback $K : X \to U$ such that (24) is satisfied. In this case for every $v \in V$ there exists a vector w such that $(A + BK)v = w$. Therefore, $Av = w - BKv + \operatorname{Im} B$ and the condition (20) is satisfied. This completes the proof.

Remark 1. The set of K matrices may be found from the formula [5].

$$K = U\left[V^{\mathrm{T}}V\right]^{-1}V^{\mathrm{T}} + Z\left\{I_n - V\left[V^{\mathrm{T}}v\right]^{-1}V^{\mathrm{T}}\right. \tag{28}$$

where $V = \left[v_1 \ldots v_p\right] \in \Re^{n \times p}$ is a full-column rank matrix, $U = \left[u_1 \ldots u_p\right]$ and Z is an arbitrary $m \times n$ matrix.

Proof. From (26) we have

$$KV = U \tag{29}$$

The matrix V has full-column rank and the matrix $[V^{\mathrm{T}}V]$ is nonsingular. It is easy to verify that the matrix (28) satisfies (29).

The class (A,B)-invariant linear subspaces of X will be denoted by $J(A, B, X)$.

Lemma 2. The class (A,B)-invariant linear subspaces of X is closed under the addition of the linear subspaces, i.e. if $V_1, V_2 \in J(A, B, X)$ then $(V_1 + V_2) \in J(A, B, X)$.

Proof. Let $V_1, V_2 \in J(A, B, X)$. Using (4.1a) we obtain $A(V_1 + V_2) = AV_1 + AV_2 \subset AV_1 + AV_2 + \operatorname{Im}B$.

Therefore, we have $(V_1 + V_2) \in J(A, B, X)$.

Definition 5. Let V be the family of linear subspaces of X. The element $V^* \subset \underline{V}$ of the family $\underline{V}$ is called maximal if $V^* \in \underline{V}$ and $V \in \underline{V}$ then $V \in V^*$, i.e.

$$V^* = \sup\{V : V \in \underline{V}\}. \tag{30}$$

Lemma 3. Let V be a nonempty family of the linear subspaces of X closed under the addition. The family V contains the maximal element V^*.

Proof. Let $V_1, V_2 \in V$. Then $V_1 \subset V_1 + V_2 \subset V_1 + V_2 + V_3 \subset \ldots$ and $\dim(V_1 + V_2 + \ldots + V_i) \geq \dim(V_1 + V_2 + \ldots + V_{i-1})$ for $i = 2, 3, \ldots$ since $\dim X < \infty$. Therefore, there exists k such that $\dim(V_1 + V_2 + \ldots + V_{k+1}) = \dim(V_1 + V_2 + \ldots + V_k)$ and $V^* = V_1 + V_2 + \ldots + V_k \subset \underline{V}$ since $\underline{V}$ is closed under the addition.

Other properties of (A,B)-invariant linear subspaces are given [5, 15, 16].

5 Descriptor Fractional Linear Systems

Consider the descriptor fractional linear continuous-time system

$$E\frac{d^\alpha x(t)}{dt^\alpha} = Ax(t) + Bu(t), \qquad 0 < \alpha < 1 \tag{31}$$

where $x(t) \in \Re^n$, $u(t) \in \Re^m$, $y(t) \in \Re^p$ are the state, input and output vectors and $E, A \in \Re^{n\times n}$, $B \in \Re^{n\times m}$, $C \in \Re^{p\times n}$ and the fractional derivative is defined by (1c).

It is assumed that

$$\text{rank } E = r < n \tag{32}$$

and

$$\text{rank } [E, A, B] = n. \tag{33}$$

Without loss of generality we may assume

$$E = \begin{bmatrix} I_r & 0 \\ 0 & 0 \end{bmatrix}, \quad A = \begin{bmatrix} A_1 & A_2 \\ A_3 & A_4 \end{bmatrix}, \quad b = \begin{bmatrix} B_1 \\ B_2 \end{bmatrix}, \quad x = \begin{bmatrix} x_1 \\ x_2 \end{bmatrix},$$
$$A_1 \in \Re^{r\times r}, \quad A_4 \in \Re^{(n-r)\times(n-r)}, \quad B_1 \in \Re^{r\times m}, \quad B_2 \in \Re^{(n-r)\times m},$$
$$x_1 \in \Re^r, \quad x_2 \in \Re^{n-r}. \tag{34}$$

and

$$\det A = 0. \tag{35}$$

If $\det A \neq 0$ then from (31) and (34) we have

$$\frac{d^\alpha x_1(t)}{dt^\alpha} = A_1 x_1(t) + A_2 x_2(t) + B_1 u(t) \tag{36a}$$

$$0 = A_3x_1(t) + A_4x_2(t) + B_2u(t) \tag{36b}$$

and from (36b)

$$x_2(t) = -A_4^{-1}A_3x_1(t) - A_4^{-1}B_2u(t) \tag{37}$$

Substituting (37) into (36a) we obtain

$$\frac{d^\alpha x_1(t)}{dt^\alpha} = \left(A_1 - A_2A_4^{-1}A_3\right)x_1(t) + \left(B_1 - A_2A_4^{-1}A_3\right)u(t) \tag{38}$$

Definition 6. The linear subspace $V \subset X$ is called (A,E)-invariant linear subspace if

$$AV \subset EV. \tag{39}$$

For given E and A there exists many different (A,E)-invariant linear subspaces. The family (A,E)-invariant linear subspaces will be denoted by $\underline{V}$, i.e.

$$\underline{V} := \{V : AV \subset EV, \quad V \subset X\}. \tag{40}$$

The family (40) has the maximal element denoted by $V^* = \sup \underline{V}$.
The maximal element can be computed by the use of the following procedure.

$$V_0 := X, \quad V_{k+1}A^{-1}V_k \text{ for } k = 0, 1, \ldots \tag{41}$$

where $A^{-1}V_k := \{x \in X : Ax \in V_k\}$.
It is easy to show that (41) is increasing and $V_q = V_{q+1} = V$ for $q \leq \dim X$.

Theorem 6. The Eq. (31) has the solution x for any u if and only if

$$\text{Im } B \subset EV^*, \quad X(0) \in V^* \tag{42}$$

The solution x is unique if and only if

$$\ker E \cap V^* = 0 \tag{43}$$

Proof. Necessity: Assume that the Eq. (31) has a solution for any input u. Then from (31) we obtain

$$Ax(t) + Bu(t) \subset \text{Im}E \tag{44}$$

Note that (44) is satisfied for any u only if $\text{Im}B \subset \text{Im}E$. Taking into account that $Ax(t) \in Ex(t)$ for $x(t) \in V^*$ form (31) we obtain (42).

Sufficiency: Let $V \in \Re^{nl}$ has full column rank and $\text{Im}V = V^*$. Then from (39) and (42) it follows that there exist matrices $\overline{A} \in \Re^{ll}$ and $\overline{B} \in \Re^{lm}$ such that

$$AV = EVA, \quad B = EVB \tag{45}$$

Note that $x(t) \in V^*$ if and only if

$$X = Vz(t) \text{ for some } z(t) \in \Re^l. \tag{46}$$

Substituting (46) and (45) into (31) we obtain

$$EV\frac{d^\alpha z(t)}{dt^\alpha} = AVz(t) + Bu(t) = EV\left(\overline{A}z(t) + \overline{B}u(t)\right). \tag{47}$$

The equation

$$\frac{d^\alpha z(t)}{dt^\alpha} = \overline{A}z(t) + \overline{B}u(t). \tag{48}$$

has a solution for any input u and an initial condition $z(0)$. Knowing the solution z we may find the solution x of the Eq. (31) for any input u and the initial condition $x(0) = Vz(0) \in V^*$. Therefore, (42) is the sufficient condition for the existence of the solution to the Eq. (31) for any u. Note that $x(t) = Vz(t)$ is the unique solution of (31) if and only if

$$\ker EV = 0. \tag{49}$$

The condition (49) is equivalent to the condition (43).

6 Concluding Remarks

The A-invariant and (A,B)-invariant linear subspaces have been introduced for fractional linear continuous-time systems. Necessary and sufficient conditions for the controllability and observability of the fractional linear systems have been given (Theorems 2 and 3). The basic properties of A-invariant and (A,B)-invariant linear subspaces have been extended to fractional standard and descriptor linear systems (Theorems 4, 5, 6). The presented considerations can be applied to investigate for example the observability of fractional linear systems with unknown inputs and the disturbance decoupling [5, 15].

The considerations can be extended to fractional linear discrete-time systems. An open problem is an extension of these considerations to the different orders fractional linear systems.

Acknowledgements. This work was supported by National Science Centre in Poland under work No. 2017/27/B/ST7/02443.

References

1. Bru, R., Coll, C., Sanchez, E.: About positively discrete-time singular systems. In: System and Control: Theory and Applications. Electrical and Computer Engineering Series, pp. 44–48. World Scientific and Engineering Society, Athens (2000)

2. Bru, R., Coll, C., Sanchez, E.: Structural properties of positive linear time-invariant difference-algebraic equations. Linear Algebra App. **349**, 1–10 (2002)
3. Dai, L.: Singular Control Systems. Lectures Notes in Control and Information Sciences, Springer, Berlin (1989)
4. Guang-Ren, D.: Analysis and Design of Descriptor Linear Systems. Springer, New York (2010)
5. Kaczorek, T.: Linear Control Systems, vol. 2, Research Studies Press J. Wiley, New York (1992)
6. Kaczorek, T.: Positive linear systems with different fractional orders. Bull. Pol. Acad. Sci. Tech. **58**(3), 453–458 (2010)
7. Kaczorek, T.: Selected Problems of Fractional System Theory. Springer, Berlin (2011)
8. Kaczorek, T.: Positive linear systems consisting of N subsystems with different fractional orders. IEEE Trans. Circuits Syst. **58**(7), 1203–1210 (2011)
9. Kaczorek, T., Rogowski, K.: Fractional Linear Systems and Electrical Circuits. Springer, Cham (2015)
10. Ostalczyk, P.: Discrete Fractional Calculus. World Scientific, River Edgle (2016)
11. Podlubny, I.: Fractional Differential Equations. Academic Press, San Diego (1999)
12. Rogowski, A.: Practical and asymptotic stability for a class of delayed fractional discrete-time linear systems. Bull. Pol. Acad. Sci. Techn. **67**(3), 509–515 (2019)
13. Sajewski, Ł.: Decentralized stabilization of descriptor fractional positive continuous-time linear systems with delays. In: 22nd International Conference on Methods and Models in Automation and Robotics, Międzyzdroje, Poland, pp. 482–487 (2017)
14. Sajewski, Ł: Stabilization of positive descriptor fractional discrete-time linear systems with two different fractional orders by decentralized controller. Bull. Pol. Acad. Sci. Techn. **65**(5), 709–714 (2017)
15. Wonham, W.M.: Linear Multivariable Control: A Geometric Approach. Springer, New York (1997)
16. Wonhan, W.M.: Geometric state-space theory in linear multivariable control: a status report. Automatica **15**, 5–12 (1979)
17. Virnik, E.: Stability analysis of positive descriptor systems. Linear Algebra Appl. **429**, 2640–2659 (2008)

The Initial Problem for a Discrete, Scalar Fractional Order System

Krzysztof Oprzędkiewicz(✉)

Department of Automatic Control and Robotics, Faculty of Electrical Engineering, Automatic Control, Informatics and Biomedical Engineering, AGH University of Science and Technology, al. A Mickiewicza 30, 30-059 Krakow, Poland
kop@agh.edu.pl

Abstract. In the paper the initial problem for a scalar, discrete, fractional order system is addressed. The fractional operator is expressed using Continuous Fraction Expansion approximation. Results are illustrated by simulations.

Keywords: Fractional order systems · Fractional order state equation · Fractional order initial problem · CFE approximation

1 Abbreviations

In the paper the following abbreviations are used:

MIMO - Multiple Input Mutliple Output,
IO - Integer Order,
FO - Fractional Order,
CFE - Continuous Fraction Expansion,
IIR - Infinite Impulse Response,
FIR - Finite Impulse Response,
PSE - Power Series Expansion.

2 Introduction

The modeling of processes and phenomena hard to describe using another mathematical tools is one of the main areas of application of non-integer order calculus. Fractional models of different physical phenomena have been presented by many Authors, e.g. [2,4–6,10,16,17].

Initial problems for FO discrete systems has been discussed for example by [9]. The fractional approach allows to analyse the response of a dynamic system to an initial function, defined at a time interval before an initial moment. This is significant difference to integer order systems, where an initial condition has the form of a single value. Such an approach allows to better describe a reality.

R. Szewczyk et al. (Eds.): AUTOMATION 2022, AISC 1427, pp. 13–23, 2022.
https://doi.org/10.1007/978-3-031-03502-9_2

This paper presents an analysis of a new, discrete, state-space model of a scalar, discrete fractional order system. The use of the CFE approximation to solving a state equation allows to keep small size of a model with its good accuracy as well as it allows to analyse the response of the system to different initial functions. The proposed results can be easily adopoted to analysis of many real systems, possible to express in the Jordan canonical form with diagonal state operator as well as scalar oscillatory systems, e.g. Spring-Mass Damper systems.

The paper is organized as follows: at the beginning elementary ideas from fractional calculus as well as the CFE approximation and its implementation to solve FO state equation are recalled. Next the considered system and its proposed, CFE-based state space model are discussed. Finally illustrative simulations are presented.

3 Preliminaries

3.1 Elementary Ideas

Elementary ideas from fractional calculus can be found in many books, for example: [4,7,13] or [16]. Here only some definitions necessary to explain of main results will be given.

Firstly the fractional-order, integro-differential operator is given (see for example [4,8,16]):

Definition 1. *(The elementary fractional order operator) The fractional-order integro-differential operator is defined as follows:*

$$
{}_aD_t^\alpha f(t) = \begin{cases} \frac{d^\alpha f(t)}{dt^\alpha} & \alpha > 0 \\ f(t) & \alpha = 0 \\ \int_a^t f(\tau)(d\tau)^\alpha & \alpha < 0 \end{cases} . \tag{1}
$$

where a and t denote time limits for operator calculation, $\alpha \in \mathbb{R}$ denotes the non integer order of the operation.

Next remember an idea of Gamma Euler function [8]:

Definition 2. *The Gamma function*

$$
\Gamma(x) = \int_0^\infty t^{x-1} e^{-t} dt. \tag{2}
$$

Furthermore the Mittag-Leffler function will be recalled. It is a non-integer order generalization of exponential function $e^{\lambda t}$ and it plays crucial rule in solution of FO state equation. It is defined as underneath:

Definition 3. *The one parameter Mittag-Leffler function*

$$E_\alpha(x) = \sum_{k=0}^{\infty} \frac{x^k}{\Gamma(k\alpha+1)}. \tag{3}$$

The fractional-order, integro-differential operator (1) can be described by different definitions, given by Grünwald and Letnikov (GL Definition), Riemann and Liouville (RL Definition) and Caputo (C Definition). Only C definition will be employed in this paper. It is as follows [8]:

Definition 4. *The Caputo Definition of the FO operator*

$${}_0^C D_t^\alpha f(t) = \frac{1}{\Gamma(N-\alpha)} \int_0^\infty \frac{f^{(N)}(\tau)}{(t-\tau)^{\alpha+1-N}} d\tau. \tag{4}$$

where $N-1 < \alpha < N$ denotes the non-integer order of operation and $\Gamma(..)$ is the complete Gamma function expressed by (2).

For the Caputo operator the Laplace transform can be defined [7]:

Definition 5. *The Laplace transform for Caputo operator*

$$\begin{aligned} &\mathcal{L}({}_0^C D_t^\alpha f(t)) = s^\alpha F(s), \quad \alpha < 0 \\ &\mathcal{L}({}_0^C D_t^\alpha f(t)) = s^\alpha F(s) - \sum_{k=0}^{n-1} s^{\alpha-k-1} {}_0 D_t^k f(0), \\ &\alpha > 0, \quad n-1 < \alpha \le n \in N. \end{aligned} \tag{5}$$

A fractional-order linear MIMO state space system, employing C definition is described as follows:

$$\begin{aligned} {}_0^C D_t^\alpha x(t) &= Ax(t) + Bu(t) \\ y(t) &= Cx(t) \end{aligned} . \tag{6}$$

where $\alpha \in (0,1)$ denotes the fractional order of the state equation, $x(t) \in \mathbb{R}^N$, $u(t) \in \mathbb{R}^L$, $y(t) \in \mathbb{R}^P$ are the state, control and output vectors respectively, A, B, C are the state, control and output matrices respectively.

3.2 The CFE Approximation

An implementation of basic operator s^α at each digital platform (PLC, microcontroller) requires to employ its integer-order, finite-length, discrete-time approximator. The typical approximators are based on PSE and CFE approximations. They allow to express a non integer-order element in the form of digital FIR or IIR filters. The PSE approximator is based directly on discrete-time version of the GL Definition and it takes the form of a FIR filter containg zeros only. However, its digital, high quality implementation requires to apply a long memory buffer (high order of the filter).

The CFE approximator has the form of an IIR filter containing both poles and zeros. It is faster covergent and easier to implement because its order is relatively low, typically not higher than 5. It is obtained via discretization of elementary fractional order element s^α. This can be done using so called generating function $s \approx \omega(z^{-1})$. The new operator raised to the power α has the following form (see for example [3,14]):

$$\begin{array}{l}(\omega(z^{-1}))^\alpha = g_h CFE\{\left(\frac{1-z^{-1}}{1+az^{-1}}\right)^\alpha\}_{M,M} \\ = \frac{P_{\alpha M}(z^{-1})}{Q_{\alpha M}(z^{-1})} = g_h \frac{CFE_N(z^{-1},\alpha)}{CFE_D(z^{-1},\alpha)} \\ = g_h \frac{\sum_{m=0}^{M} w_m z^{-m}}{\sum_{m=0}^{M} v_m z^{-m}}\end{array} . \tag{7}$$

where M is the order of approximation, g_h is the coefficient depending on sample time and type of approximation:

$$g_h = \left(\frac{1+a}{h}\right)^\alpha . \tag{8}$$

In (8) h is the sample time and a is the coefficient depending on approximation type. For $a = 0$ and $a = 1$ we obtain the Euler and Tustin approximations respectively. For $a \in (0,1)$ we arrive at the Al-Alaoui-based approximation, which is a linear combination of the Euler and Tustin approaches. Note that in this case the parameter a in Eq. (7) is equal to $a = \frac{1-\gamma}{1+\gamma}$, with γ being the Al-Alaoui weighting coefficient [1,18]. Numerical values of coefficients w_m and v_m and various values of the parameter a can be calculated with the use of the MATLAB function written by I. Petras and available at [15]. If the Tustin approximation is considered ($a = 1$) then $CFE_D(z^{-1},\alpha) = CFE_N(z^{-1},-\alpha)$ and the polynomial $CFE_D(z^{-1},\alpha)$ can be given in the direct form (see [3]).

3.3 The CFE Based Method of Solution a FO State Equation

The method of memory-effective solution a fractional state equation was proposed and analysed by [12] and applied to modeling of heat transfer process by [11]. Its idea consists in replacing the continuous operator s^α in the Laplace transform of the FO state equation by its discrete CFE approximant expressed by (7). The CFE approximant is a function of discrete complex variable z^{-1}. This allows to directly pass to the discrete time domain. Then the solution of the state equation takes the following form:

$$\sum_{m=0}^{M} E_m x^+(k-m) = \sum_{m=0}^{M} F_m u^+(k-m) + \sum_{m=-M}^{0} x_0(m). \tag{9}$$

where matrices E_m and F_m are defined as follows:

$$\begin{cases} E_m = g_h w_m I_{N\times N} - v_m A. \\ F_m = v_m B. \\ m = 0, 1, ..., M. \end{cases} \tag{10}$$

In (10) g_h is described by (8), w_m and v_m denote coefficients of CFE approximant. From (9) the state vector $x^+(k)$ can be directly calculated as follows:

$$\begin{aligned} x^+(k) &= -E_0^{-1} \sum_{m=1}^{M} E_m x^+(k-m) \\ &+ E_0^{-1} \sum_{m=0}^{M} F_m u^+(k-m) + E_0^{-1} \sum_{m=-M}^{0} x_0(m). \end{aligned} \tag{11}$$

The Eq. (11) allows to directly solve the discrete-time FO state equation using CFE. The solution has the form of the M-th order difference equation, requiring to know M previous steps of state and control signals.

The existence of the solution (11) is determined by the non-singularity of the matrix E_0^{-1}. In general this can be expressed as:

$$\det E_0 \neq 0 \iff \det\left(g_h w_0 I - v_0 A\right) \neq 0. \tag{12}$$

For the scalar case: $A = a_{11}$ and Tustin approximation, when $w_0 = v_0 = 1$ the general condition (12) reduces to:

$$\det E_0 \neq 0 \iff g_h \neq a_{11}. \tag{13}$$

Notice that the initial condition has the form of a discrete function M steps long, starting M steps before an initial step. This is one of crucial advantages of FO modeling over "classic" approach. It allows to take into account a behaviour of a modeled system not only in an initial moment, but also M steps before. Both initial state and control functions can be considered. An impact of different initial functions on trajectories of the considered system will be shown in simulations. The deep analysis of initial problems for discrete FO systems is presented in [9].

The state Eq. (9) with M delays can also be written in the equivalent extended form without delay:

$$\begin{cases} x_q^+(k+1) = A_q^+ x_q^+(k) + B_q^+ u_q^+(k) \\ y_q^+(k) = C_q^+ x_q^+(k) \end{cases}. \tag{14}$$

where:

$$x_q^+(k) = \begin{bmatrix} x_1^+(k) \\ x_2^+(k) \\ \dots \\ x_M^+(k) \end{bmatrix}_{MN\times 1}. \tag{15}$$

$$\begin{cases} x_1^+(k) = x(k) \\ x_2^+(k) = x(k-1) \\ \dots \\ x_M^+(k) = x(k+1-M) \end{cases} . \quad (16)$$

$$u_q^+(k) = \begin{bmatrix} u_1^+(k) \\ u_2^+(k) \\ \dots \\ u_{M+1}^+(k) \end{bmatrix}_{M+1\times 1} . \quad (17)$$

$$\begin{cases} u_1^+(k) = u(k) \\ u_2^+(k) = u(k-1) \\ \dots \\ u_{M+1}^+(k) = u(k-M) \end{cases} . \quad (18)$$

$$A_q^+ = \begin{bmatrix} -E_0^{-1}E_1, ..., -E_0^{-1}E_M \\ I_{N\times N}, 0, 0, ..., 0 \\ 0, I_{N\times N}, 0, ..., 0 \\ .., .., .., .., .. \\ 0, ..., I_{N\times N}, 0 \end{bmatrix}_{MN\times MN} . \quad (19)$$

$$B_q^+ = \begin{bmatrix} E_0^{-1}F_0, ..., E_0^{-1}F_M \\ 0, 0, 0, ..., 0 \\ .., .., .., .., .. \\ 0, 0, 0, ..., 0 \end{bmatrix}_{MN\times M+1} . \quad (20)$$

The output matrix C is as follows:

$$C_q^+ = \left[C^+, 0, ..., 0\right]_{N\times MN} . \quad (21)$$

The initial conditions for state and control (14) also turn to the extended form:

$$x_{q0}^+ = \begin{bmatrix} x^+(0) \\ x^+(-1) \\ \dots \\ x^+(-M+1) \end{bmatrix}_{MN\times 1} . \quad (22)$$

$$u_{q0}^+ = \begin{bmatrix} u^+(0) \\ u^+(-1) \\ \dots \\ u^+(-M+1) \end{bmatrix}_{M\times 1} . \quad (23)$$

In (22), (23) index $-m$ denotes m-th time instant before initial one. This allows to take in consideration "a history" of process dynamics before start of modeling. The summarized size of the proposed discrete, FO model equals to NM. This size is significantly smaller than a size of model using PSE approximation. In our case the maximum size of discrete CFE model equals to 10.

Next, notice the extended system (14)–(21) is the IO discrete system due to its "fractionality" is hidden in its parameters. It allows to analyze its properties using IO approach. Typical numerical tools available for example in MATLAB can be employed too.

4 The Considered Scalar, Free FO System

4.1 The Time-Continuous Model

Consider an elementary, free, scalar FO system described as follows:

$$\begin{cases} D^{\alpha}x(t) = [a_{11}]x(t), \\ y(t) = x(t). \end{cases} \tag{24}$$

where $0.0 < \alpha \leq 2.0$ is the fractional order and a_{11} is the eigenvalue of the system.

The above elementary FO model is used to construct of the proposed discrete CFE model presented in the next subsection.

4.2 The Discrete CFE State Space Model

Due to the free system without control is analyzed, the direct solution (11) reduces to the following form:

$$x^{+}(k) = -E_0^{-1}\sum_{m=1}^{M} E_m x^{+}(k-m) + E_0^{-1}\sum_{m=-M}^{0} x_0(m) \tag{25}$$

where:

$$E_m = [g_h w_m - a_{11}v_m]\,,\ m = 0, 1, ..., M. \tag{26}$$

During further analysis consider different inital functions having the same value x_0 in the step $k = 0$. Denote these functions by $x_{01-4}(m)$. They are following:

$$\begin{aligned} x_{01}(m) &= \begin{cases} x_0,\ m = 0, \\ 0,\ m = -4, ..., -1, \end{cases} \\ x_{02}(m) &= x_0,\ m = -4, ..., 0, \\ x_{03}(m) &= \frac{M+m}{M}x_0,\ m = -4, ..., 0, \\ x_{04}(m) &= \frac{M-m}{M}x_0,\ m = -4, ..., 0. \end{aligned} \tag{27}$$

The function x_{01} is the same as in IO case, the function x_{02} is constant in the whole interval, the function x_{02} is decreasing and the function x_{04} is increasing.

Consequently the solution (25) takes different forms:

$$x_1^{+}(k) = -E_0^{-1}\sum_{m=1}^{M} E_m x^{+}(k-m) + E_0^{-1}x_0. \tag{28}$$

$$x_1^+(k) = -E_0^{-1} \sum_{m=1}^{M} E_m x^+(k-m) + M E_0^{-1} x_0. \tag{29}$$

$$x_1^+(k) = -E_0^{-1} \sum_{m=1}^{M} E_m x^+(k-m) + E_0^{-1} \sum_{m=-M}^{0} \frac{M+m}{M} x_0. \tag{30}$$

$$x_1^+(k) = -E_0^{-1} \sum_{m=1}^{M} E_m x^+(k-m) + E_0^{-1} \sum_{m=-M}^{0} \frac{M-m}{M} x_0. \tag{31}$$

The extended Eq. (14) turns to the form as beneath:

$$\begin{cases} x_q^+(k+1) = A_q^+ x_q^+(k) \\ y_q^+(k) = C_q^+ x_q^+(k) \end{cases}. \tag{32}$$

Assume that the order M of CFE approximation equals to 5. Then the state matrix (19) takes the following form:

$$A_q^+ = \begin{bmatrix} q_1 & q_2 & q_3 & q_4 & q_5 \\ 1 & 0 & 0 & 0 & 0 \\ 0 & 1 & 0 & 0 & 0 \\ 0 & 0 & 1 & 0 & 0 \\ 0 & 0 & 0 & 1 & 0 \end{bmatrix}_{5\times 5}. \tag{33}$$

where:

$$q_m = -\frac{g_h w_m - a_{11} v_m}{g_h w_0 - a_{11} v_0}, \; m = 1, ..., 5. \tag{34}$$

The characteristic polynomial of the system (32) with respect to (33) is as follows:

$$w(\lambda) = \det(\lambda I - A) = \lambda^5 - q_1\lambda^4 + q_2\lambda^3 - q_3\lambda^2 - q_4\lambda - q_5. \tag{35}$$

The discrete FO system (24) will be practically stable if and only if the extended system (32)–(35) is asymptotically stable. It is the IO discrete system and its stability can be tested using known methods.

The initial condition (22) reduces to:

$$x_{q0}^+ = \begin{bmatrix} x^+(0) \\ x^+(-1) \\ x^+(-2) \\ x^+(-3) \\ x^+(-4) \end{bmatrix}_{5\times 1}. \tag{36}$$

With respect to (27) and $M = 5$ the initial condition (36) is as follows:

$$x_{q01}^+ = x_0 \begin{bmatrix} 1.0 \\ 0.0 \\ 0.0 \\ 0.0 \\ 0.0 \end{bmatrix} x_{q02}^+ = x_0 \begin{bmatrix} 1.0 \\ 1.0 \\ 1.0 \\ 1.0 \\ 1.0 \end{bmatrix} x_{q03}^+ = x_0 \begin{bmatrix} 1.0 \\ 0.8 \\ 0.6 \\ 0.4 \\ 0.2 \end{bmatrix} x_{q03}^+ = x_0 \begin{bmatrix} 1.0 \\ 1.2 \\ 1.4 \\ 1.6 \\ 1.8 \end{bmatrix} \tag{37}$$

The extended system (32)–(37) approximates the FO scalar system (24) and it can be used to simulate the FO system with the use of standard MATLAB tools. This is presented in the next section.

5 Simulations

Simulations were done with the use of extended model (32)–(37) and MATLAB function *initial*. Parameters of the CFE model employed to simulations are given in the Table 1. The experiments were done for two values of coefficient a: $a = 0.0$ (Euler method) and $a = 1.0$ (Tustin method). Free responses of the system to initial functions x_{01-4} and both experiments are given in Figs. 1 and 2.

Table 1. The parameters of the simulated system and CFE model.

Parameter	α	a_{11}	$h[s]$	a	x_0
Experiment 1	0.5	-1	0.1	0.0	1.0
Experiment 2	0.5	-1	0.1	1.0	1.0

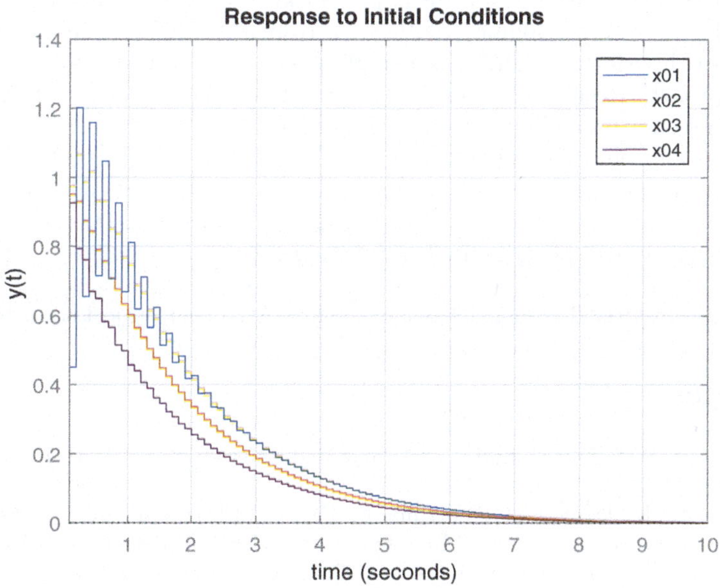

Fig. 1. The response to initial conditions, Tustin approximation ($a = 1.0$).

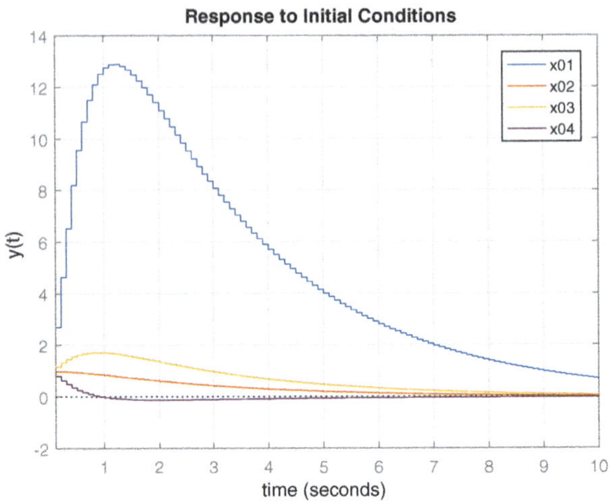

Fig. 2. The response to initial conditions, Euler approximation ($a = 0.0$).

6 Conclusions

The main final conclusion from this paper is that the FO approach allows to analyze a behaviour of a dynamic system with respect to different initial functions. This better describes a reality than single initial value taken into consideration in IO approach. Results of simulations show that the trajectory of the system strongly depends on a shape of initial function.

Next, it can be observed that the value of coefficient a in the CFE approximation has a strong impact to the shape of the trajectory of the system.

The presented results require deeper theoretical analysis. It is recently being done and results will be presented in future works.

Acknowledgments. This paper was sponsored by AGH UST project no 16.16.120.773.

References

1. Al-Alaoui, M.A.: Novel digital integrator and differentiator. Electron. Lett. **29**(4), 376–378 (1993)
2. Caponetto, R., Dongola, G., Fortuna, L., Petras, I.: Fractional order systems: modeling and control applications. In: Chua, L.O. (ed.) World Scientific Series on Nonlinear Science, pp. 1–178. University of California, Berkeley (2010)
3. Chen, Y.Q., Moore, K.L.: Discretization schemes for fractional-order differentiators and integrators. IEEE Trans. Circuits Syst. I: Fundam. Theory Appl. **49**(3), 263–269 (2002)
4. Das, S.: Functional Fractional Calculus for System Identyfication and Control. Springer, Berlin (2010)
5. Dzieliński, A., Sierociuk, D., Sarwas, G.: Some applications of fractional order calculus. Bull. Pol. Acad. Sci. Tech. Sci. **58**(4), 583–592 (2010)

6. Gal, C.G., Warma, M.: Elliptic and parabolic equations with fractional diffusion and dynamic boundary conditions. Evol. Equ. Control Theory **5**(1), 61–103 (2016)
7. Kaczorek, T.: Selected Problems of Fractional Systems Theory. Springer, Berlin (2011)
8. Kaczorek, T., Rogowski, K.: Fractional Linear Systems and Electrical Circuits. Bialystok University of Technology, Bialystok (2014)
9. Mozyrska, D., Pawluszewicz, E.: Fractional discrete-time linear control systems with initialisation. Int. J. Control **1**(1), 1–7 (2011)
10. Obrączka, A.: Control of heat processes with the use of non-integer models. Ph.D. thesis, AGH University, Krakow, Poland (2014)
11. Oprzędkiewicz, K., Mitkowski, W.: A memory efficient non integer order discrete time state space model of a heat transfer process. Int. J. Appl. Math. Comput. Sci. **28**(4), 649–659 (2018)
12. Oprzędkiewicz, K., Stanisławski, R., Gawin, E., Mitkowski, W.: A new algorithm for a CFE approximated solution of a discrete-time non integer-order state equation. Bull. Pol. Acad. Sci. Tech. Sci. **65**(4), 429–437 (2017)
13. Ostalczyk, P.: Discrete Fractional Calculus. Applications in Control and Image Processing. World Scientific, Singapore (2016)
14. Petras, I.: Fractional order feedback control of a DC motor. J. Electr. Eng. **60**(3), 117–128 (2009)
15. Petras, I.: (2009). http://people.tuke.sk/igor.podlubny/usu/matlab/petras/dfod2.m
16. Podlubny, I.: Fractional Differential Equations. Academic Press, San Diego (1999)
17. Sierociuk, D., et al.: Diffusion process modeling by using fractional-order models. Appl. Math. Comput. **257**(1), 2–11 (2015)
18. Stanisławski, R., Latawiec, K., Łukaniszyn, M.: A comparative analysis of Laguerre-based approximators to the Grünwald-Letnikov fractional-order difference. Math. Probl. Eng. **2015**(1), 1–10 (2015)

Variability in the Height of Layers for Robotised WAAM Process

Julia Wilk[1,2](✉), Norbert Prokopiuk[1], and Piotr Falkowski[1,2](✉)

[1] Institute of Aeronautics and Applied Mechanics, Faculty of Power and Aeronautical Engineering, Warsaw University of Technology, Warsaw, Poland
julia.wilk2.stud@pw.edu.pl, julia.wilk@piap.lukasiewicz.gov.pl
[2] ŁUKASIEWICZ Research Network – Industrial Research Institute of Automation and Measurements PIAP, Warsaw, Poland
piotr.falkowski@piap.lukasiewicz.gov.pl

Abstract. The dynamic development of industrial needs leads to the creation of innovative methods of manufacturing. In recent years, additive manufacturing, especially with the use of polymer material has arose in popularity. In heavy industry, there is a seek for manufacturing objects with high mechanical and temperature properties; met by WAAM technology. The following paper addresses the challenge of investigating geometry proprieties of welds. Moreover, the measurements are the base to develop automatic methods for creating objects with predictable shapes. For this purpose, welding trials with various parameters were conducted. These subsequently allowed finding optimal settings for additive manufacturing with the use of MIG-MAG technology. The study concerned an analysis of the geometry of the welds dependent on the height of the printouts with the WAAM method. The dependence between the number of layers and the general height of the object was modelled. There was a tendency to melt the previous layers of welds in the case of no breaks between the layers applied. On the contrary, a linear relationship between the number of layers applied and the height of the weld appeared while pausing between laying down subsequent layers. Hence, this proves that proper cooling is necessary for additive manufacturing with the use of this welding technology.

Keywords: Additive manufacturing · Industry 4.0 · Process modelling · Robotic welding · WAAM

1 Introduction

Advanced manufacturing draws from innovative technologies to enhance manufacturing processes and improve the quality of products [1]. Additive manufacturing (AM), often called 3D printing is one of its top-grossing branches, in which development is eagerly invested. It is based on laying down material layer-by-layer based on CAD models to create real, three-dimensional models [2]. What is more, AM stays in line with the Industry 4.0 principles; among others, manufacturing in a smart, innovative cost-effective way, and focused on clients [3]. There is no doubt, that the most rapidly

R. Szewczyk et al. (Eds.): AUTOMATION 2022, AISC 1427, pp. 24–34, 2022.
https://doi.org/10.1007/978-3-031-03502-9_3

developing and popular method of additive manufacturing is 3D printing with the use of thermoplastic polymer materials [4]. In the search of AM technology that may be useful in operation within high-temperature conditions and under high dynamic loads, the idea of metal printing came up. Wire Arc Additive Manufacturing (WAAM) is a relatively new approach to manufacturing. Despite this, it has already found use in fields such as aerospace, automotive, defence, naval or nuclear energy - mainly for prototyping and creating models [5]. It is based on using welding wire as the building material and electric arc as a source of heat [6]. Usually, this involves special WAAM machines but nowadays also robots connected with welding equipment [1, 7–11]. This type of AM generally uses welding Cold Metal Transfer (CMT) technology [12–15] but also plasma welding (PAW) [16] and TIG [2, 16]. WAAM manufacturing enables the fast building of complicated, complex geometries with high precision [1, 14] in a stable, and well-predicted way. However, this requires an accurate model of a process. Hence, a deep understanding of its principles is needed.

Objects manufactured with this approach are relatively precise after appropriate finishing [16]. Moreover, such technology is cost-effective, especially while compared to other AM solutions. The operational space of the robotised production device may be almost unlimited in terms of size, which makes it particularly useful for large objects [6, 17, 18]. On top of these, there is a wide range of applicable materials, including aluminium [8, 11, 19], stainless steel [20] and nickel [21] or even titanium under specific conditions. Available literature presents recent results of researches on material properties such as hardness, tensile, toughness [13], stresses [22], and the microstructure of the samples [12, 23]; based on X-Ray [9, 17], electromagnetic or thermal analyses [17]. The geometric properties are tested for matching the weld profile to a known curve such as parabola or cosine function [24]. Both the reduction of energy by adjusting the voltage and current [13, 25], as well as the direction of trails [2, 26], are also in the spot of interest. In terms of economic and social aspects, costs of manufacturing and environmental impact are the key factors investigated to be lowered [7].

The purpose of this paper is to optimize the robot's dynamics parameters, mainly the position of the welding torch and the welding speed, in order to achieve the intended weld geometries in terms of their dimensional accuracy. Its outcomes shall contribute to building more exact simulation models for computer-aided manufacturing (CAM) software dedicated to robotic applications. Original findings from the investigation of the correlation between velocity and vertical shift of the tool may enhance the capabilities of slicers. As a result, the built objects will have the constant thickness of the layers. Also, their lower layers ones will neither be over-melted nor the new ones will be laid down from a too big distance. Presented data are used to draw general conclusions on robotised WAAM process applicable for various setups.

3D metal printing tests using MIG technology were conducted in order to adjust these parameters for additive manufacturing with metal use. In the beginning, the quality of welds was examined dependent on the height of the welding torch, next – on the velocity of welding. Consequently, the optimal torch position and welding velocity were selected. This enables running the target experiments, which is the investigation of the correlation between the layers' thickness and the number of layers applied.

2 Methodology

To conduct welding tests the experimental setup shown in Fig. 1(a) was used, which includes a KUKA KR 16-2F with DKP400 V1 table equipped with a welding module and a Fronius TransPuls Synergic 3200 MIG-MAG welder.

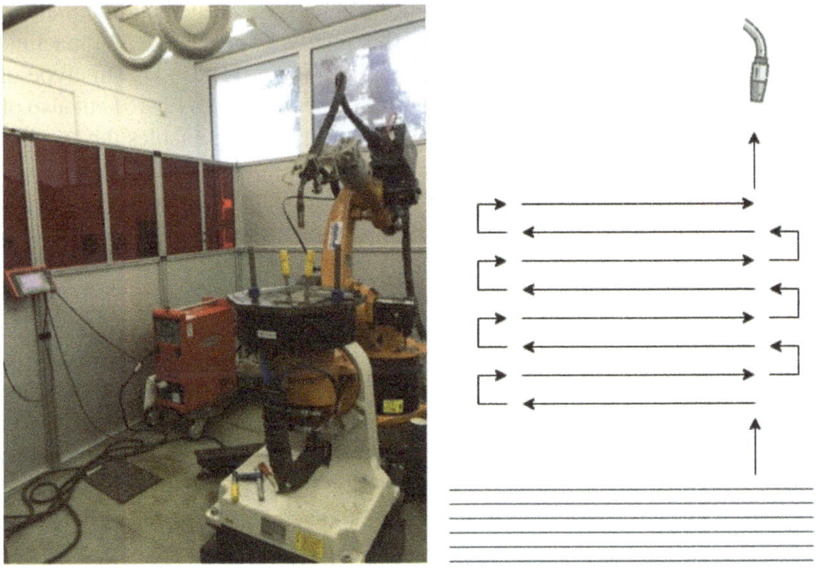

(a) (b)

Fig. 1. (a) Experimental setup, (b) The layering strategy

There were three kinds of test carried out – the dependence of the height of the weld on the position of the welding torch, the dependence of the height of the weld on the welding speed and the study of the variability of the height of a single layer depending on the number of applied layers.

The WAAM process was conducted with a wire of chemical properties presented in Table 1, and a mixture of shielding gases was used in a given proportion: argon 82% and carbon dioxide 18%.

Table 1. Chemical properties of welding wire

Material [%]	C	Si	Mn	Cr	Ni	Mo
Wire	0.09	0.80	1.80	0.36	2.23	0.55

Two tests of welding torch positioning and its transitional velocity were carried out in order to define optimal robot's settings for the final investigation. Every test was realised by laying down single trails on a steel plate of overall dimensions 400 mm × 160 mm × 10 mm.

Both the first tests were conducted for twenty different settings, except one torch position trial was run for nineteen only. Every weld has 70 mm in length.

For the first set of tests, the position of the torch was increased by 0.2 mm for each subsequent weld (the torch was raised by a given height according to the Z-axis). The starting position of the torch was almost at the edge of the sheet. Twenty welds were placed, so the position of the tool characteristic point for the last weld was 4 mm distant from the first.

In the next attempt, the position was increased by 0.5 mm, and thus, the difference in the height of the torch changed by 10 mm within the trials.

Afterwards, the welding speed tests were conducted at the equal vertical torch position. The velocities used for the subsequent welds were correspondingly: 0.2, 0.3, 0.4, 0.5, 0.6, 0.7, 0.8, 0.9, 1.0, 1.1, 1.2, 1.3, 1.4, 1.5, 1.6, 1.7, 1.8, 1.9, and 2.0 [m/min]. These cover the complete range of the robot's dynamics capabilities. The welding parameters used are presented in Table 2.

Table 2. Welding parameters

Welding parameters	
Voltage	16.9 V
Current	130 A
Wire feed speed	3.1 m/min
Welding type	Standard
Gas supply	15 l/min

The outcomes of the trials were 3D-scanned with the use of GOM ATOS II. Further analyses were carried in the ATOS Professional 2018 software. As to construct a mesh enabling numerical computations, constructional planes were defined to obtain measured values in the corresponding nodes. First of all, hundred cross-sections were created through all the welds across their width. The sections were symmetrical regarding the centre plane of the welds and distanced by 0.5 mm one from another. Due to the accumulation of material in the ends of the welds, these areas were not analysed. In addition, at the theoretical centre of every weld (understood as the section including the welding trail and perpendicular to the building plate surface), the main cross-section was created and ten additional parallel ones were added symmetrically. Also, a plane coinciding with the plane of the top surface of the metal sheet was constructed as well.

Using the Python scripting capabilities offered by the software [27, 28] for both height of the torch and speed tests, the heights of the welds on each of the hundred sections at eleven horizontal coordinates (corresponding to the planes described beforehand) were measured. This was aimed at fitting the curve reflecting the shape of the surface geometry. The most accurate was a cubic spline, which was approximated with the MATLAB software and used to precisely estimate the maximum of the weld's height in a section.

After analysing the results, optimal parameters of the torch height and welding speed were selected, and the main experimental phase was carried out. The test was carried out, where twenty single-trail walls were built, each of them with the number of layers increased by one regarding the previous one. The layering method is shown in Fig. 1(b). The welds' height analysis was performed analogically to the mentioned before.

3 Results and Discussion

Figure 2 presents the results of the speed tests and the welding torch position analysis. The upper bar plots represent the average maximum weld height for a given test and the standard deviation of this value for different heights of the torch above the building plate. The black dotted line was added as the mean value for all the analysed samples. As almost every value fits this within their uncertainty of measurement range, that torch vertical position was assumed to have a negligible impact on the height of the welds.

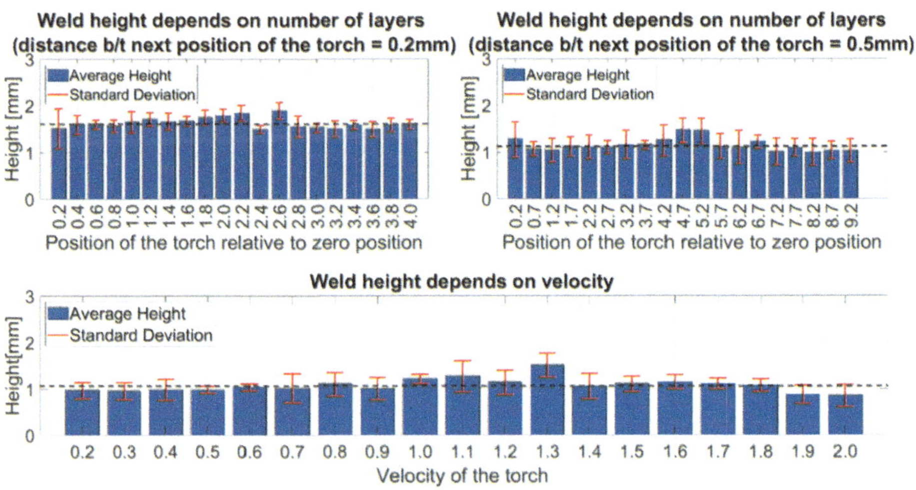

Fig. 2. Results of the first two experiments

The value of the average height of the weld has a slightly increasing linear trend. However, it starts to decrease from a velocity of 1.3 m/min. This is caused to the loss of quality while increasing the velocity of a robot. Discontinuities appear along the entire length of the weld which affects the measured geometrical parameters, and hence, the final outcomes. Therefore, the velocities resulting in such irregular surfaces are strongly advised against being used for WAAM manufacturing.

The obtained outcomes show that the height of the layer does not change significantly with a change of the speed and the position of the torch in relation to their average value. However, it can be concluded that some settings guarantee more stable and predictable geometry results. Hence, it was mainly the analysis of standard deviations that influenced the selection of optimal parameters.

Thence, the parameters for further welding tests were chosen. Taking into account the lowest standard deviation for velocity and the minimal value (marked), it was decided to set to it 0.6 m/min. After defining the average value of medians from each experiment it was decided that the torch should be placed 1.26 mm above the building plate and lifted by this distance after every layer which corresponds precisely to its value.

Initially, it was decided not to wait between laying down the layers, so to minimise additional material accumulated at the ends of the trails. However, the analysis of the results and fitting a flattening trend line, proved that this method only allows for the application of a limited, small number of layers, approximated as no more than 30. The outcome of these is illustrated in Fig. 3 with the estimation for a larger number of layers. The fitted blue exponential curve is suspected to flatten due to the too high object's temperature during the WAAM process.

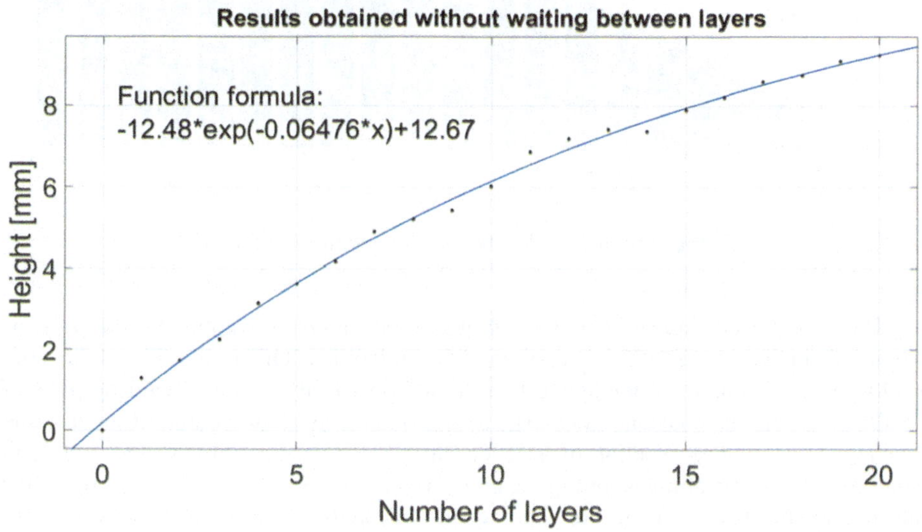

Fig. 3. Results obtained without waiting between layers

Consequently, the robot was stopped for 20 s between applying every next layer to prevent the print from collapsing due to too much heat input [29].

Figure 4 shows the values obtained for the process with waiting between layers. The red trend line, obtained with the use of the 'fit' function in MATLAB, proves the linear correlation between the number of layers and the total height of the object. As the estimated height for no layers is equal to zero, the constructed model corresponds to the physical one. However, the directional coefficient is not exactly equal to 1.26. Therefore, a correction of the torch lifting height should be introduced, so as to avoid a collision of the weld with the robot or distancing the torch from the top layer.

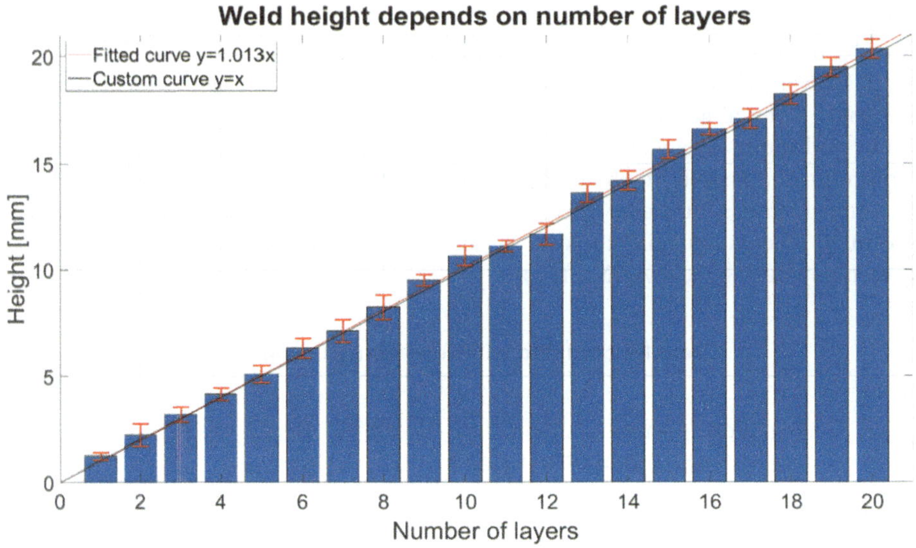

Fig. 4. Final results obtained for additive welding

The dependence of the height of the printout on the number of applied welds remains linear only in the case of sufficient cooling of the built object. Otherwise, this relationship is close to exponential, which suggests a limited possibility of manufacturing parts of small sections in the constant mode. The results and the available literature suggest that mechanical cooling is essential. Its absence causes irregularities at the welds' ends, and excessive material melting as in Fig. 5a compared to Fig. 5b. Nevertheless, a controlled and predictable fusing of the previous layers is desirable as it reduces pre-stress and bounds the layers.

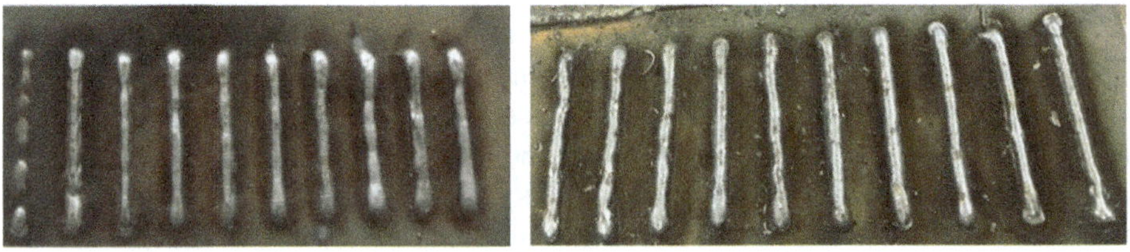

(a) **(b)**

Fig. 5. Surface appearance (a) without waiting and (b) with waiting between the layers

The WAAM technology with MIG-MAG may be used in the future for the production of retail components with specialized, custom shapes. It could be also applied for repair or feature addition. Due to the possible large operation space, WAAM devices are suitable for manufacturing large objects; particularly for automotive, marine, nuclear, and aerospace industries. This may be used for parts of wings, turbines, and metal connectors, among others. Moreover, such manufactured objects can be used as a new way of artistic expression in art or in the industries focused on aesthetics, e.g. architecture.

The demonstrated analysis methodology can be used for qualitative research of manufactured objects, as well as in the study of their mechanical properties in order to model and hence, improvement of the AM process.

4 Summary

The paper determines the optimal values of welding speed and welding torch lifting between successive layers. It turns out that these parameters do not have a significant impact on the average value of the weld height. However, it is possible to select ones to guarantee more predictable geometry, thanks to the analysis of the standard deviation. Based on the performed calculations, the parameters were selected and the correlation between the WAAM print height and the number of layers was investigated. Such experiments were repeated for continuous printing and with waiting between laying down the layers.

Using the trend lines constructed for the experimental outcomes, complete modelling of the geometrical profile and developing an algorithm constantly adjusting the torch lifting is possible. Thanks to these the manufactured objects may be built more accurately and keep their expected mechanical properties. Also, the single trails should be homogenous, hence, the layers would remain relatively flat within the printing process. Due to this, the Z-axis differences would be minimal even for a large number of layers, and the quality of the top surface would be enhanced. Therefore, with a deep understanding of the process principles, it would be available to set the basic welding parameters in such a way as to be able to predict future results with high accuracy.

The original outcomes of the research may contribute to manufacturing parts in a more controlled way without the application of advanced sensory systems (e.g. involving vision control [30] or neural networks [31]). The accurately investigated process enables the exact modelling of the built objects. Therefore, such constructed models may be used for further simulations, including finite elements (FEM) optimisation [32]. With this approach, the geometries could be automatically modified to decrease their mass while keeping the strength parameters. Consequently, the production and operation costs may be decreased so to maximise the enterprises' profits. As the WAAM method is particularly useful for manufacturing large geometries, obtained results may be especially important in the field of architecture [33], space [6], and heavy industry [34] mainly for mechanical engineers, AM specialists, and CAM software developers.

In future research, it is planned to investigate the impact of the trail length on the height of the layer. As with the time to cool down the risk of collapsing the object decreases, the time of laying down a single trail may eliminate the need of waiting between the layers. Due to the observed problem of high heat transfer during printing, it is necessary to implement a method of cooling the object. With this, it should be possible to carry out printing without pauses between the application of layers, even for small objects. Due to the existing need for constant cooling of the material, the suggested approach is to mount cooling on the robot's flange with the use of fluid while shielding the arc [35, 36]. For future applications, it will be extremely valuable to develop a method of robot-table collaboration to produce geometries of any complexity [37] or the wider use of 3D scanning for more complex analyses [38].

Acknowledgements. The paper is based on the results of the "Development of ultra-lightweight design for volatile adapters attaching satellites to a rocket and with readiness their to produce by use Additive Manufacturing", SPACE_Adaptors project (LIDER/45/0181/L-11/19/NCBR/2020), financed in 2021–2023, in the scope of scientific research and development works, by the National Center for Research and Development.

References

1. Köhler, M., Fiebig, S., Hensel, J., Dilger, K.: Wire and arc additive manufacturing of aluminum components. Metals (Basel) **9**(5) (2019). https://doi.org/10.3390/met9050608
2. Dinovitzer, M., Chen, X., Laliberte, J., Huang, X., Frei, H.: Effect of wire and arc additive manufacturing (WAAM) process parameters on bead geometry and microstructure. Addit. Manuf. **26**, 138–146 (2019). https://doi.org/10.1016/j.addma.2018.12.013
3. Vaidya, S., Ambad, P., Bhosle, S.: Industry 4.0 - a glimpse. Procedia Manuf. **20**, 233–238 (2018). https://doi.org/10.1016/j.promfg.2018.02.034
4. Zastrow, M.: The new 3D printing. Nature **578**, 20–24 (2020)
5. Rodrigues, T.A., Duarte, V., Miranda, R.M., Santos, T.G., Oliveira, J.P.: Current status and perspectives on wire and arc additive manufacturing (WAAM). Materials (Basel) **12**(7) (2019). https://doi.org/10.3390/ma12071121
6. Knezović, N., Topić, A.: Wire and arc additive manufacturing (WAAM) – a new advance in manufacturing. In: Karabegović, I. (ed.) New Technologies, Development and Application, pp. 65–71. Springer, Cham (2019). https://doi.org/10.1007/978-3-319-90893-9_7
7. Priarone, P.C., Pagone, E., Martina, F., Catalano, A.R., Settineri, L.: Multi-criteria environmental and economic impact assessment of wire arc additive manufacturing. CIRP Ann. **69**(1), 37–40 (2020). https://doi.org/10.1016/j.cirp.2020.04.010
8. Horgar, A., Fostervoll, H., Nyhus, B., Ren, X., Eriksson, M., Akselsen, O.M.: Additive manufacturing using WAAM with AA5183 wire. J. Mater. Process. Technol. **259**, 68–74 (2018). https://doi.org/10.1016/j.jmatprotec.2018.04.014
9. Chabot, A., Laroche, N., Carcreff, E., Rauch, M., Hascoët, J.Y.: Towards defect monitoring for metallic additive manufacturing components using phased array ultrasonic testing. J. Intell. Manuf. **31**(5), 1191–1201 (2020). https://doi.org/10.1007/s10845-019-01505-9
10. Hassel, T., Carstensen, T.: Properties and anisotropy behaviour of a nickel base alloy material produced by robot-based wire and arc additive manufacturing. Weld. World **64**(11), 1921–1931 (2020). https://doi.org/10.1007/s40194-020-00971-7
11. Gierth, M., Henckell, P., Ali, Y., Scholl, J., Bergmann, J.P.: Wire arc additive manufacturing (WAAM) of aluminum alloy AlMg5Mn with energy-reduced gas metal arc welding (GMAW). Materials (Basel) **13**(12), 1–22 (2020). https://doi.org/10.3390/ma13122671
12. Prado-Cerqueira, J.L., et al.: Analysis of favorable process conditions for the manufacturing of thin-wall pieces of mild steel obtained by wire and arc additive manufacturing (WAAM). Materials (Basel) **11**(8) (2018). https://doi.org/10.3390/ma11081449
13. Eriksson, M., et al.: Additive manufacture of superduplex stainless steel using WAAM. In: MATEC Web of Conferences, vol. 188, August 2018. https://doi.org/10.1051/matecconf/201818803014

14. Prado-Cerqueira, J.L., Diéguez, J.L., Camacho, A.M.: Preliminary development of a wire and arc additive manufacturing system (WAAM). Procedia Manuf. **13**, 895–902 (2017). https://doi.org/10.1016/j.promfg.2017.09.154
15. Stinson, H., Ward, R., Quinn, J., Mcgarrigle, C.: Comparison of properties and bead geometry in MIG and CMT single layer samples for WAAM applications (2021)
16. Ivántabernero, A.P., Álvarez, P., Suárez, A.: Study on arc welding processes for high deposition rate additive manufacturing. Procedia CIRP **68**, 358–362 (2018). https://doi.org/10.1016/j.procir.2017.12.095
17. Lopez, A., Bacelar, R., Pires, I., Santos, T.G., Sousa, J.P., Quintino, L.: Non-destructive testing application of radiography and ultrasound for wire and arc additive manufacturing. Addit. Manuf. **21**, 298–306 (2018). https://doi.org/10.1016/j.addma.2018.03.020
18. Jin, W., Zhang, C., Jin, S., Tian, Y., Wellmann, D., Liu, W.: Wire arc additive manufacturing of stainless steels: a review. Appl. Sci. **10**(5), 1563 (2020). https://doi.org/10.3390/app10051563
19. Li, Y., Yu, S., Chen, Y., Yu, R., Shi, Y.: Wire and arc additive manufacturing of aluminum alloy lattice structure. J. Manuf. Process. **50**, 510–519 (2020). https://doi.org/10.1016/j.jmapro.2019.12.049
20. Wang, L., Xue, J., Wang, Q.: Correlation between arc mode, microstructure, and mechanical properties during wire arc additive manufacturing of 316L stainless steel. Mater. Sci. Eng. A **751**, 183–190 (2019). https://doi.org/10.1016/j.msea.2019.02.078
21. Dhinakaran, V., Ajith, J., Fathima Yasin Fahmidha, A., Jagadeesha, T., Sathish, T., Stalin, B.: Wire arc additive manufacturing (WAAM) process of nickel based superalloys-a review. Mater. Today: Proc. **21**, 920–925 (2020). https://doi.org/10.1016/j.matpr.2019.08.159
22. Szost, B.A., et al.: A comparative study of additive manufacturing techniques: residual stress and microstructural analysis of CLAD and WAAM printed Ti-6Al-4V components. Mater. Des. **89**, 559–567 (2016). https://doi.org/10.1016/j.matdes.2015.09.115
23. Rodrigues, T.A., et al.: In-situ strengthening of a high strength low alloy steel during wire and arc additive manufacturing (WAAM). Addit. Manuf. **34**, 101200 (2020). https://doi.org/10.1016/j.addma.2020.101200
24. Ding, D., Pan, Z., Cuiuri, D., Li, H.: A multi-bead overlapping model for robotic wire and arc additive manufacturing (WAAM). Robot. Comput. Integr. Manuf. **31**, 101–110 (2015). https://doi.org/10.1016/j.rcim.2014.08.008
25. Henckell, P., Gierth, M., Ali, Y., Reimann, J., Bergmann, J.P.: Reduction of energy input in wire arc additive manufacturing (WAAM) with gas metal arc welding (GMAW). Materials (Basel) **13**(11) (2020). https://doi.org/10.3390/ma13112491
26. Bandari, Y.K., Charrett, T.O.H., Michel, F., Ding, J., Williams, S.W., Tatam, R.P.: Compensation strategies for robotic motion errors for additive manufacturing (AM)
27. Kurc, K., et al.: Robotic machining in correlation with a 3D scanner. Mech. Mech. Eng. **24**(1), 36–41 (2020). https://doi.org/10.2478/mme-2020-0003
28. Cheluszka, P.: Automation of coordinate measurements of mining machines working units with 3D scanning (in Polish). Pomiary Autom. Robot. **20**(3), 33–42 (2016). https://doi.org/10.14313/PAR
29. Liberini, M., et al.: Selection of optimal process parameters for wire arc additive manufacturing. Procedia CIRP **62**, 470–474 (2017). https://doi.org/10.1016/j.procir.2016.06.124
30. Xiong, J., Zhang, Y., Pi, Y.: Control of deposition height in WAAM using visual inspection of previous and current layers. J. Intell. Manuf. **32**(8), 2209–2217 (2021). https://doi.org/10.1007/s10845-020-01634-6
31. Wang, Y., et al.: Coordinated monitoring and control method of deposited layer width and reinforcement in WAAM process. J. Manuf. Process. **71**, 306–316 (2021). https://doi.org/10.1016/j.jmapro.2021.09.033

32. Kovšca, D., Starman, B., Ščetinec, A., Klobčar, D., Mole, N.: Advanced computational modelling of metallic wire-arc additive manufacturing. **13**, 1–11 (2021)
33. Müller, J., et al.: Design and parameter identification of wire and arc additively manufactured (WAAM) steel bars for use in construction. Metals (Basel) **9**(7) (2019). https://doi.org/10.3390/met9070725
34. Singh, S.R., Khanna, P.: Wire arc additive manufacturing (WAAM): a new process to shape engineering materials. Mater. Today: Proc. **44**, 118–128 (2021). https://doi.org/10.1016/j.matpr.2020.08.030
35. Kozamernik, N., Bračun, D., Klobčar, D.: WAAM system with interpass temperature control and forced cooling for near-net-shape printing of small metal components. Int. J. Adv. Manuf. Technol. **110**(7–8), 1955–1968 (2020). https://doi.org/10.1007/s00170-020-05958-8
36. Li, F., Chen, S., Shi, J., Zhao, Y., Tian, H.: Thermoelectric cooling-aided bead geometry regulation in wire and arc-based additive manufacturing of thin-walled structures. Appl. Sci. **8**(2), 207 (2018). https://doi.org/10.3390/app8020207
37. Rauch, M., Dorado, J.P., Hascoet, J.: A novel method for additive manufacturing of complex shape curved parts by using variable height layers. J. Mach. Eng. **21**(3), 80–91 (2021)
38. Laghi, V., Palermo, M., Gasparini, G., Girelli, V.A., Trombetti, T.: Geometrical characterization of wire-and-arc additive manufactured steel element. Adv. Mater. Lett. **10**(10), 695–699 (2019). https://doi.org/10.5185/amlett.2019.0019

Synthesis of a State Feedback Controller for an Averaging Tank with Variable Filling

Michał Kolankowski(✉) and Robert Piotrowski

Faculty of Electrical and Control Engineering, Gdańsk University of Technology, Narutowicza 11/12, 80-233 Gdańsk, Poland
michkola1999@gmail.com, robert.piotrowski@pg.edu.pl

Abstract. In paper, a nonlinear averaging tank with variable filling is considered. The main purpose of this research work was the modelling and control system synthesis of an averaging tank. The control objectives included ensuring stability and zero steady-state error of the system and achieving settling time as short as possible, while maintaining a minimal overshoot. In order to achieve the intended purpose, firstly a mathematical model of the control plant was derived. The model was adapted to the form required to design the control system by linearization and reduction of its dimensions, resulting in two system variants. A state feedback and output feedback with integral action control system of the averaging tank was designed using a linear-quadratic regulator (LQR) and optimization of weights. The developed control system was tested using MATLAB environment. Finally, based on simulation results, performance of the control system in different variants and effects of optimization were assessed.

Keywords: Control system · Integral control · Mathematical model · State feedback controller · Tank with variable filling

1 Introduction

An averaging tank with variable filling is a tank containing a substance with variable component concentration. The concentration is assumed to be even in the entire volume of substance. The considered tank is with variable filling and thus the volume of substance contained in the tank is variable and the system's performance is not affected by it not being constant. Averaging in the tank may be achieved by mixing.

Averaging tanks are widely used, primarily in wastewater treatment for stabilizing the composition of wastewater. It is important, because the technological parameters of the treatment process are determined based on the average composition of wastewater. Moreover, the averaging tank ensures a steady flow of wastewater to the further stages of the treatment process despite the input flow of wastewater to the tank being variable [1].

In [2] PID controllers were applied for tank control. Multivariable PID was proposed for the control of a tank with heating [3]. In [4] a Smith predictor for a system composed of three tank was implemented. In [5] the control of a two-tank system was achieved using

R. Szewczyk et al. (Eds.): AUTOMATION 2022, AISC 1427, pp. 35–44, 2022.
https://doi.org/10.1007/978-3-031-03502-9_4

a state feedback structure. In [6, 7] fuzzy PID control systems for a vertical two-tank system and a single tank were designed.

The remainder of this paper is organized as follows. The derivation and implementation of the mathematical model of the averaging tank are described in Sect. 2. The design of the control system is illustrated in Sect. 3. In Sect. 4, the results of control are discussed. The last section presents the conclusions.

2 Mathematical Models of an Averaging Tank with Variable Filling

A substance of variable component concentration flows into the averaging tank (see Fig. 1). A substance of component concentration equal to the average concentration in the tank flows out of the tank. It is assumed that both inflow and outflow are forced. Input (1–3) and output (4–5) variables of the plant are described in Table 1. The considered control plant is a multiple-input–multiple-output (MIMO) plant with 3 input variables and 2 output variables.

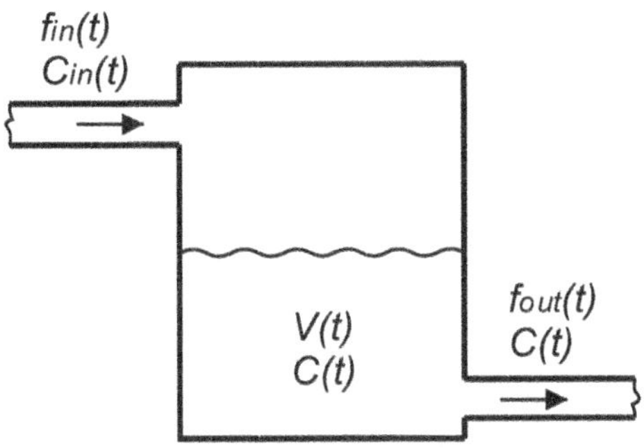

Fig. 1. Scheme of the averaging tank

Table 1. Symbols of variables and their units

No.	Description	Symbol	Unit
1	Inflow rate of subs. with component concentration $C_{in}(t)$	f_{in}	m^3/s
2	Outflow rate of subs. with component concentration $C(t)$	f_{out}	m^3/s
3	Concentration of the component in the subst. flowing into the tank	C_{in}	$kmol/m^3$
4	Volume of substance of concentration $C(t)$ in tank	V	m^3
5	Averaged component concentration in the tank	C	$kmol/m^3$

2.1 Nonlinear Model of Tank

It is assumed that there are no disturbances and that mixing is instantaneous and complete, lumped-parameter model can thus be applied. Mass conservation law (1) and impurity conservation law (2) were applied in the modelling process.

$$\frac{dV(t)}{dt} = f_{in}(t) - f_{out}(t) \tag{1}$$

$$\frac{d(V(t)C(t))}{dt} = f_{in}(t) \cdot C_{in}(t) - f_{out}(t) \cdot C(t) \tag{2}$$

The result of product *V(t)* · *C(t)* in Eq. (2) is:

$$\frac{dV(t)}{t}C(t) + V(t)\frac{dC(t)}{dt} = f_{in}(t) \cdot C_{in}(t) - f_{out}(t) \cdot C(t) \tag{3}$$

Substituting *V(t)* from Eq. (1) in (3), we get:

$$(f_{in}(t) - f_{out}(t)C(t) + V(t)\frac{dC(t)}{t} = f_{in}(t) \cdot C_{in}(t) - f_{out}(t) \cdot C(t) \tag{4}$$

After transformation, the Eq. (4) takes the form:

$$V(t)\frac{dC(t)}{dt} = f_{in}(t)C_{in}(t) - f_{in}(t)C(t) \tag{5}$$

The final model of the plant takes the form (Eqs. (5) and (1)):

$$\begin{cases} V(t)\frac{dC(t)}{dt} = f_{in}(t)(C_{in}(t) - C(t)) \\ \dot{V}(t) = f_{in}(t) - f_{out}(t) \end{cases} \tag{6}$$

The model is continuous, dynamic, nonlinear, deterministic, stationary and a lumped-parameter model.

2.2 Linearization of Tank Model

Many control system structures and analysis methods are intended for linear plants exclusively. Thus linearization of the control plant model was necessary. It was performed by using a Taylor expansion in the neighbourhood of the equilibrium point ($f_{in0}, f_{out0}, C_{in0}, C_0, V_0$). The final result linearized model takes the form:

$$\begin{cases} V_0 \cdot \Delta\frac{dC(t)}{dt} = f_{in0}(\Delta C_{in}(t) - \Delta C(t)) + (C_{in0} - C_0)\Delta f_{in}(t) \\ \Delta\frac{dV(t)}{dt}(t) = \Delta f_{in}(t) - \Delta f_{out}(t) \end{cases} \tag{7}$$

In the equations of the linearized model the coordinates of the equilibrium point (around which the linearization was performed) appear as coefficients. The linearized model was converted to a state-space representation.

State variables:

$$\Delta \boldsymbol{x}(t) = \begin{bmatrix} \Delta x_1(t) \\ \Delta x_2(t) \end{bmatrix} = \begin{bmatrix} \Delta C(t) \\ \Delta V(t) \end{bmatrix} \tag{8}$$

Input variables:

$$\Delta \boldsymbol{u}(t) = \begin{bmatrix} \Delta u_1(t) \\ \Delta u_2(t) \\ \Delta u_3(t) \end{bmatrix} = \begin{bmatrix} \Delta f_{in}(t) \\ \Delta f_{out}(t) \\ \Delta C_{in}(t) \end{bmatrix} \tag{9}$$

Output variables:

$$\Delta \boldsymbol{y}(t) = \begin{bmatrix} \Delta y_1(t) \\ \Delta y_2(t) \end{bmatrix} = \begin{bmatrix} \Delta C(t) \\ \Delta V(t) \end{bmatrix} \tag{10}$$

The coordinates of the equilibrium point in state-space are as follows:

$$(f_{in0}, f_{out0}, C_{in0}, C_0, V_0) = \left(u_{1,0}, u_{2,0}, u_{3,0}, x_{1,0}, x_{2,0}\right) \tag{11}$$

In Sect. 3, two versions of the plant model in state-space representation were applied:

A. Assumption: $f_{out}(t) = const$ – the input vector was changed:

$$\Delta \boldsymbol{u}(t) = \begin{bmatrix} \Delta u_1(t) \\ \Delta u_3(t) \end{bmatrix} \tag{12}$$

The result was the following linearized model of the system:

$$\Delta \dot{\boldsymbol{x}}(t) = \boldsymbol{A}\Delta \boldsymbol{x}(t) + \boldsymbol{B}\Delta \boldsymbol{u}(t) \tag{13}$$

$$\Delta \boldsymbol{y}(t) = \boldsymbol{C}\Delta \boldsymbol{x}(t) + \boldsymbol{D}\Delta \boldsymbol{u}(t) \tag{14}$$

where:

$$\boldsymbol{A} = \begin{bmatrix} \frac{-u_{1,0}}{x_{2,0}} & 0 \\ 0 & 0 \end{bmatrix}; \quad \boldsymbol{B} = \begin{bmatrix} \frac{u_{3,0}-x_{1,0}}{x_{2,0}} & \frac{u_{1,0}}{x_{2,0}} \\ 1 & 0 \end{bmatrix}; \quad \boldsymbol{C} = \begin{bmatrix} 1 & 0 \\ 0 & 1 \end{bmatrix}; \quad \boldsymbol{D} = \begin{bmatrix} 0 & 0 \\ 0 & 0 \end{bmatrix} \tag{15}$$

B. Assumption: $f_{in}(t) = const$ – the input vector and the matrices $\boldsymbol{A}$, $\boldsymbol{B}$ were changed:

$$\Delta \boldsymbol{u}(t) = \begin{bmatrix} \Delta u_2(t) \\ \Delta u_3(t) \end{bmatrix} \tag{16}$$

$$\boldsymbol{A} = \begin{bmatrix} \frac{-u_{1,0}}{x_{2,0}} & 0 \\ 0 & 0 \end{bmatrix}; \quad \boldsymbol{B} = \begin{bmatrix} 0 & \frac{u_{1,0}}{x_{2,0}} \\ -1 & 0 \end{bmatrix}; \quad \boldsymbol{C} = \begin{bmatrix} 1 & 0 \\ 0 & 1 \end{bmatrix}; \quad \boldsymbol{D} = \begin{bmatrix} 0 & 0 \\ 0 & 0 \end{bmatrix} \tag{17}$$

A static equilibrium point was selected for the linearized model: (f_{in0}, f_{out0}, C_{in0}, C_0, V_0) = (0.2 m^3/s; 0.2 m^3/s; 5 kmol/m^3; 5 kmol/m^3; 2 m^3) and the initial values of variables $C(t)$ and $V(t)$ were assumed as: $C_p = 5$ kmol/m^3; $V_p = 2$ m^3.

3 Design of State Feedback Control System

In order to design a control system structure with output feedback and state feedback, it was necessary to reduce the number of control inputs of the plant, because the number of control inputs must not exceed the number of state variables of the control plant. It was accomplished in two ways: by assumption of $f_{out}(t) = f_{out0}$ and by assumption of $f_{in}(t) = f_{in0}$.

Assumption of $C_{in}(t) = C_{in0}$ is inexpedient for this model, because it would prevent the component concentration $C(t)$ from changing, as a substance of only one component concentration flows into the tank.

Integral Control (IC) structure is a control structure with state feedback and output feedback with integral action (see Fig. 2).

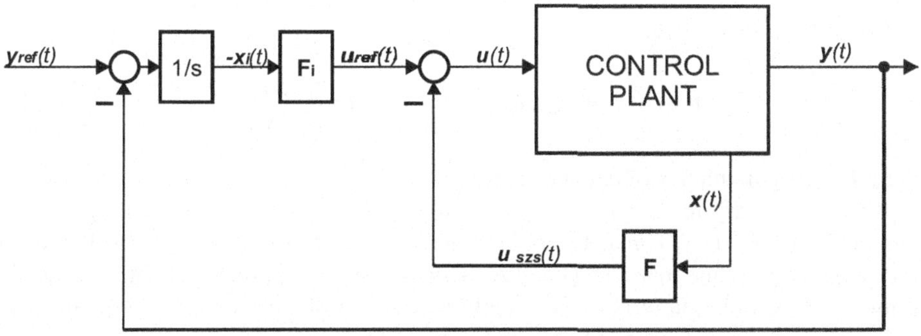

Fig. 2. Block diagram of Integral Control sytem

where: $\boldsymbol{y}_{ref}(t)$, $\boldsymbol{x}_i(t)$, $\boldsymbol{u}_{ref}(t)$, $\boldsymbol{u}_{szs}(t)$, $\boldsymbol{F}_i$, $\boldsymbol{F}$ are reference value vector, integral state variable vector, the component of control variable vector influenced by output feedback, the component of control variable vector influenced by state feedback, gain in output feedback loop matrix, gain in state feedback loop matrix, respectively.

In Integral Control design a linearized model of the tank was applied ((13, 14) and additionally (12, 15) in the $f_{out}(t) = const$ system and (16, 17) in the $f_{in}(t) = const$ system). In order to calculate $\boldsymbol{F}_i$ and $\boldsymbol{F}$ matrices, Linear Quadratic Regulator (LQR) was used [8].

In order to formulate a description of an expanded plant, it is necessary to introduce a deviation integral state variable $\Delta \boldsymbol{x}_i(t)$. It consists of integrated errors which allow for output variable following its reference value. Then the equations of the expanded plant take form (18) and (19). The state vector of the expanded plant takes form of Eq. (20) or in its expanded form (21).

$$\Delta \dot{\boldsymbol{x}}(t) = \boldsymbol{A}\Delta \boldsymbol{x}(t) + \boldsymbol{B}\Delta \boldsymbol{u}(t) \tag{18}$$

$$\Delta \dot{\boldsymbol{x}}_i(t) = \boldsymbol{C}\Delta \boldsymbol{x}(t) + \boldsymbol{M}\Delta \boldsymbol{u}(t) - \boldsymbol{y}_{ref}(t) \tag{19}$$

$$\Delta \boldsymbol{x}_{exp}(t) = \begin{bmatrix} \Delta \boldsymbol{x}(t) \\ \Delta \boldsymbol{x}_i(t) \end{bmatrix} = \begin{bmatrix} \Delta x_1(t) \\ \Delta x_2(t) \\ \Delta x_{i1}(t) \\ \Delta x_{i2}(t) \end{bmatrix} \tag{20}$$

$$\Delta \dot{\boldsymbol{x}}_{exp} = \boldsymbol{A}_{exp} \Delta \boldsymbol{x}_{exp}(t) + \boldsymbol{B}_{exp} \Delta \boldsymbol{u}(t) - \begin{bmatrix} 0 \\ 1 \end{bmatrix} y_{ref}(t) \tag{21}$$

where $\boldsymbol{M}$ is direct input on output impact matrix.

State matrix $\boldsymbol{A}_{exp}$, input matrix $\boldsymbol{B}_{exp}$ and $\boldsymbol{M}$ for the expanded plant are:

$$\boldsymbol{A}_{exp} = \begin{bmatrix} \boldsymbol{A} & 0 \\ \boldsymbol{C} & 0 \end{bmatrix}; \quad \boldsymbol{B}_{exp} = \begin{bmatrix} \boldsymbol{B} \\ \boldsymbol{M} \end{bmatrix}; \quad \boldsymbol{M} = \begin{bmatrix} 0\,0\,0 \\ 0\,0\,0 \end{bmatrix} \tag{22}$$

Control law takes form:

$$\Delta \boldsymbol{u}(t) = -\boldsymbol{F}_{exp} \Delta \boldsymbol{x}_{exp}(t) = -\begin{bmatrix} \boldsymbol{F} & \boldsymbol{F}_i \end{bmatrix} \begin{bmatrix} \Delta \boldsymbol{x}(t) \\ \Delta \boldsymbol{x}_i(t) \end{bmatrix} \tag{23}$$

where $\boldsymbol{F}_{exp}$ is gain matrix of expanded plant.

It was necessary to design a control system in two configurations: with $f_{out}(t) = const$ and with $f_{in}(t) = const$. In both versions the control system is controllable. In order to manually tune matrices $\boldsymbol{Q}$ and $\boldsymbol{R}$, which served the purpose of initial values in optimization, simulation tests of the weight matrices' influence on control quality and performance of control system were conducted.

For the $f_{out}(t) = const$ system values of matrices $\boldsymbol{Q}$, $\boldsymbol{R}$, $\boldsymbol{F}$ and $\boldsymbol{F}_i$ were achieved as follows:

$$\boldsymbol{Q} = \begin{bmatrix} 2 & 0 & 0 & 0 \\ 0 & 3 & 0 & 0 \\ 0 & 0 & 10 & 0 \\ 0 & 0 & 0 & 10 \end{bmatrix}; \quad \boldsymbol{R} = \begin{bmatrix} 1 & 0 \\ 0 & 0.1 \end{bmatrix} \tag{24a}$$

$$\boldsymbol{F} = \begin{bmatrix} 0 & 3.0536 \\ 13.8661 & 0 \end{bmatrix}; \quad \boldsymbol{F}_i = \begin{bmatrix} 0 & 3.0536 \\ 10 & 0 \end{bmatrix} \tag{24b}$$

For the $f_{in}(t) = const$ system the results were:

$$\mathbf{Q} = \begin{bmatrix} 3 & 0 & 0 & 0 \\ 0 & 1 & 0 & 0 \\ 0 & 0 & 10 & 0 \\ 0 & 0 & 0 & 3 \end{bmatrix}; \quad \mathbf{R} = \begin{bmatrix} 2 & 0 \\ 0 & 0.1 \end{bmatrix} \tag{25a}$$

$$\mathbf{F} = \begin{bmatrix} 0 & -1.7174 \\ 14.1987 & 0 \end{bmatrix}; \quad \boldsymbol{F}_i = \begin{bmatrix} 0 & -1.2247 \\ 10 & 0 \end{bmatrix} \tag{25b}$$

Then, in order to find optimal $\boldsymbol{Q}$ and $\boldsymbol{R}$ weight matrices and, in consequence, optimal $\boldsymbol{F}$ and $\boldsymbol{F}_i$ gain matrices, optimization was performed using integral performance indices of control variables $C(t)$ and $V(t)$:

$$IE = \int_0^\infty e(t)dt \tag{26a}$$

$$ISE = \int_0^\infty e^2(t)dt \tag{26b}$$

$$IAE = \int_0^\infty |e(t)|dt \tag{26c}$$

$$ITAE = \int_0^\infty t|e(t)|dt \tag{26d}$$

$$ISEG = \int_0^\infty \left(e^2(t) + a\dot{e}^2(t)\right)dt;\ a = 0.5 \tag{26e}$$

where $e(t)$ is control error.

Decision variables: $\boldsymbol{Q}$ and $\boldsymbol{R}$ ($Q_1, Q_2, Q_3, Q_4, R_1, R_2$).

Constraints:

$$Q_1 \geq 0;\quad Q_2 \geq 0;\quad Q_3 \geq 0;\quad Q_4 \geq 0;\quad R_1 > 0;\quad R_2 > 0 \tag{27}$$

Optimization was performed using the method of minimizing the maximum out of a set of objective functions [9].

For the $f_{out}(t) = const$ system the lowest integral performance index values for both of the control variables and the best control quality in terms of settling time and overshoot were achieved for the ITAE index. The optimal values for $\boldsymbol{Q}$, $\boldsymbol{R}$, $\boldsymbol{F}$ and $\boldsymbol{F}_i$ matrices are:

$$\boldsymbol{Q} = \begin{bmatrix} 10^{-6} & 0 & 0 & 0 \\ 0 & 0.1309 & 0 & 0 \\ 0 & 0 & 12.04 & 0 \\ 0 & 0 & 0 & 10.88 \end{bmatrix};\ \boldsymbol{R} = \begin{bmatrix} 10^{-6} & 0 \\ 0 & 10^{-6} \end{bmatrix} \tag{28a}$$

$$\boldsymbol{F} = \begin{bmatrix} 0 & 370.7667 \\ 262.4497 & 0 \end{bmatrix};\ \boldsymbol{F}_i = \begin{bmatrix} 0 & 3298.0 \\ 3470.2 & 0 \end{bmatrix} \tag{28b}$$

Analogically to the $f_{out}(t) = const$ system, for the $f_{in}(t) = const$ system the lowest integral performance index values were achieved for the ISE index. The optimal values for $\boldsymbol{Q}$, $\boldsymbol{R}$, $\boldsymbol{F}$ and $\boldsymbol{F}_i$ matrices are:

$$\boldsymbol{Q} = \begin{bmatrix} 10^{-6} & 0 & 0 & 0 \\ 0 & 0.3995 & 0 & 0 \\ 0 & 0 & 0.1092 & 0 \\ 0 & 0 & 0 & 4.528 \end{bmatrix};\ \boldsymbol{R} = \begin{bmatrix} 0.0113 & 0 \\ 0 & 10^{-6} \end{bmatrix} \tag{29a}$$

$$\boldsymbol{F} = \begin{bmatrix} 0 & -8.6885 \\ 256.1044 & 0 \end{bmatrix};\ \boldsymbol{F}_i = \begin{bmatrix} 0 & -20.0 \\ 3305.0 & 0 \end{bmatrix} \tag{29b}$$

4 Control Results

The graphs of process and control values for the best values of $\boldsymbol{Q}$ and $\boldsymbol{R}$ matrices, which had been selected, tuned manually (24a, 25a) and achieved in the process of optimization (28a, 29a) for the $f_{out}(t) = const$ system and for the $f_{in}(t) = const$ system were compared. By far the best results were achieved for the version with constant substance outflow – settleng time of variables $C(t)$ (Fig. 3a) and $V(t)$ (Fig. 3b) is significantly shorter and the control signal $C_{in}(t)$ (Fig. 3e) is significantly less demanding in comparison to the control with constant inflow. It is due to the fact that in the nonlinear model of the averaging tank the variable $f_{in}(t)$ influences the rate of change of concentration $C(t)$ both directly and through the change of volume $V(t)$, while the variable $f_{out}(t)$ influences only the change of $V(t)$ which then influences $C(t)$. The advantage is especially prominent in Fig. 3b, which presents an approx. 3 times shorter settling time than in case of $f_{in}(t)$ = const. For both versions of the control system the optimization improved the quality of control. It is indicated by the values of integral performance indices as well (see Table 2). The indices are in all instances lower for the $f_{out}(t) = const$ system than for the $f_{in}(t) = const$ system and they are lower for optimal $\boldsymbol{Q}$ and $\boldsymbol{R}$ matrices compared to the results of manual tuning (with the exception of ISEG). The only reservation is that the maximum values of control variables – $f_{in}(t)$ in the $f_{out}(t) = const$ system (Fig. 3c), $f_{out}(t)$ in the $f_{in}(t) = const$ system (Fig. 3d) and $C_{in}(t)$ in the $f_{out}(t) = const$ system (Fig. 3e) was increased visibly.

The results of Integral Control were compared with the results achieved using a control system structure with two independent PID controllers (optimized using the same method). Integral Control provided no overshoot and, above all, significantly less demanding control signals. It may be due to the fact that PID controllers, in their basic form, are designed for linear SISO control, whereas in this case the control plant is multidimensional and nonlinear.

Table 2. Integral quality indicators for $f_{out}(t) = const$ system and for $f_{in}(t) = const$ system

Performance index	Control value	$f_{out}(t) =$ *const; hand*	$f_{out}(t) =$ *const; optimal*	$f_{in}(t) =$ *const; hand*	$f_{in}(t) = const;$ *optimal*
IAE	C	0.398	0.019	0.415	0.022
	V	0.098	0.011	0.149	0.043
ISE	C	0.064	0.003	0.068	0.004
	V	0.006	0.001	0.010	0.003
ITAE	C	0.481	0.001	0.510	0.002
	V	0.067	0.001	0.161	0.014
IE	C	0.372	0.019	0.380	0.019
	V	0.097	0.011	0.140	0.043
ISEG	C	156.073	156.200	156.078	156.202
	V	24.969	24.981	24.972	24.968

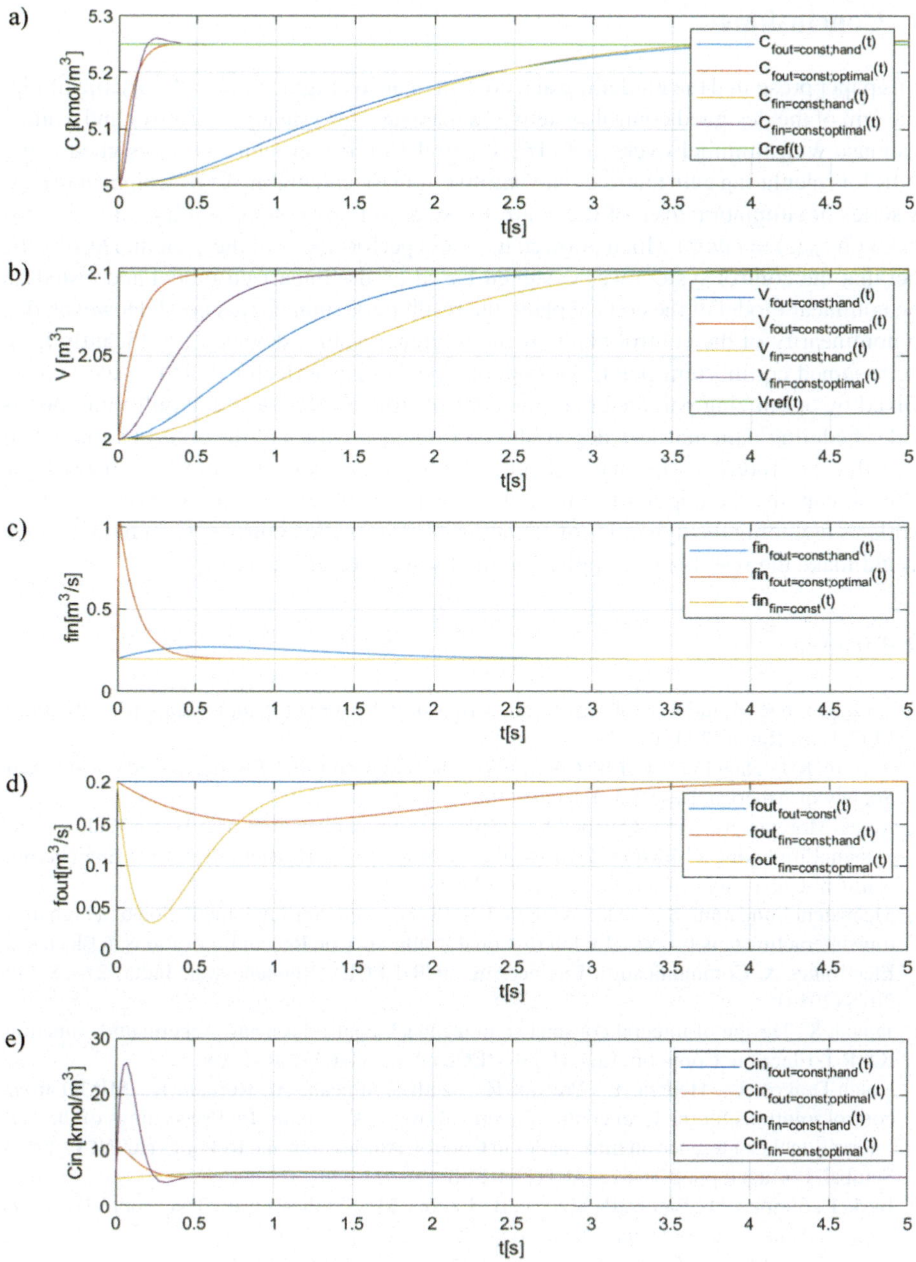

Fig. 3. Control results for $f_{out}(t) = const$ system and for $f_{in}(t) = const$ system

5 Conclusions

This paper presented the modelling and control of an averaging tank with variable filling. The aim of the designed control system was to ensure zero steady-state error and a quick response with minimal overshoot. The Integral Control structure was designed using LQR for calculating gain matrices and optimization for calculating LQR weight matrices. A series of simulation tests of the control system in two versions (with $f_{out}(t) = const$ and with $f_{in}(t) = const$) which allowed to assess performance of the system. Despite the fact that the control system was designed based on the linearized model and tested on the nonlinear model of the control plant, the result performance was good. However, due to nonlinearity of the control plant, as the reference values deviated significantly from the assumed equilibrium point, the systems' performance declined. This issue may be solved by optimizing parameters of the control structure for several equilibrium points and scheduling gain matrices depending on reference values of the control variables. It is a subject of future works on the system. Nonlinear control algorithms, e.g. fuzzy logic control, can also be applied to this control plant. Another issue to be raised in future works is including the dynamics of actuating devices in the control system model which would make the results more applicable to physical averaging tanks.

References

1. Spellman, F.R.: Handbook of Water and Wastewater Treatment Plant Operations. CRC Press LLC, Boca Raton (2003)
2. Astrom, K.J., Hagglund, T.: PID Controllers: Theory, Design and Tuning, 2nd edn. Instrument Society of America, Research Triangle (1995)
3. Rojas–Moreno, A., Parra–Quispe, A.: Design and Implementation of a Water Tank Control System Employing a MIMO PID Controller, Faculty of Electrical and Electronic Engineering, National University of Engineering Lima (2008)
4. Meenatchi Sundaram, S., Venkateswaran, P.R.: Smith predictor implementation of a high dead time interacting tank process. In: International Conference on Recent Innovations in Electrical, Electronics & Communication Engineering (ICRIEECE), Bhubaneswar, India, 27–28 July 2018 (2018)
5. Janani, S.: Design of integral constant state feedback controller using Ackermann's function. IOSR J. Electron. Commun. Eng. (IOSR-JECE) **9**(1), 58–63 (2014). Ver. I
6. Bojan-Dragos, C., Hedrea, E., Precup, R., Szedlak-Stinean, A., Roman, R.: MIMO fuzzy control solutions for the level control of vertical two tank systems. In: Proceedings of the 16th International Conference on Informatics in Control, Automation and Robotics (ICINCO 2019), Setúbal, Portugal, pp. 810–817. SCITEPRESS (2019)
7. Berk, P., Stajnko, D., Vindis, P., Mursec, B., Lakota, M.: Synthesis water level control by fuzzy logic. J. Achievements Mater. Manuf. Eng. **45**(2), 204–210 (2011)
8. Naidu, D.S.: Optimal Control Systems. CRC Press, Boca Raton (2003)
9. Murray, R.M.: Optimization-Based Control. California Institute of Technology, Pasadena (2008)

Discrete Time Sliding Mode Control with State Constraints and Limited Control Signal

Marek Jaskuła and Piotr Leśniewski(✉)

Institute of Automatic Control, Łodź University of Technology,
18 Stefanowskiego St., 90-537 Łodź, Poland
{marek.jaskula,piotr.lesniewski}@p.lodz.pl

Abstract. In this paper we discuss the issue of constraining the state variables and the control input in the sliding mode control approach. The considered linear system is perturbed by unknown but bounded external disturbances. In addition, the model uncertainties can vary in time. The sliding mode controller is designed via the reaching law technique and the convergence rate is selected to ensure the fastest and monotonic approaching of the representative point to the predefined switching hyperplane. Moreover, the state constraints are considered in form of linear combinations and the duration of the reaching phase is bounded from above. Results are presented in lemmas that are formally stated and proved.

Keywords: Sliding mode control · Discrete time system · Control signal limitation · Reaching law method · Linear combinations · State constraints

1 Introduction

Presently, a rising number of variable structure controllers is being applied to various types of uncertain systems, even with changing parameters. Although, the perturbed plants are challenging to deal with, the sliding mode control approach proved to be an effective way of regulation, due to its computational efficiency and robustness [3,11,17,20–22,25]. Nevertheless, there are research fields that require further study, such as maintaining the state constraints in the reaching phase or designing the optimal time-varying switching hyperplanes. First and fundamental released papers, considering sliding mode control methodology, were published by Emelyanov [10], Draženowić [9] and Utkin [23]. The results, obtained for the continuous time systems, were quickly extended also onto the discrete time objects [1,2,8,12,16,19,24]. Soon after, the theoretical considerations found their practical application, for example in power electronics.

The main concept of the sliding mode approach is to initially drive the representative point of the system to the predefined switching hyperplane, that

R. Szewczyk et al. (Eds.): AUTOMATION 2022, AISC 1427, pp. 45–53, 2022.
https://doi.org/10.1007/978-3-031-03502-9_5

represents the designer preferences. Subsequently, the regulation method forces the sliding motion, up to the moment when the desired state is reached.

The process of designing the sliding mode controller should begin with the proper selection of the switching hyperplane. Next, the control input can be specified by proving the stability of the sliding motion or predefining the evolution of the sliding variable by the proposed reaching law [4–7,13–15]. In our paper we use the second approach, because it ensures additional control over the state behavior in the reaching phase, where the state can most likely violate the constraints. The results presented in this work extend our earlier outcomes [18], by taking into account the constraints in forms of state variables linear combinations. Moreover, we enhance the convergence rate speed by considering the uncertainties influence at every time instant, rather then on the whole admissible set. Research conclusions are formulated as lemmas and formally proved.

2 Controller Design

To begin with, we consider the discrete time linear system, perturbed by unknown external disturbances. What is more, we also take into account the time varying parameter uncertainties. As a consequence, the plant model is given as follows

$$\boldsymbol{x}(k+1) = \left[\boldsymbol{A} + \widetilde{\boldsymbol{A}}(k)\right]\boldsymbol{x}(k) + \boldsymbol{b}u(k) + \boldsymbol{d}(k). \tag{1}$$

The $n \times 1$ state vector is denoted by $\boldsymbol{x}(k)$, $\boldsymbol{A}$ represents the $n \times n$ state matrix, $\widetilde{\boldsymbol{A}}(k)$ describes the time changing uncertainties and $u(k)$ is a scalar control input. Let us notice that the external disturbances $\boldsymbol{d}(k) = [d_1(k), \ldots, d_n(k)]^T$ do not fulfill the matching conditions. Nevertheless, they need to be constrained in order to guarantee that the control signal can overcome their impact: $\max|d_i(k)| = D_i$ for every $i \in \{1, \ldots, n\}$. Due to the same reason the uncertainties $\widetilde{a}_{ij}(k)$ (elements of matrix $\widetilde{\boldsymbol{A}}(k)$) are limited in following manner: $|\widetilde{a}_{ij}(k)| \leq v_{ij}$, $\boldsymbol{V} = [v_{ij}]$.

Next, we define the sliding manifold $\boldsymbol{c}^T\boldsymbol{x}(k) = 0$. Vector $\boldsymbol{c}^T = [c_1, \ldots, c_n]$ needs to be selected to fulfill condition $\boldsymbol{c}^T\boldsymbol{b} \neq 0$. Thus, we receive the sliding variable as $s(k) = \boldsymbol{c}^T\boldsymbol{x}(k)$. Let us outline that after the reaching phase we do not require the representative point to cross the sliding hyperplane in each step, although it must remain in the neighborhood of this manifold.

The controller design process starts with formulating the following reaching law

$$s(k+1) = s(k) - K\mathrm{sgn}[s(k)] + \boldsymbol{c}^T\boldsymbol{d}(k) + \boldsymbol{c}^T\widetilde{\boldsymbol{A}}(k)\boldsymbol{x}(k), \tag{2}$$

where K is called the convergence rate. Using (1) and the above approach law we calculate the control signal obtaining

$$u(k) = -\left(\boldsymbol{c}^T\boldsymbol{b}\right)^{-1}\left\{\boldsymbol{c}^T\boldsymbol{A}\boldsymbol{x}(k) - s(k) + K\mathrm{sgn}[s(k)]\right\}. \tag{3}$$

The designed controller ensures a stable quasi-sliding motion after the time instant when the neighborhood of the switching manifold is reached. Nevertheless, in order to guarantee the finite time reaching phase, we need to satisfy the inequality $|s(k)| - |s(k+1)| \geq \lambda > 0$ until s converges to a certain threshold. As a consequence, we have to assure that

$$K - D_{\max} - V_{\max} \geq \lambda > 0, \tag{4}$$

where $D_{\max} = [|c_1|, \ldots, |c_n|]\,\boldsymbol{D}$, $V_{\max} = [|c_1|, \ldots, |c_n|]\,\boldsymbol{V}\,[|x_1(t)|, \ldots, |x_n(t)|]^T$ and $\boldsymbol{D} = [D_1, \ldots, D_n]^T$.

3 Constraints and Control Strategy

In this section we introduce the respective convergence rates: for the control signal limitation and state variables constraints (in the form of linear combinations). Moreover, we will demonstrate the general control strategy, providing methodology of selecting the appropriate convergence rate at every time instant.

First of all, we consider the problem of restricting the control input. Hence, we require that $-r_u \leq u(k) \leq r_u$, where r_u is a strictly positive constant. Taking into account our previous results [18], we get the biggest possible convergence rate that maintains the limitation as

$$K_u = \left|\boldsymbol{c}^T \boldsymbol{b}\right| r_u - \operatorname{sgn}[s(k)]\,\boldsymbol{c}^T \boldsymbol{A}\boldsymbol{x}(k) + |s(k)|\,. \tag{5}$$

By the use of the convergence rate K_u the controller will generate a maximal or minimal admissible control signal value. On the other hand, the lowest possible convergence rate in admissible range of u is

$$K_{\min_u} = -\left|\boldsymbol{c}^T \boldsymbol{b}\right| r_u - \operatorname{sgn}[s(k)]\,\boldsymbol{c}^T \boldsymbol{A}\boldsymbol{x}(k) + |s(k)|\,. \tag{6}$$

Let us move to the problem of constraining the state variables. We demand that for the whole regulation time the condition $\alpha_i(\boldsymbol{x}) \leq r_i$ holds, where α_i represents the certain linear combination, i.e. $\alpha_i(\boldsymbol{x}) = \alpha_{i1}x_1 + \cdots + \alpha_{in}x_n, \alpha_{ij} \in \Re$. To clarify the following calculations we introduce $\alpha_{ie} = [\alpha_{i1}, \ldots, \alpha_{in}]$. Thus, we can express $\alpha_i(\boldsymbol{x}(k+1))$ as

$$\begin{aligned}\alpha_i(\boldsymbol{x}(k+1)) = &\alpha_{ie}\left[\boldsymbol{A} + \widetilde{\boldsymbol{A}}(k)\right]\boldsymbol{x}(k) - \\ &\alpha_i(\boldsymbol{b})\left(\boldsymbol{c}^T\boldsymbol{b}\right)^{-1}\left\{\boldsymbol{c}^T\boldsymbol{A}\boldsymbol{x}(k) - s(k) + K\operatorname{sgn}[s(k)]\right\} + \alpha_i(\boldsymbol{d}(k))\,.\end{aligned} \tag{7}$$

Subsequently, we compute the highest possible impacts of system uncertainties and external disturbances in the direction of the hyperplane connected to the given constraint, obtaining

$$V_{\alpha_i} = [|\alpha_{i1}|, \ldots, |\alpha_{in}|]\,\boldsymbol{V}\,[|x_1(k)|, \ldots, |x_n(k)|]^T, \tag{8}$$

$$D_{\alpha_i} = [|\alpha_{i1}|, \ldots, |\alpha_{in}|]\,\boldsymbol{D}. \tag{9}$$

Taking into account (7), (8) and (9) from $\alpha_i(\boldsymbol{x}(k+1)) \leq r_i$ we derive (assuming that $\alpha_i(\boldsymbol{b})\, \boldsymbol{c}^T \boldsymbol{b} s(t)$ is negative) the biggest possible convergence rate, getting

$$K_{\alpha_i} = \left|[\alpha_i(\boldsymbol{b})]^{-1} \boldsymbol{c}^T \boldsymbol{b}\right| [r_i - D_{\alpha_i} - V_{\alpha_i} - \alpha_{ie} \boldsymbol{A}\boldsymbol{x}(k)] \\ - \operatorname{sgn}[s(t)]\, \boldsymbol{c}^T \boldsymbol{A}\boldsymbol{x}(k) + |s(k)| . \tag{10}$$

However, in such a case there is no lower bound of the convergence rate value that we can apply. Now, let us consider the situation when $\alpha_i(\boldsymbol{b})\, \boldsymbol{c}^T \boldsymbol{b} s(t)$ is positive. Then there is no maximal convergence rate from the i-th constraint point of view, but the lowest possible convergence rate that ensures $\alpha_i(\boldsymbol{x}(k+1)) \leq r_i$ exists and is equal to

$$K_{\min_{\alpha_i}} = - \left|[\alpha_i(\boldsymbol{b})]^{-1} \boldsymbol{c}^T \boldsymbol{b}\right| [r_i - D_{\alpha_i} - V_{\alpha_i} - \alpha_{ie} \boldsymbol{A}\boldsymbol{x}(k)] \\ - \operatorname{sgn}[s(k)]\, \boldsymbol{c}^T \boldsymbol{A}\boldsymbol{x}(k) + |s(k)| . \tag{11}$$

We can observe from the above formulas that is it crucial to assume that $\alpha_i(\boldsymbol{b})$ does not equal zero. Else, the control signal would not have any impact on the satisfaction of the constraint.

Implementation of the convergence rate K_{α_i} will effect in driving the state near the hyperplane connected to the i-th constraint. It is not likely to obtain $\alpha_i(\boldsymbol{b}) = r_i$ due to the unknown impact of external disturbances and time changing model uncertainties. The convergence rate is designed in such a manner to reach the limitation only when the perturbations are the most intensive.

The general strategy of selecting the convergence rate at each time instant is given as follows

$$K = \min\{K_u, K_{\alpha_{i_1}}, \ldots, K_{\alpha_{i_l}}, |s(k)|\}, \tag{12}$$

where $i_1, \ldots, i_l$ are indexes, for which the maximal convergence rates exist. The above methodology includes the absolute value of the sliding variable, since it is important to reach the certain neighborhood of the switching manifold and it would be undesirable to cross the hyperplane. Of course due to the impact of external disturbances and model uncertainties the manifold can be exceeded, however only by a small, known a priori distance.

In the end, let us outline that the initial point $\boldsymbol{x}_0 = \boldsymbol{x}(0)$ belongs to the admissible set. Therefore, if the monotonic convergence of the representative point to the switching hyperplane is obtained, then the state will remain in

$$\mathbf{X}_0 = \{\boldsymbol{x} \in \Re^n : \forall_{i \in \{1,\ldots,n\}} |x_i| \leq r_i, \left|\boldsymbol{c}^T \boldsymbol{x}\right| \leq \left|\boldsymbol{c}^T \boldsymbol{x}_0\right|\}, \tag{13}$$

for the whole regulation process.

4 Sufficient Conditions

This section focus on formally stating and proving lemmas that are crucial to guarantee the fastest, monotonic and terminal convergence of sliding variable to the certain threshold, simultaneously ensuring all given constraints.

First of all, we need to verify if convergence rates K_u and K_{α_i} are large enough to overcome even the most intensive impact of uncertain terms. Moreover, we demand that the quasi-sliding mode band can be reached without the violation of any of the considered restrictions. Therefore, we require

$$K_u - D_{\max} - V_{\max} \geq \lambda > 0, \tag{14}$$
$$K_{\min_u} \leq |s(k)|, \tag{15}$$
$$K_{\alpha_i} - D_{\max} - V_{\max} \geq \lambda > 0, \tag{16}$$
$$K_{\min_{\alpha_j}} \leq |s(k)|, \tag{17}$$

for i, for which the maximal convergence rates exist, and j, for which the minimal convergence rates exist.

Lemma 1. *In order to ensure (14) and (15) it is sufficient that for* $\boldsymbol{x}(k) \in \boldsymbol{X}_0$

$$\left|\boldsymbol{c}^T \boldsymbol{b}\right| r_u \geq sgn\left[s(k)\right] \boldsymbol{c}^T \boldsymbol{A}\boldsymbol{x}(k) - |s(k)| + D_{\max} + V_{\max} + \lambda, \tag{18}$$
$$\left|\boldsymbol{c}^T \boldsymbol{b}\right| r_u \geq -sgn\left[s(k)\right] \boldsymbol{c}^T \boldsymbol{A}\boldsymbol{x}(k). \tag{19}$$

Proof. Above inequalities can be derived directly by using K_u, $K_{\min_u}$ definitions, into conditions (14), (15).

Lemma 2. *In order to ensure (16) and (17) it is sufficient that for* $\boldsymbol{x}(k) \in \boldsymbol{X}_0$

$$\left|\left[\alpha_i(\boldsymbol{b})\right]^{-1} \boldsymbol{c}^T \boldsymbol{b}\right| \left[r_i - D_{\alpha_i} - V_{\alpha_i} - \alpha_{ie}\boldsymbol{A}\boldsymbol{x}(k)\right] \geq sgn\left[s(k)\right] \boldsymbol{c}^T \boldsymbol{A}\boldsymbol{x}(k) - |s(k)| + D_{\max} + V_{\max} + \lambda, \tag{20}$$
$$\left|\left[\alpha_j(\boldsymbol{b})\right]^{-1} \boldsymbol{c}^T \boldsymbol{b}\right| \left[r_j - D_{\alpha_j} - V_{\alpha_j} - \alpha_{je}\boldsymbol{A}\boldsymbol{x}(k)\right] \geq -sgn\left[s(k)\right] \boldsymbol{c}^T \boldsymbol{A}\boldsymbol{x}(k). \tag{21}$$

Proof. By substituting formulas for K_{α_i}, $K_{\min_{\alpha_j}}$ in (20) and (21) we get the conditions (16), (17).

Next, we need to guarantee that keeping one of the constraints would not result in violating the other one. In other words, we must ensure that convergence rates $K_{\min_u}$, $K_{\min_{\alpha_j}}$ are lower or equal then both K_u, K_{α_i}. From the convergence rates definition we get that $K_{\min_u} \leq K_u$, thus it is sufficient to require

$$K_{\min_u} \leq K_{\alpha_i}, \tag{22}$$
$$K_{\min_{\alpha_i}} \leq K_u, \tag{23}$$
$$K_{\min_{\alpha_j}} \leq K_{\alpha_i}, \tag{24}$$

for i, for which maximal convergence rates exist, and j, for which minimal convergence rates exist.

Lemma 3. *In order to ensure (22) and (23) it is sufficient that for* $\boldsymbol{x}(k) \in \boldsymbol{X}_0$

$$r_u + |\alpha_i(\boldsymbol{b})|^{-1} \left[r_i - D_{\alpha_i} - V_{\alpha_i} - \alpha_{ie}\boldsymbol{A}\boldsymbol{x}(k)\right] \geq 0. \tag{25}$$

Proof. To begin with, let us notice that substituting the respective convergence rates formulas into (14), (15) and then subtracting $|s(k)| - \text{sgn}[s(t)]\, \boldsymbol{c}^T \boldsymbol{A}\boldsymbol{x}(k)$ leads to same inequality. After moving all terms to one side and dividing by $|\boldsymbol{c}^T \boldsymbol{b}|$ we obtain condition (25).

Lemma 4. *In order to ensure (24) it is sufficient that for* $\boldsymbol{x}(k) \in \boldsymbol{X}_0$

$$\begin{aligned} &|\alpha_i(\boldsymbol{b})|^{-1}\left[r_i - D_{\alpha_i} - V_{\alpha_i} - \alpha_{ie}\boldsymbol{A}\boldsymbol{x}(k)\right] \\ &+ |\alpha_j(\boldsymbol{b})|^{-1}\left[r_j - D_{\alpha_j} - V_{\alpha_j} - \alpha_{je}\boldsymbol{A}\boldsymbol{x}(k)\right] \geq 0. \end{aligned} \tag{26}$$

Proof. First of all, let us substitute the definitions of $K_{\min_{\alpha_j}}, K_{\alpha_i}$ into (24). Then we subtract $|s(k)| - \text{sgn}[s(t)]\, \boldsymbol{c}^T \boldsymbol{A}\boldsymbol{x}(k)$ and divide both sides by $|\boldsymbol{c}^T \boldsymbol{b}|$. As a consequence we obtain (26).

5 Results and Discussion

The simulation scenario demonstrates the properties of the presented control strategy. Firstly, let us introduce the plant parameters, external disturbances and time changing model uncertainties:

$$\boldsymbol{A} = \begin{bmatrix} 1 & 0.5 \\ 0.5 & -1 \end{bmatrix}, \qquad \boldsymbol{b} = \begin{bmatrix} 0.5 \\ 1 \end{bmatrix},$$

$$\widetilde{\boldsymbol{A}}(k) = \begin{bmatrix} 0.01\sin(k) & 0 \\ 0 & 0.02\sin(k) \end{bmatrix} \qquad \boldsymbol{d} = \begin{bmatrix} 0.1(-1)^k \\ 0.2\sin(20k) \end{bmatrix}.$$

The switching line is described by equation $x_1 + 2x_2 = 0$ and the initial state is $\boldsymbol{x}_0 = [65, 10]^T$. Our goal is to drive the representative point to the switching line monotonically and in finite time, simultaneously maintaining the following constraints: $x_1(k) - x_2(k) \leq 55$, $x_1(k) - 5x_2(k) \leq 75$, $x_1(k) \geq -50$, $x_2(k) \leq 10$, $|u(k)| \leq 40$. Therefore, the initial point lies in the corner or the admissible set. Let us also notice that all conditions in Lemmas from the last section hold, so from the theoretical point of view the strategy can be successfully applied.

Figure 1 presents the state trajectory. The region bounded by the blue lines is the admissible set. We can observe that the state variable limitations hold for the whole regulation process. Moreover, at first the representative point moves along the constraint that is represented by $x_1(k) - x_2(k) = 55$ and then reaches the other restriction: $x_1(k) - 5x_2(k) = 75$. This was caused by the change in convergence rate: from K_{α_1} to K_{α_2}. After that, the switching line is attained and the quasi-sliding motion begins. Let us notice, that due to the impact of unknown external disturbances and system uncertainties, the state does not precisely reach the constraints. This statement can be verified from Fig. 2, which represents the state trajectory in case when perturbations are disabled. Further, Fig. 3 shows the evolution of the control signal. The admissible range of the input was not fully utilized, because other constraints were more restrictive from the convergence rate point of view. From Fig. 4 we can see that the monotonic and finite time convergence was successfully achieved, same as the stable quasi-sliding motion.

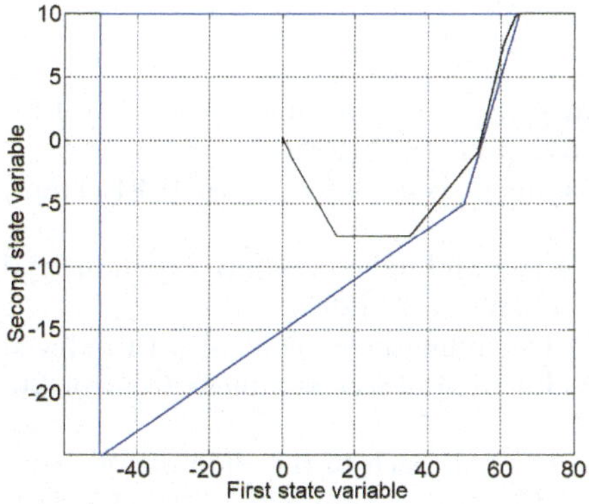

Fig. 1. State trajectory (perturbed)

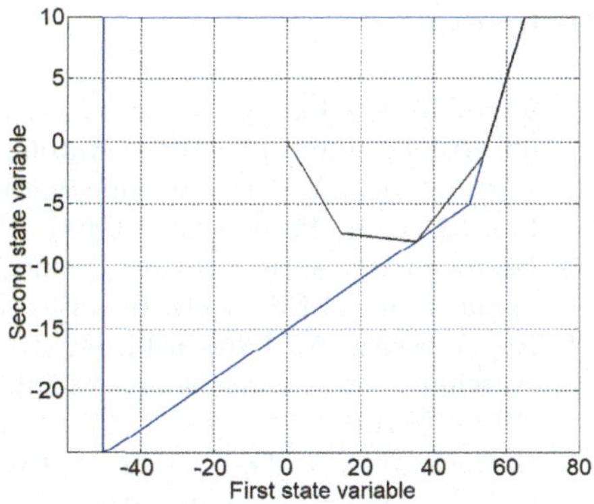

Fig. 2. State trajectory (not perturbed)

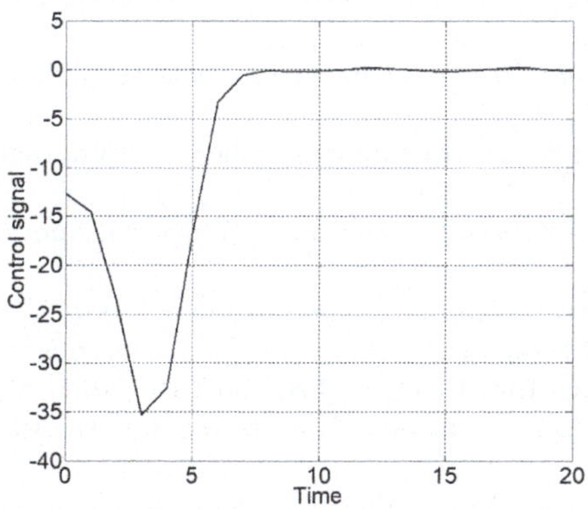

Fig. 3. Control signal

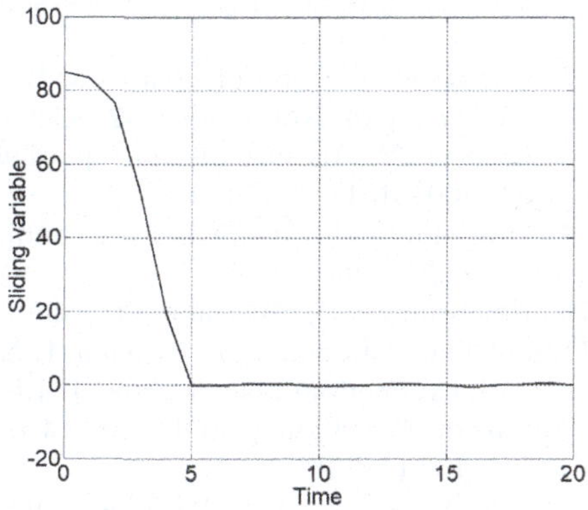

Fig. 4. Sliding variable

6 Conclusions

In this paper the discrete time sliding mode control approach was presented via the reaching law technique. The introduced control strategy was dedicated to solve the problem of constraining the control input and state variables from the very beginning of the regulation process. The proposed sliding mode controller results in the fastest, monotonic and terminal convergence of the representative point to the predefined switching hyperplane. Moreover, the control signal was kept in the admissible range and the state constrains, in the form of linear combinations, were not violated. Theoretical background was encapsulated in lemmas, which were formally proved. Our further goal is to consider nonlinear constraints and time changing constraints.

References

1. Bartolini, G., Pisano, A., Usai, E.: Digital second-order sliding mode control for uncertain nonlinear systems. Automatica **37**, 1371–1377 (2001)
2. Bartoszewicz, A.: Discrete-time quasi-sliding-mode control strategies. IEEE Trans. Ind. Electron. **45**, 633–637 (1998)
3. Bartoszewicz, A., Kaynak, O., Utkin, V.I. (eds.): Sliding mode control in industrial applications. IEEE Trans. Ind. Electron. **55**, 3805–4103 (2008)
4. Bartoszewicz, A., Latosiński, P.: Discrete time sliding mode control with reduced switching - a new reaching law approach. Int. J. Robust Nonlinear Control **26**, 47–68 (2016)
5. Bartoszewicz, A., Leśniewski, P.: Reaching law approach to the sliding mode control of periodic review inventory systems. IEEE Trans. Autom. Sci. Eng. **11**, 810–817 (2014)
6. Bartoszewicz, A., Leśniewski, P.: New switching and nonswitching type reaching laws for SMC of discrete time systems. IEEE Trans. Control Syst. Technol. **24**, 670–677 (2016)
7. Chakrabarty, S., Bandyopadhyay, B.: A generalized reaching law for discrete sliding mode control. Automatica **52**, 83–86 (2015)
8. Chan, C.Y.: Robust discrete quasi-sliding mode tracking controller. Automatica **31**, 1509–1511 (1995)
9. Draženović, B.: The invariance conditions in variable structure systems. Automatica **5**, 287–295 (1969)
10. Emelyanov, S.V.: Variable Structure Control Systems. Nauka, Moscow (1967)
11. Fallaha, C.J., Saad, M., Kanaan, H.Y., Al-Haddad, K.: Sliding-mode robot control with exponential reaching law. IEEE Trans. Ind. Electron. **58**, 600–610 (2011)
12. Furuta, K.: Sliding mode control of a discrete system. Syst. Control Lett. **14**, 145–152 (1990)
13. Gao, W., Hung, J.C.: Variable structure control of nonlinear systems: a new approach. IEEE Trans. Ind. Electron. **40**, 45–55 (1993)
14. Gao, W., Wang, Y., Homaifa, A.: Discrete-time variable structure control systems. IEEE Trans. Ind. Electron. **42**, 117–122 (1995)
15. Golo, G., Milosavljević, Č.: Robust discrete-time chattering free sliding mode control Syst. Control Lett. **41**, 19–28 (2000)
16. Hui, S., Żak, S.: On discrete-time variable structure sliding mode control. Automatica **38**, 283–288 (1999)
17. Janardhanan, S., Bandyopadhyay, B.: Multirate output feedback based robust quasi-sliding mode control of discrete-time systems. IEEE Trans. Autom. Control **52**, 499–503 (2007)
18. Jaskuła, M., Leśniewski, P.: State and control signal limitations in discrete time sliding mode control with time varying system's uncertainties. In: Szewczyk, R., Zieliński, C., Kaliczyńska, M. (eds.) AUTOMATION 2021. AISC, vol. 1390, pp. 117–127. Springer, Cham (2021). https://doi.org/10.1007/978-3-030-74893-7_12
19. Milosavljević, Č.: General conditions for the existence of a quasisliding mode on the switching hyperplane in discrete variable structure systems. Autom. Remote Control **46**, 307–314 (1985)
20. Qu, Q., Zhang, H., Yu, R., Xiao, G.: Constrained robust optimal sliding mode control for uncertain nonlinear systems using ADP approach. In: 6th Data Driven Control and Learning Systems (DDCLS), pp. 208–213 (2017)

21. Sarpturk, S.Z., Istefanopulos, Y., Kanyak, O.: On the stability of discrete-time sliding mode control systems. IEEE Trans. Autom. Control **32**, 930–932 (1987)
22. Schreibeis, J., Wulff, K., Reger, J., Moreno, J.: Lyapunov-stability for the sliding-mode control of a turbocharger subject to state constraints. In: 43rd Annual Conference of the IEEE Industrial Electronics Society, IECON 2017, pp. 4068–4073 (2017)
23. Utkin, V.I.: Variable structure systems with sliding modes. IEEE Trans. Autom. Control **22**, 212–222 (1977)
24. Utkin, V.I., Drakunow, S.V.: On discrete-time sliding mode control. In: IFAC Conference on Nonlinear Control, pp. 484–489 (1989)
25. Veluvolu, K.C., Soh, Y.C., Cao, W.: Robust discrete-time nonlinear sliding mode state estimation of uncertain nonlinear systems. Int. J. Robust Nonlinear Control **17**, 803–828 (2007)

PID and FOPID Controllers Combinations During Control of 3D Crane Optimized with GWO Algorithm

Jakub Żegleń-Włodarczyk(✉) and Klaudia Wajda

Department of Automatics and Robotics, AGH University, Kraków, Poland
zeglenjakub@gmail.com

Abstract. The article presents 3D crane control of all three axes (XYZ) and both angles (α and β). PID controllers and their fractional counterparts - FOPID controllers, which have two additional parameters, were used. Previous research focused on a direct comparison of the two types of controllers. This time, the authors used various combinations to check which part of the system influences the quality of control the most. The model and simulations were implemented in the Matlab/Simulink environment. Grey-Wolf Optimizer was used to optimize the controller coefficients. Several combinations of PID and FOPID controllers were compared (e.g. 2 PID controllers on the XY axes and 3 FOPID controllers on the Z axis and α and β angles).

Keywords: 3D crane · PID · FOPID · Optimization · GWO · ORA

1 Introduction

Due to the increasing access to the Internet, more and more people decide to shop online. The result is increased movement of items by courier and shipping companies that use various types of warehouses and ports. In the first phase, items are often packed into large containers. In order to reload payload as quickly as possible, it is necessary to prepare an appropriate control for the cranes that can cope with weather conditions (e.g. rain, wind) and the imperfections of the system.

Many types of controllers are in use, ranging from PID control [1], Model Predictive Control (MPC) [2], to neural networks [3] and fractional controller variants (e.g. FOPID [4]). Each of the XYZ axes requires a separate control. Additionally, the transferred load should oscillate as little as possible - both angles should eventually reach a stable value of 0.

Five controllers are needed in the crane control system. Therefore, in order to determine the coefficients, it is best to use optimization algorithms. There are many, including those inspired by nature: Grey-Wolf Optimizer (GWO) [5,11], Firefly Algorithm (FA) [6], Particle Swarm Optimization (PSA) [6], Artificial

R. Szewczyk et al. (Eds.): AUTOMATION 2022, AISC 1427, pp. 54–63, 2022.
https://doi.org/10.1007/978-3-031-03502-9_6

Bee Colony (ABC) [6]. In this article the authots decided to choose the GWO algorithm.

Fractional calculus is not a new idea. One of the first mentions can be found in letter from L'Hopital to Leibniz from 1695. [7] The development of this field resulted in creation of many fractional equivalents of known controllers: PID [4], adaptive control [8], Kalman Filter [9].

The above topics allow to define the problem of 3D crane control with the use of FOPID regulators optimized by GWO algorithm. Other papers indicate that FOPID control is better than PID [10,12]. However, the authors decided to check which part of the 3D crane affects the control quality the most. For this purpose, various control options were used, e.g. 2 PID controllers with 3 FOPID controllers.

The following section provide basic information about the FOPID controller. The next chapter discusses the GWO algorithm. Later, 3D crane model was presented. The next part presents the results. At the end the results are presented together with a discussion.

2 FOPID Controller

The FOPID controller is an extension of the classic PID version. In this case the integral and the derivative part can be of a non-integer order. It manifests itself through two additional parameters - lambda and mi. Ultimately, the transfer function of FOPID controller takes the following form:

$$G_c(s) = K_P + \frac{K_I}{s^\lambda} + K_D s^\mu \tag{1}$$

where:

K_P - coefficient of proportional action
K_I - coefficient of integral action
K_D - coefficient of derivative action
λ - order of integral
μ - order of differentiation

In order to use the FOPID controller, the non-integer order of s should be used. One example is the Oustaloup Recursive Approximation which was used in this paper [13]. This method is done in the frequency range of ω:

$$s^\lambda \approx C_o \Pi \frac{s + \omega_k'}{s + \omega_k} \tag{2}$$

where each coefficient is described by:

$$\omega_k' = \omega_b \frac{\omega_h}{\omega_b}^{\frac{k+N+0.5(1-r)}{2N+1}}, \omega_k = \omega_b \frac{\omega_h}{\omega_b}^{\frac{k+N+0.5(1-r)}{2N+1}}, C_o = \frac{\omega_h}{\omega_b}^{\frac{-r}{2}} \Pi \frac{\omega_k}{s + \omega_k'}. \tag{3}$$

where (ω_b , ω_h) is the fixed pulsation interval and N is the fixed degree of the approximation.

3 Grey Wolf Optimizer

Grey Wolf Optimizer, as a name suggest, is inspired by grey wolves. It is a meta-heuristic optimization method. In the pack of wolves there is a strict hierarchy: the alfa wolf, beta and delta individuals and the weakest units - omega. Additionally, a pack of wolves has a predefined hunting scenario:

1. Tracking and approaching a victim
2. Circling and harassing
3. Attacking the victim

$$D = |C * X_v(t) - X(t)| \quad C = 2 * r_2 \tag{4}$$

$$X(t+1) = X_v(t) - A * D \quad A = 2a * r_1 - a \tag{5}$$

$$X(t+1) = \frac{X_\alpha + X_\beta + X_\delta}{3}, \tag{6}$$

where X_v is the location of prey, X is the position of wolf, t is the current iteration, r_1, $r_2 \in [0,1]$ are random numbers, X_α, X_β and X_δ specify the location vector of the alpha, beta and delta, the a decrease in each iteration from 2 to 0. Reaching the maximum number of iterations is the end criterion. This is the point at which the algorithm ends its operation.

4 Mathematical Model of the 3d Crane

3D crane is a system often used in industry that can be classified as highly nonlinear MIMO (Multiple Input Multiple Output) [17]. The crane works in three dimensions:

- X and Y axes by moving the cart,
- Z axis by lengthening or shortening the line with the payload.

A laboratory example of such a setup is a 3D crane prepared by the INTECO company - shown in Fig. 1. Such a system was equipped with three DC motors and five encoders - three for measuring the position in the XYZ axes and two for measuring the angle in the appropriate 2D planes. The scheme of the system with illustrated coordinates and forces is shown in Fig. 2, where [15,16]:

- α denotes the angle between the x axis and the lift-line;
- β denotes the angle between the negative direction on the z axis and the projection of the lift-line onto the yz plane;
- y_w denotes the distance of the rail with the cart from the center of the construction frame;
- x_w denotes the distance of the cart from the center of the rail;
- R denotes the length of the lift-line.

Fig. 1. 3D crane from INTECO

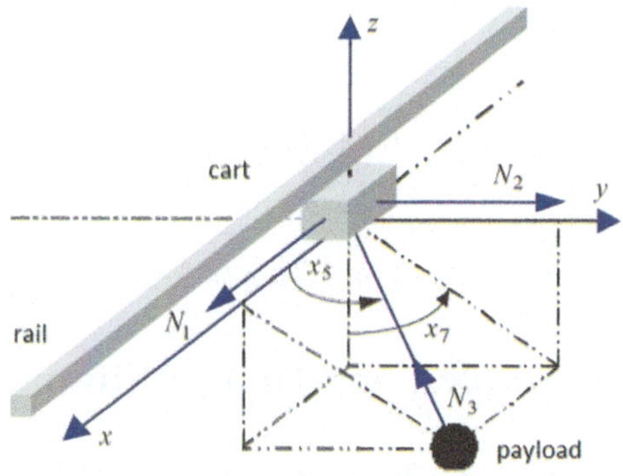

Fig. 2. 3D crane diagram [15]

The other variables needed to describe the system are:

m_c - mass of the payload
m_w - mass of the cart
m_s - mass of the moving rail
x_c, y_c, z_c - coordinates of the payload
S - reaction force in the lift-line acting on the cart
F_x - force driving cart
F_y - force driving the rail with cart
F_R - force controlling the length of the lift-line
T_x, T_y, T_R - friction forces
u_1 - control on the x axis
u_2 - control on the y axis
u_3 - control on the z axis

Using them, the following system of differential equations describing the 3D gantry system can be obtained:

$$\begin{cases} x_1 = x_w \\ x_2 = \dot{x}_1 = \dot{x}_w \\ x_3 = y_w \\ x_4 = \dot{x}_3 = \dot{y}_w \\ x_5 = \alpha \\ x_6 = \dot{x}_5 = \dot{\alpha} \\ x_7 = \beta \\ x_8 = \dot{x}_7 = \dot{\beta} \\ x_9 = R \\ x_{10} = \dot{x}_9 = \dot{R} \end{cases} \quad (7) \qquad \begin{cases} x_1 = x_2 \\ x_2 = N_1 + \mu_1 c_5 N_3 \\ x_3 = x_4 \\ x_4 = N_2 + \mu_2 s_5 s_7 N_3 \\ x_5 = x_6 \\ x_6 = (s_5 N_1 - c_5 s_7 N_2 + (\mu_1 - \mu_2 s_7^2) c_5 s_5 N_3 + V_5)/x_9 \\ x_7 = x_8 \\ x_8 = -(c_7 N_2 + \mu_2 s_5 c_7 s_7 N_3 + V_6)/(s_5 x_9) \\ x_9 = x_{10} \\ x_{10} = -c_5 N_1 - s_5 s_7 N_2 - (1 + \mu_1 c_5^2 + \mu_2 s_5^2 s_7^2) N_3 + V_7 \end{cases} \quad (8)$$

where:

$$\mu_1 = m_c/m_w \qquad V_6 = 2x_8(c_5s_6x_9 + s_5x_{10}) + gs_7$$
$$\mu_2 = m_c/(n_m + m_s) \qquad V_7 = s_5^2x_8^2x_9 + gs_5c_7 + x_6^2x_9$$
$$s_n \equiv \sin x_n \qquad N_1 = u_1 - k_1\dot{x}_w = u_1 - k_1x_2$$
$$c_n \equiv \cos x_n \qquad N_2 = u_2 - k_2\dot{y}_w = u_2 - k_2x_4$$
$$V_5 = c_5s_5x_8^2x_9 - 2x_{10}x_6 + gc_5c_7 \qquad N_3 = u_3 + k_3\dot{R} = u_3 + k_3x_{10}$$

5 Implementation of Simulation

The aim of this article is to compare different combinations of PID and FOPID controllers for a 3D crane. For this purpose, it was necessary to prepare an appropriate simulation that would allow finding appropriate coefficients for all variants. Figure 3 shows the prepared project in the Matlab/Simulink environment for the variant with five FOPID controllers. Of course, each of them can be converted to a classic PID. The orange part symbolizes the 3D crane model.

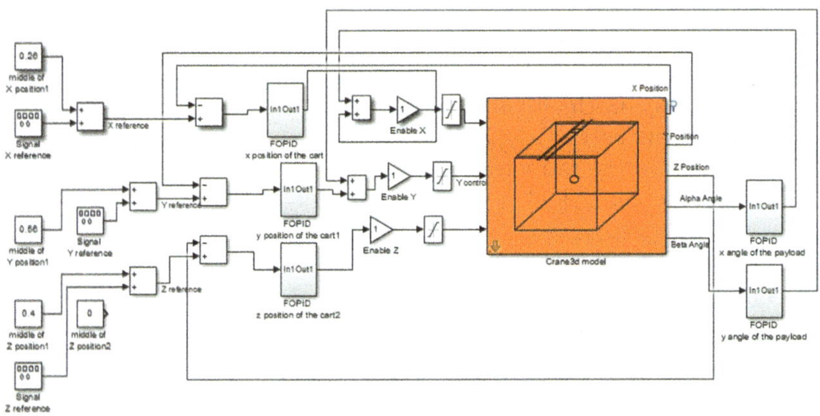

Fig. 3. Simulation model of the 3D crane with FOPID controllers

GWO algorithm is searching for unknown coefficients. The initial range of them is shown below:

$$\begin{aligned} K_P, &= [0:50] \\ K_I, &= [0:50] \\ K_D &= [0:50]. \\ \lambda &= [0:1] \\ \mu &= [0:1] \end{aligned} \tag{9}$$

The algorithm itself also requires the definition of additional parameters presented in the Table 1.

Table 1. Initial parameters for GWO algorithm.

Parameters	Values
Number of wolves in herd	5
Maximum number of iterations	30

In order to estimate the quality of the system operation, the following cost function was used:

$$J = \int_0^\infty (|e_x(t)| + |e_y(t)| + |e_z(t)| + |\alpha(t)| + |\beta(t)|)\, dt. \tag{10}$$

where:

$e_x(t)$ - difference between the set point and the actual value on the X axis
$e_y(t)$ - difference between the set point and the actual value on the Y axis
$e_z(t)$ - difference between the set point and the actual value on the Z axis
$\alpha(t)$ - X angle of the payload
$\beta(t)$ - Y angle of the payload

6 Results of Simulation

The coefficients for each controller were selected after multiple simulations were conducted. Ultimately, it was decided to prepare three types of control:

- Variant 1 - X/Y/Z axes and α/β angles - FOPID controllers (Table 2)
- Variant 2 - X/Y axes - PID controllers, Z axis and α/β angles - FOPID controllers (Table 2)
- Variant 3 - α/β angles - PID controllers, X/Y/Z axes - FOPID controllers (Table 2)

Figure 4 shows a comparison of the behavior of the X axis of all variants. The second and third control versions reach the set value at a similar time. The only more noticeable difference is the occurrence of the overshoot for the third variant. The version with five FOPID regulators performs surprisingly the worst. Approaching the set point is slower. There is even a small offset between the set point and steady state (albeit small - around 0.004).

In the case of the Y axis (Fig. 5), there are much greater differences between each of the control versions. The only thing in common is the presence of overshoot. This time variant 1 with five FOPID controllers behaves in the best way

Table 2. Example of coefficients for all variants

Variant 1	P	I	λ	D	μ
X position	30.2	43.3	−0.99	13.2	0.47
Y position	4.98	15.03	−0.99	20.6	0.67
Z position	19.78	11.9	−0.0096	40.23	0.0100
X angle	3.50	19.9	−0.017	38.8	−0.023
Y angle	22.3	3.73	−0.99	58.1	0.048
Variant 2	P	I	λ	D	μ
X position	50.4	1.29	—	44.3	—
Y position	69.5	5.04	—	39.01	—
Z position	27.49	17.58	−0.02	8.37	0.83
X angle	42.4	6.99	−0.32	73.39	0.24
Y angle	15.45	2.85	−0.26	101.31	0.39
Variant 3	P	I	λ	D	μ
X position	21.28	24.2	−0.99	20.06	0.19
Y position	20.4	18.9	−0.129	26.09	0.12
Z position	37.007	27.4	−0.225	29.02	0.78
X angle	66.04	8.14	—	2.13	—
Y angle	27.6	30.3	—	0.19	—

- the set point is quickly reached. The stabilization for version 2 is very similar, but the final approach to the set value is very slow. Variant 3 performs the worst, with oscillations appearing for half of the steady state period, but they eventually expire.

Figure 6 shows the behavior of the Z axis, with a FOPID controller in each version. The set point for the Z axis is achieved in a very similar way for all control versions. After a very aggressive approach to the set point, it suddenly slows down. Only in the case of the third version of the control the course of the graph is slightly rounded. In all variants, the set point is reached very quickly.

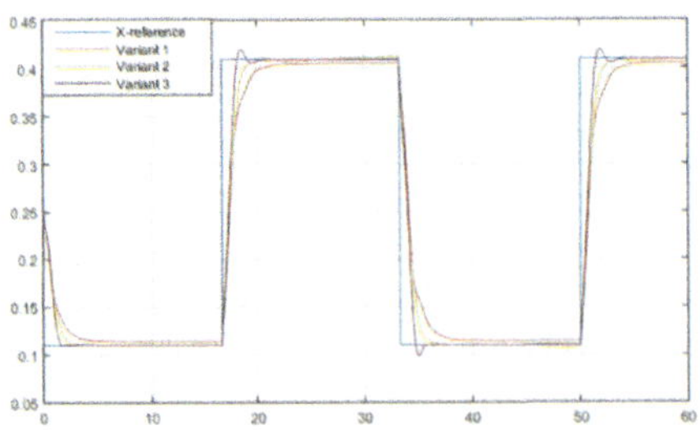

Fig. 4. 3D crane response for X axis

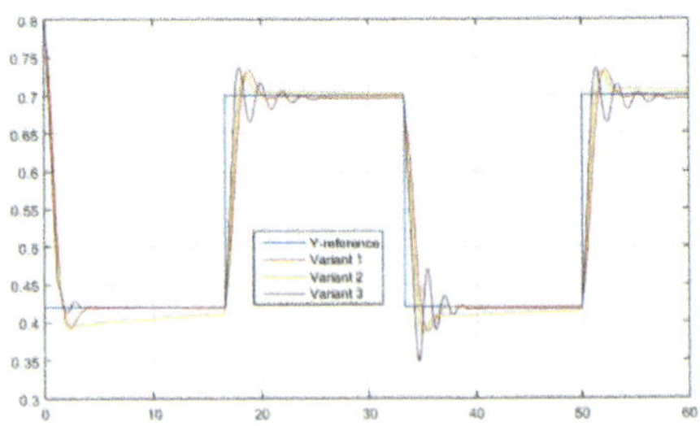

Fig. 5. 3D crane response for Y axis

The behavior of the alpha angle is shown in the Fig. 7. One should pay attention to the behavior of variant 2 - the initial change of angle is always the smallest and there is only one major overshoot. The set point is reached very quickly. Version 1 is a bit worse, there are more oscillations, but the stabilization time is very similar to the previous case. With variant 3, the largest initial swings can be seen. Additionally, oscillations around a set point last the longest.

In the case of the beta angle (Fig. 8), one can see a change in behavior between variants 1 and 2 in relation to the angle alpha - this time version 1 has the smallest initial angle change and the approach to the set value itself occurs in a softer way compared to version 2. However, variant 2 stabilizes slightly faster although the changes are very small). The most noticeable is the occurrence of large oscillations for variant 3. Eventually, the set point is reached, but it takes much longer time than in the previous cases.

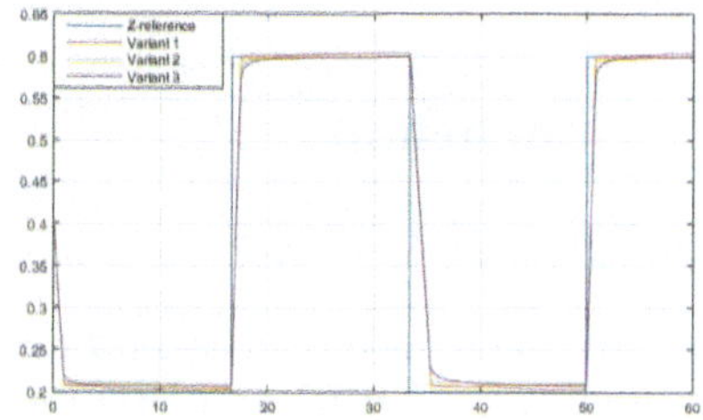

Fig. 6. 3D crane response for Z axis

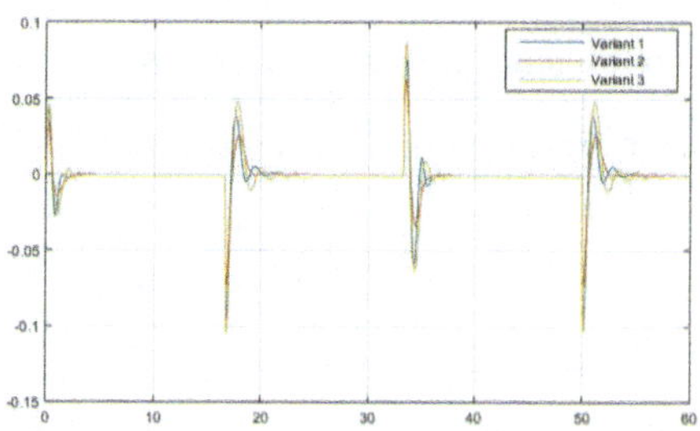

Fig. 7. 3D crane response for α angle

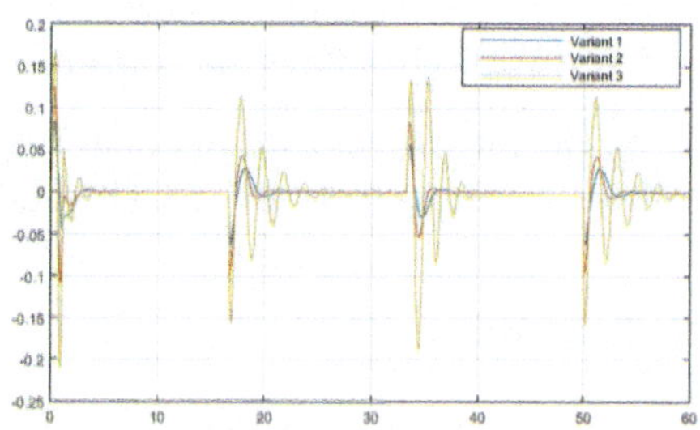

Fig. 8. 3D crane response for β angle

Table 3 contains the values of the cost function for all three variants. The smallest is for the version with five FOPID controllers. However, only a slightly worse result was achieved by version 2, which had classic PID controllers on the X and Y axes. This possibility could already be indicated by the coefficients of variant 1, where for the X and Y axes the value of the λ coefficient was -0.99. It means that integral part is almost of the integer order. Additionally, due to the occurring friction, the control of these axes is relatively simple. Variant 3 is much more different than the rest. Most likely it is due to the fact that the task of extinguishing appearing oscillations is more difficult, and in this case classic PID controllers were used for this purpose.

Table 3. Value of cost function for every variant

	Cost function
Variant 1	8226
Variant 2	8311
Variant 3	8756

7 Conclusion

The following conclusions can be drawn from the article:

- three versions of the control for the 3D crane were prepared, differing in the number of FOPID and PID regulators
- more accurate control can be achieved with FOPID controllers
- if the optimized λ and μ values for FOPID controllers are -0.99 or 0.99 respectively, then it is worth checking the behavior of the system with the classic PID controller in this place (the control quality is likely to be worse, albeit not much, but it will save a lot of computing power)
- less complex controllers can be used to control the easier parts of the system (e.g. the position of the cart)
- Grey Wolf Optimizer has optimized the coefficients for all controllers

In the future, the authors plan to compare the results on a real system, which will be 3D crane from INTECO (as shown in Fig. 1).

References

1. Samin, R.E., Jie, L.M., Zawawi, M.A.: PID implementation of heating tank in mini automation plant using Programmable Logic Controller (PLC). In: International Conference on Electrical, Control and Computer Engineering 2011 (InECCE), pp. 515–519 (2011)
2. Cortes, F., Linares, D., Patino, D., Melo, K.: A distributed model predictive control (D-MPC) for modular robots in chain configuration. In: IX Latin American Robotics Symposium and IEEE Colombian Conference on Automatic Control, pp. 1–6. IEEE (2011)
3. Dubois, O., Nicolas, J., Billat, A.: Adaptive neural network control of the temperature in an oven. In: IEE Colloquium on Advances in Neural Networks for Control and Systems, pp. 8/1–8/3 (1994)
4. Puchalski, B., Rutkowski, T.A., Duzinkiewicz, K.: Implementation of the FOPID algorithm in the PLC controller - PWR thermal power control case study. In: 23rd International Conference on Methods and Models in Automation and Robotics (MMAR), pp. 229–234 (2018)
5. Anbumani, K., Ranihemamalini, R., Pechinathan, G.: GWO based tuning of PID controller for a heat exchanger process. In: Third International Conference on Sensing, Signal Processing and Security (ICSSS), pp. 417–421 (2017)

6. Agarwal, S., Singh, A.P., Anand, N.: Evaluation performance study of Firefly algorithm, particle swarm optimization and artificial bee colony algorithm for non-linear mathematical optimization functions. In: Fourth International Conference on Computing, Communications and Networking Technologies (ICCCNT), pp. 1–8 (2013)
7. Machado, J.T., Kiryakova, V., Mainardi, F.: Recent history of fractional calculus. Commun. Nonlinear Sci. Numer. Simul. **16**(3), 1140–1153 (2011)
8. Veisi, A., Delavari, H.: Adaptive fractional order control of photovoltaic power generation system with disturbance observer. In: 7th International Conference on Control, Instrumentation and Automation (ICCIA), pp. 1–5 (2021)
9. Feng, P., Lu, L., Xue, D.: Compensation for network data dropouts based on modified fractional-order Kalman filter. In: The 27th Chinese Control and Decision Conference (2015 CCDC), pp. 348–352 (2015)
10. Żegleń-Włodarczyk, J., Dziedzic, K.: Optimization of the FOPID parameters of the 3D crane control system by using GWO. In: 25th International Conference on Methods and Models in Automation and Robotics (MMAR 2021), pp. 13–18 (2021)
11. Dziedzic, K.: Identification of fractional order transfer function model using biologically inspired algorithms. In: Szewczyk, R., Zieliński, C., Kaliczyńska, M. (eds.) AUTOMATION 2019. AISC, vol. 920, pp. 47–57. Springer, Heidelberg (2019). https://doi.org/10.1007/978-3-030-13273-6_5
12. Żegleń-Włodarczyk, J., Dziedzic, K.: Optimization of the FOPID parameters of the 3D crane control system by using GWO. Paper Accepted to be Published in Proceedings of the International Conference on Fractional Differentiation and its Applications (ICFDA'21) (2021)
13. Oustaloup, A., Levron, F., Mathieu, B., Nanot, F.M.: Frequency-band complex noninteger differentiator: characterization and synthesis. IEEE Trans. Circuits Syst. I Fundam. Theory Appl. **47**(1), 25–39 (2000)
14. Mirjalili, S., Mirjalili, S.M., Lewis, A.: Grey wolf optimizer. Adv. Eng. Softw. **69**, 46–61 (2014). https://doi.org/10.1016/j.advengsoft.2013.12.007
15. Pauluk, M., Korytowski, A., Turnau, A., Szymkat, M.: Time optimal control of 3D crane. In: MMAR 2001 : Proceedings of the 7th IEEE International Conference on Methods and Models in Automation and Robotics: Miedzyzdroje 28–31 August 2001, Robotics; Marine automation; Control engineering; Identification and Signal Processing, vol. 2, pp. 927–932 (2001)
16. INTECO: 3D Crane User's Manual. http://www.inteco.com.pl/Docs/3DCrane_um.pdf. Accessed 31 Mar 2021
17. INTECO: 3D Crane - Laboratory model of industrial gantry crane controlled from PC. http://www.inteco.com.pl/products/3d-crane/. Accessed 24 Oct 2021

Multi Depot UAVs Routing Subject to Changing Weather and Time Windows Variation

Grzegorz Radzki[1], Grzegorz Bocewicz[1](✉), Jarosław Wikarek[2], Peter Nielsen[3], and Zbigniew Banaszak[1]

[1] Faculty of Electronics and Computer Science, Koszalin University of Technology, Koszalin, Poland
{grzegorz.bocewicz,zbigniew.banaszak}@tu.koszalin.pl
[2] Department of Information Systems, Kielce University of Technology, Kielce, Poland
[3] Department of Materials and Production, Aalborg University, Aalborg, Denmark
peter@mp.aau.dk

Abstract. The presented problem being an extension of well know Vehicle Routing Problem (VRP) concerns the routes planning of the UAVs fleet carrying out the deliveries to spatially dispersed impatient customers in a highly dynamic environment within a specified timeframe. The problem under consideration integrates the Capacitated VRP, Split Delivery VRP, Heterogeneous VRP, Multi Depot VRP, Multi-visit VRP as well as VRP with Time Windows. The developed model allows for proactive and reactive delivery missions planning in terms of the Constraint Satisfaction Problem. The sufficient conditions allowing to eliminate unfeasible solutions and thus allow to speed up the calculations have been implemented in decision support procedure aimed at UAVs online routing. The experiments carried out showed the usefulness of the proposed approach in planning of UAVs' mission performed in dynamic environment.

Keywords: UAVs fleet mission planning · Vehicle routing problem · Re-routing

1 Introduction

In this paper we consider a planning of UAVs' fleet mission problem with dynamic and unpredictable environment constraints [3, 10, 12, 14]. Typical disruptions in deliveries by UAVs, may be caused by ad hoc changes to the deliveries ordered and/or delivery receipt periods as well as changing weather conditions, which affect the energy consumption of UAVs and result in their shorter range [11]. Therefore, UAVs' missions should take into account the possibility of implementing both proactive and reactive behavior. In this context our study focuses on reactive planning of deliveries, that are resistant to sudden atmospheric changes in wind direction and intensity as well as unforeseen changes in the delivery schedules caused by cancellation of ordered deliveries or notification of new orders. The considered problem is NP-hard and belongs to the class of Vehicle Routing Problems (VRP) [2] and can be seen as integration of the Capacitated VRP (CVRP) [9], Split Delivery VRP (SDVRP) [4], Heterogeneous VRP (HVRP) [5, 6], Multi Depot

R. Szewczyk et al. (Eds.): AUTOMATION 2022, AISC 1427, pp. 64–74, 2022.
https://doi.org/10.1007/978-3-031-03502-9_7

VRP (MDVRP) [1], Multi-Visit VRP (MVVRP) [7], and VRP with Time Windows (VRPTW) [13]. In other words it can be seen as extension of the basic VRP about the possibility of taking into account the different capacities and costs of the vehicles as well as the collection of depots. Additional extensions assume that the time windows determine the delivery times of services and more than one vehicle is allowed to service a customer, so that a customer's demand can be split among several vehicles on different routes. The originality of paper results from merging the proactive and reactive planning of missions of UAVs' fleet in the framework of above mentioned integrated problem. In this context the conducted research can be seen as a continuation of the previously obtained results [8, 11]. The developed model allows planning of delivery missions in terms of the Constraint Satisfaction Problem (CSP) easily implemented in a constraint programming environment, e.g., IBM ILOG. The vast majority of publications on the subject usually focuses on technical capabilities of the available UAVs fleet and/or the restrictions resulting from the distributions. The restrictions related to dynamic change of the delivery time windows are usually omitted. This may lead to unnecessary search for routes to customers whose service in the changed delivery time windows prevents the UAV from returning to base safely. Indicated research gaps have become inspiration for extension of the declarative model by sufficient condition the fulfilment of which allows to limit the space of potential solutions and the time of their search. This condition allow to foreseen the UAV's battery depletion before completing its mission in online mode. In this context the main objective of paper boils down to: (*i*) development of a declarative model of the problem of routing the UAVs fleet that integrates the CVRP, SDVRP, HVRP, MDVRP, MVVRP, and VRPTW, (*ii*) extension of the proposed declarative model with conditions allowing 30% reduction of solution space, and (*iii*) conducting computational experiments confirming the usefulness of the proposed approach for planning of UAVs fleet missions.

The remaining of this paper is organized as follows. Section 2 presents the methodology used in the approach to UAVs mission contingency planning. Section 3 describes a mathematical formulation the declarative model for reactive planning of deliveries by UAVs fleet in terms of the CSP. Section 4 presents experiments conducted and elaborates results obtained. Finally, Sect. 5 provides the conclusions followed by the description of future research.

2 Approach to UAVs Fleet Dynamic Routing

We consider a distribution network, that is modelled by the graph $G = (N, E)$ where $N = \{N_1, \ldots, N_\lambda, \ldots, N_n\}$ signifies the set of $n = |N|$ nodes and $E = \{(N_\beta, N_\lambda) | \beta, \lambda \in \{1, \ldots, n\}, \beta \neq \lambda\}$ signifies the set of edges determining the possible connections between nodes. Set $NB \subset N$ contains the nodes representing the depots and set of $ND \subset N$ contains the nodes representing delivery points: $NB \cup ND = N$, $NB \cap ND = \emptyset$. Given a fleet of UAVs $\mathcal{U} = \{U_1, \ldots, U_k, \ldots, U_K\}$ that delivers goods to the points ND. The fleet $\mathcal{U}$ is spread in between the depots NB, the subset $\mathcal{U}^\lambda \subseteq \mathcal{U}$ determines the UAVs which are allocated at depot $N_\lambda \in NB$. To each delivery point $N_\lambda \in ND$ an ordered quantity of goods $z_\lambda \in \mathbb{N}$ (taken from depots) should be transported. Deliveries are made as part of the mission S, which consists of sub-missions

${}^{l}S$ (i.e., delivery plans that include a single course of UAVs: depot-delivery points-depot). Z denotes a sequence $Z = (z_1, \ldots, z_n)$ where $z_\lambda = 0$ for each $N_\lambda \in NB$. The all required goods should be delivered in the given horizon time H. Amount of goods delivered during one sub-mission ${}^{l}S$ by the U_k to the delivery point $N_\lambda \in ND$ is determined by the variable ${}^{l}c_\lambda^k \in \mathbb{N}$. ${}^{l}C = \left({}^{l}c_1^1, \ldots, {}^{l}c_1^K, \ldots, {}^{l}c_n^1, \ldots, {}^{l}c_n^K\right)$ is the sequence determining the payload weight delivered by fleet $\mathcal{U}$. Variable Q_k denotes the payload capacity of U_k. Moreover each U_k is described by technical parameters: battery capacity CAP, airspeed va, drag coefficient C_D, front surface A, UAV's width b. Time spent on take-off and landing U_k on delivery point N_λ is indicated by variable $w_\lambda \in \mathbb{N}$. ${}^{l}\mathcal{U} \subseteq \mathcal{U}$ denotes a set of UAVs used during sub-mission ${}^{l}S$. The moment when the $U_k \in {}^{l}\mathcal{U}$ arrives to the delivery point N_λ during sub-mission ${}^{l}S$ is indicated by variable ${}^{l}y_\lambda^k \in \mathbb{N}$. Therefore the sequence ${}^{l}Y = \left({}^{l}y_1^1, \ldots, {}^{l}y_1^K, \ldots, {}^{l}y_n^1, \ldots, {}^{l}y_n^K\right)$ is called the fleet. In addition, each delivery point $N_\lambda \in ND$ is associated with the delivery time window $dX_\lambda = \left[dx_\lambda^{down}; dx_\lambda^{up}\right]$ specifying the time interval in which the goods should be delivered to N_λ. Moreover we assume that the variable $t_{\beta,\lambda} \in \mathbb{N}$ determines traveling time between nodes N_β,N_λ, where: $\left(N_\beta, N_\lambda\right) \in E$, and routes of $U_k \in {}^{l}\mathcal{U}$ during sub-mission ${}^{l}S$ are represented by sequences: ${}^{l}\pi_k = \left(N_{k_1}, \ldots, N_{k_i}, N_{k_{i+1}}, \ldots, N_{k_\mu}\right)$, where: $k_i \in \{1, .., n\}$, $\left(N_{k_i}, N_{k_{i+1}}\right) \in E$. ${}^{l}\Pi$ denotes a sequence of UAVs routes executed during sub-mission ${}^{l}S$: ${}^{l}\Pi = \left({}^{l}\pi_1, \ldots, {}^{l}\pi_k, \ldots, {}^{l}\pi_K\right)$. Delivery plan of one UAVs sub-mission ${}^{l}S$ is defined as a sequence: ${}^{l}S = \left({}^{l}\mathcal{U}, {}^{l}\Pi, {}^{l}Y, {}^{l}C\right)$. It is assumed that a plan of UAVs sub-mission ${}^{l}S$ is implemented in accordance with the forecasted conditions known for each sub-mission ${}^{l}S$. The forecasted weather conditions are described by the set $\mathbb{F}$ of pairs composed of direction θ and wind speed vw $(\theta, vw) \in \mathcal{F}$, i.e. $\mathbb{F}$ is defined as follows: $\mathbb{F} = \{(\theta, vw) | \theta \in [0^\circ, 360^\circ), vw \in [0, \mathcal{F}(\theta)]\}$. $\mathcal{F}(\theta)$ is a function whose values determine the maximum forecasted wind speed for the given direction θ. The weather conditions determine the admissibility of the adopted sub-mission's plan ${}^{l}S$, i.e. they determine whether the batteries of UAVs will not be prematurely discharged. A function $\Upsilon_{k,l}(\theta)$ determines the borderline wind speed, which guarantees the successful completion of delivery plan by the U_k during sub-mission ${}^{l}S$ in the distribution network G: $\Upsilon_{k,l}(\theta) = \max \Gamma_{k,l}(\theta)$, where: $\Gamma_{k,l}(\theta)$ – set of wind speed's values vw for a given direction θ, for which the batteries of U_k is not discharged. The sub-mission's plan ${}^{l}S$ is assumed (Radzki et al. 2021) to be resistant to the forecast weather conditions $\mathbb{F}$ if the boundary wind $\Upsilon_{k,l}(\theta)$ of all $U_k \in {}^{l}\mathcal{U}$ in any direction θ does not exceed the forecasted value $Z(\theta)$: ${}_{U_k \in {}^{l}\mathcal{U}} \Upsilon_{k,l}(\theta) \geq \mathcal{F}(\theta)$.

The implementation of the designated mission S may be subject to various disturbances IS. Among them, there are sudden changes in the weather (beyond the expected $\mathcal{F}^*(\theta)$ range), order changes (increase /decrease amount of ordered deliveries Z^*), change a number of delivery points served (changing the structure of G^*) and change a delivery time windows dX^*. The UAVs' fleet, when performing the mission plan S, meets a disturbance $IS(t^*)$ - covering one of cases: $\mathcal{F}^*(\theta)$, G^*, Z^*, dX^*, at the time t^*. This situation raises the question: *Does a re-route plan exit for mission S^* that guarantees the timely deliveries (with given delivery time windows dX^*) in a given time horizon T and at acceptable battery's levels?*

The problem under consideration boils down to the search for a solution solutions taking into account split deliveries, multi-visit of customers, multi depots, resistance to the forecast weather and changing delivery time windows.

In order to illustrate the proposed problem let us consider a network from Fig. 1a) covering area 100 km^2, containing 3 depots $NB = \{N_1, N_2, N_3\}$ and 22 delivery points $ND = \{N_4, N_5, \ldots, N_{25}\}$. The goods are delivered by the fleet $\mathcal{U} = \{U_1, U_2, U_3\}$. Technical parameters of UAVs are collected in the table from Fig. 1c). The weight of individual orders z_λ are presented on the Fig. 1a) while the delivery time windows dX_λ are shown on the Fig. 2.

a)

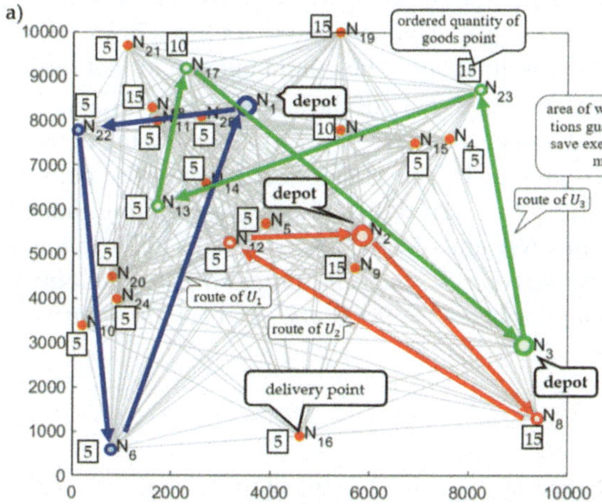

b)

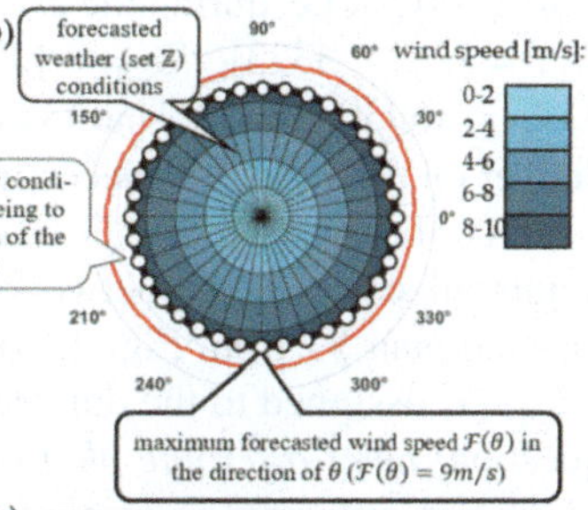

c)

technical parameters of UAVs	value	unit
payload capacity Q	25	kg
battery capacity CAP	7500	kJ
flight speed v	20	m/s
drag coefficient C_D	0.54	-
surface A of front UAV	1.2	m²
UAV width b	8.7	m

Fig. 1. Layout of the distribution network with UAVs routes highlighted

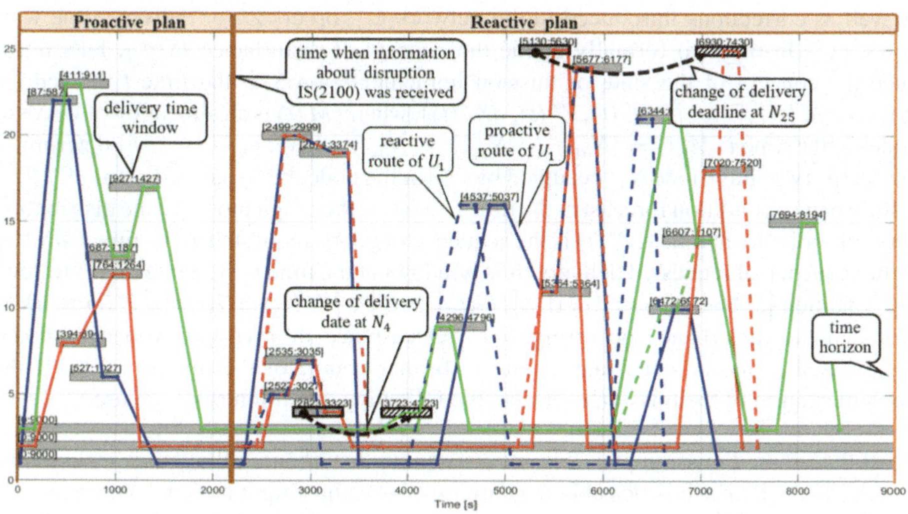

Fig. 2. Schedule of delivery missions for the fleet $\mathcal{U} = \{U_1, U_2, U_3\}$

With these assumptions, the UAVs mission S plan guaranteeing timely deliveries of the required amount of goods is sought. The example of admissible solution presented in Fig. 2 consists of 4 sub-missions: $S = ({}^{1}S, \ldots, {}^{4}S)$ following the wind speed constraints ≤ 9 m/s. An example illustrating the routes of UAVs fleet mission ${}^{1}S$: ${}^{1}\pi_1 = (\mathrm{N}_1, \mathrm{N}_{22}, \mathrm{N}_6, \mathrm{N}_1)$; ${}^{1}\pi_2 = (\mathrm{N}_2, N_8, N_{12}, N_2)$; ${}^{1}\pi_3 = (N_3, N_{23}, N_{13}, N_{17}, N_3)$; and the resistance function: $\Upsilon_{2,1}(\theta)$, is shown in Fig. 1. All goods are delivered within a time horizon not exceeding 2.5 h and deliveries are made under various weather conditions (set $\mathbb{F}$) illustrated in Fig. 1b). According to the forecast, the wind speed does not exceed $vw = 9\frac{m}{s}$ in direction $\theta = 0° - 360°$. Let us now consider the situation in which the delivery time windows on the delivery points N_4 and N_{25} changed at the time of $t^* = 2100$ [s], i.e. during the execution of the sub-mission ${}^{2}S$, the delivery time windows $dX_4 = [2823, 3323]$, $dX_{25} = [3723, 4223]$ have been changed to $dX_4^* = [7930, 8830]$, $dX_{25}^* = [6930, 7430]$. That means the mission cannot be continued due to inability to deliver on time. So, it is necessary to correct the route of the sub-mission ${}^{2}S$ being carried out, that poses a new Given a UAVs fleet $\mathcal{U} = \{U_1, U_2, U_3\}$ servicing the delivery points allocated in the network G from Fig. 1. The UAVs fleet realizes the delivery mission plan S^* from Fig. 2. At the time $t^* = 2100$ disturbance IS, resulting in dX_4^*, dX_{25}^*, i.e. assigned to the delivery points N_{18} and N_{19}, is observed. This leads to a new question: *Does a reroute plan exist for mission S^*, that guarantees the timely delivery of the ordered supplies in a given time horizon $T = 9000$ and at acceptable battery's level?*

3 CSP Formulation

The proposed reaction to the disturbance $IS(t^*)$ can be reduced to dynamic re-routing and rescheduling of previously adopted routes ${}^{l}\Pi$, schedules ${}^{l}Y$ and delivered goods ${}^{l}C$ stated in the basic proactive plan for mission by UAVs' fleet. It is a feasible adjustment of assumed ${}^{l}\Pi$, ${}^{l}Y$ and ${}^{l}C$ values to the changes in forecasted weather $\mathcal{F}^*(\theta)$, as well as corrections introduced to the network G^*, orders Z^* or delivery time windows dX^*. In order to formally define the concept of disturbance $IS(t^*)$, let's introduce the concept of the state of mission implementation S at the time t defined as: $IS(t) = (M(t), \mathcal{F}^*(\theta, t), {}^{*}G(t), Z^*(t), dX^*(t))$, where: $M(t)$ is an allocation of UAVs to nodes at the time t: $M(t) = (N_{a_1}, \ldots, N_{a_k}, \ldots, N_{a_K})$, where: $a_k \in \{1, .., n\}$ determines the (delivery points) node N_{a_k} occupied by U_k (or the node the U_k is headed to). $\mathcal{F}^*(\theta, t)$ is the weather condition forecast at the time t. ${}^{*}G(t)$ is the graph model of the distribution network structure at time t. $Z^*(t)$ is the sequence of goods requested at the time t. $dX^*(t)$ is the sequence of requested delivery time windows at the time t. The state $IS(t^*)$ following condition $[\mathcal{F}^*(\theta, t^*) \neq \mathcal{F}^*(\theta)] \vee [{}^{*}G(t^*) \neq G] \vee [Z^*(t^*) \neq Z] \vee [dX^*(t) \neq dX^*]$ is called the disturbance occurring at t^*. Occurrence of $IS(t^*)$ disturbance should be assessed in terms of its impact on the further course of the mission of S. If the implementation of the mission is at risk, the following condition rules are used:

1. If the adopted mission plan S is not resistant to disturbance $IS(t^*)$ then it should be checked whether it is possible to adapt (re-plan) adjusting it to new conditions.
2. If there are UAVs cannot continue to fly due to disturbance $IS(t^*)$, then they should be returned to the depot.

3. If the tasks of the UAVs returning to the depot cannot be taken over then it should be checked whether the reserve UAVs, can take over their responsibilities.
4. If the reserve UAVs are unable to take over the responsibilities returned to the depot then they activity should be suspended until the until the disturbance is resolved.

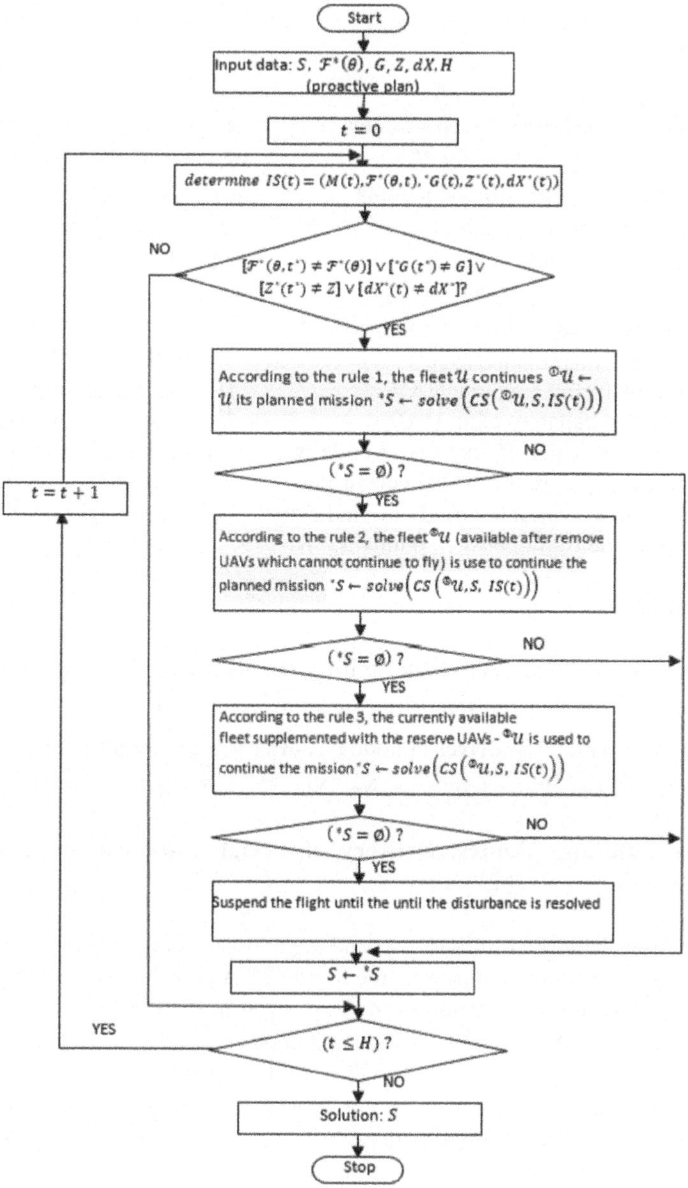

Fig. 3. Reactive mission planning method

The above rules have been used in the reactive planning method shown in Fig. 3.

During the implementation of the mission S, there is a continuous monitoring of the state of the $IS(t)$ (for $t \in \{0 \ldots H\}$). In practice, it comes down to solving constraints satisfaction problem $CS\left({}^{\circ}\mathcal{U}, S, IS(t)\right)$(where: ${}^{\circ}\mathcal{U}$– defines the fleet designated by rules 1–4) which mathematical formulation employs the following parameters, variables, sets, and constraints.

Parameters

${}^{l}G$	graph of a network for sub-mission ${}^{l}S$	z_λ	demand at node N_λ, $z_1 = 0$
dX_λ	delivery time window for N_λ	$d_{\beta,\lambda}$	distance between N_β, N_λ
w	time spent on take-off and landing	$t_{\beta,\lambda}$	travel time between N_β, N_λ
Q	maximum loading capacity	H	time horizon
$\Upsilon_{k,l}(\theta)$	weather resistance function	$\mathcal{F}(\theta)$	forecasted wind speed
$va_{\beta,\lambda}$	air speed between nodes N_β, N_λ	ep	the empty weight of a UAV
${}^{l}y_\lambda^k$	the time at which U_k arrives at node N_λ, before the disturbance $IS(t^*)$	C_D	the aerodynamic drag coefficient
A	the front-facing area of a UAV	D	an air density
CAP	the energy capacity of an UAV	b	the width of an UAV
${}^{l}y_\lambda^k$	the time at which U_k arrives at node N_λ, before the disturbance $IS(t^*)$ occurrence	$vg_{\beta,\lambda}$	ground speed between N_β, N_λ
${}^{l}x_{\beta,\lambda}^k$	the binary variable used to indicate whether the flight between points N_β, N_λ	${}^{l}c_\lambda^k$	the weight of freight delivered to N_λ by U_k,

Decision Variables

$\overline{{}^{l}x_{\beta,\lambda}^k}$ the binary variable used to indicate if U_k travels between nodes N_β, N_λ, after the disturbance $IS(t^*)$ occurrence (during sub-mission ${}^{l}S$)

$\overline{{}^{l}y_\lambda^k}$ the time at which U_k arrives at node N_λ, after the disturbance $IS(t^*)$ occurrence,

$\overline{{}^{l}\pi_k}$ the route of U_k, $\overline{{}^{l}\pi_k} = \left(N_{k_1}, \ldots, N_{k_i}, N_{k_{i+1}}, \ldots, N_{k_\mu}\right)$,

Constraints Limiting: Routes, Delivery of Freight and Energy Consumption Describing (Radzki et al. 2021):

- the relationship between routes (${}^{l}x_{\beta,\lambda}^k$) and the delivery schedule (variables ${}^{l}y_\lambda^k$) at the same time guaranteeing congestion-free flight of UAVs as well as delivery of that the ordered goods within time windows ($dx_\lambda^{up} \le {}^{l}y_\lambda^k + w \le dx_\lambda^{down}$),
- constraints limiting the amounts of delivered goods (c_i^k) along each UAV route (${}^{l}x_{\beta,\lambda}^k$) and ensuring that UAVs are not overloaded while delivery correct amounts of goods,
- constraints guaranteeing that the plan S is safe i.e. resistance functions $\Upsilon_{k,l}(\theta)$ for the fleet $\mathcal{U}$ exceed the value of the function $Z(\theta)$: $\Upsilon_{k,l}(\theta) \ge \mathcal{F}(\theta)$; $\forall\theta \in [0°, 360°)$. The constraints assume that UAVs battery consumption is specified by nonlinear function from the weight of freight, ground speed $vg_{\beta,\lambda}$ and wind speed vw.

Since the re-planning of the mission delivery plan S is the result of the disturbance $IS(t^*)$, hence the new set of sub-missions $\overline{{}^1S}, \ldots, \overline{{}^lS}, \ldots, \overline{{}^LS}$ guaranteeing timely delivery, are determined by solving the following problem (1):

$$CS\left({}^{\circ}\mathcal{U}, S, IS(t^*)\right) = \left((\mathcal{V}, \mathcal{D}), \mathcal{C}({}^{\circ}\mathcal{U}, S, IS(t^*))\right) \tag{1}$$

where: $\hat{\mathcal{V}} = \left\{\overline{{}^l\Pi}, \overline{{}^lY}, \overline{{}^lC} | l = 1 \ldots L\right\}$ – the set of decision variables: $\overline{{}^l\Pi}$ – the set of routes determining the schedule; $\overline{{}^lY}$ – schedule of the fleet ${}^{\circ}\mathcal{U}$guarantees timely service of delivery points in the case of disturbance $IS(t^*)$, and $\overline{{}^lC}$ – sequence of weights of delivered goods by the fleet ${}^{\circ}\mathcal{U}$; $\mathcal{D}$ – the finite set of decision variable domains: $\overline{{}^lx^k_{i,j}} \in$.

Solving CS problem, however, is very time-consuming, which results from the necessity to verify the inequality $\Upsilon_{k,l}(\theta) \geq \mathcal{F}(\theta)$ carried out for each increment $\theta \in [0°, 360°)$. In order to replace this inequality with an equivalent but less time consuming computationally, it was assumed that continuous $\Upsilon_{k,l}(\theta)$ function will be approximated by a discrete function represented by a finite set $\mathbb{Y}_{k,l} = \left\{\Upsilon_{k,l}(\theta_i) | i = 1 \ldots lq; \theta_i < \theta_{i+1}\right\}$, where lq is an arbitrarily taken number of samples.

Therefore, the following property holds: For any wind direction $\theta \in [0°, 360°)$ function of $\Upsilon_{k,l}(\theta)$ mission ${}^*S$ takes values no less than its discrete form $\mathbb{Y}^*_{k,l}(\theta)$: $\theta \in [0°, 360°)$, $\Upsilon_{k,l}(\theta) \geq \mathbb{Y}^*_{k,l}(\theta)$. This means that the function $\Upsilon_{k,l}(\theta)$ can be replaced by the function $\mathbb{Y}^*_{k,l}(\theta)$ and, in fact, the set of vertices in $\mathbb{Y}_{k,l}$ reducing the number of vertices as in the adopted polygon. Consequently, the considered inequality can be replaced by $\mathbb{Y}^*_{k,l}(\theta) \geq \mathcal{F}(\theta); \theta \in \left\{\theta_1, \ldots, \theta_{lq}\right\}$.

4 Computational Experiments

A Consider the network from Fig. 1, in which the UAVs fleet $\mathcal{U} = \{U_1, U_2, U_3\}$ service delivery points N_4–N_{25}. Let us assume that at the time $t^* = 2100$ [s] (see Fig. 2 – proactive plan period) an information about the delivery time windows change (for N_4 and N_{25}) was received. This disruption implies the necessity of the mission re planning (see Fig. 2 – rescheduling period). For this purpose the CS problem (1) has been used. Its implementation in the constraint programming environment IBM ILOG (Intel Core i7-M4800MQ 2.7 GHz, 32 GB RAM) has shown that the solution time for problems of a size considered does not exceed 35 s. The Fig. 2 (dashed lines of the reactive plan period), shows the plan of mission adapted to the disturbance. The rule 1 was used in assigning mission ${}^*S$, i.e., If the adopted mission plan S is not resistant to disturbance $IS(t^*)$ then it should be checked whether it is possible to adapt (re-plan) adjusting it to new conditions. In the case under consideration, none of the UAVs were turned back. The received mission is resistant to the given weather conditions (9 m/s) and guarantees timely delivery of the expected goods (also for N_4 and N_{25}).

In order to assess the scalability of proposed approach in terms of the possibility of its usage in an online mode (i.e., to solve the problem in <600 s) in decision support systems the series of quantitative experiments have been carried out. Table 1 contains the results of experiments that are conducted for $\mathcal{F}(\theta) = 9, 10, 11$ m/s.

Table 1. Results of the experiments conducted

n	K	NS	$\mathcal{F}(\theta) = 11\frac{m}{s}$ $\forall\Theta \in [0°, 360°)$		NS	$\mathcal{F}(\theta) = 12\frac{m}{s}$ $\forall\Theta \in [0°, 360°)$		NS	$\mathcal{F}(\theta) = 13\frac{m}{s}$ $\forall\Theta \in [0°, 360°)$	
			TCR [s]	TC [s]		TCR [s]	TC [s]		TCR [s]	TC [s]
40	2	7	18.76	1373.41	7	35.71	2534.7	7	27.69	2091.37
	3	6	154.16	t > 600	6	147.46	t > 600	7	541.51	t > 600
	4	5	408.95	t > 600	5	442.02	t > 600	5	698.24	t > 600
60	2	12	309.44	t > 600	12	66.42	t > 600	12	67.94	t > 600
	3	11	334.01	t > 600	×	t > 600	t > 600	9	236.8	t > 600
	4	×	t > 600	t > 600	×	t > 600	t > 600	×	t > 600	t > 600
70	2	13	411.58	t > 600	13	139.59	t > 600	14	153.53	t > 600
	3	12	756.58	t > 600	×	t > 600	t > 600	×	t > 600	t > 600
	4	×	t > 600	t > 600	×	t > 600	t > 600	×	t > 600	t > 600
80	2	16	360.98	t > 600	15	175.65	t > 600	16	502.24	t > 600
	3	×	t > 600	t > 600	×	t > 600	t > 600	×	t > 600	t > 600
	4	×	t > 600	t > 600	×	t > 600	t > 600	×	t > 600	t > 600

n – number of delivery points; K – size of fleet; *TRC*– time of computation with relaxation; *TC*– time of computation without relaxation; × – no solution allowed in time $t < 600$ s; NS - number of sub-missions.

The experiments are carried out for the network of n randomly designated collection of delivery points (on the area 10 km × 10 km) and a fleet consisting of K UAVs with the technical parameters, as shown in Table form Fig. 1c). For each of the considered variants of the network, a proactive mission plan has been set out to guarantee the delivery in the time horizon $T = 9000$ [s]. It was assumed that at the moment $t^* = 2000$ [s] there is a change in delivery time window. Results (i.e. the times with relaxation *TCR* and without relaxation *TC* determining reactive mission plans design) are presented in the Table 1. The conducted experiments show that for the distribution network of a size up to the 80 delivery points, and serviced by 2 UAVs the fleet mission plans can be effectively refined, and successfully carried out taking into account specific types of disturbances.

5 Conclusions

The class of considered cases include situations related to planning UAVs routes under conditions of weather changes exceeding previously forecasted ones and/or changes to previously agreed terms of order fulfillment. Consequently, the need to react in such situations enforces the establishment of condition-action rules that allow the designation of appropriate possible end-to-end routes enabling emergency safe completion of the mission or its continuation in a modified version. The main advantage of the proposed approach follows from the open structure of the proposed model, that allows for taking into account several other variables and restrictions [8]. Computational results show that, the permissible size of the distribution network (80 nodes and 2 UAVs) for which considered reactive planning aimed at routing design of delivery a fleet mission can be undertaken, makes the proposed approach suitable for online applications. In other words obtained computational results confirm the achievement of the assumed goals particularly the development of declarative model and sufficient conditions allowing reduction of searching space for VRP.

In our future research, we want to take into account the uncertain nature of the real world variables which are not deterministic. Thus, a fuzzy approach could be applied to the UAVs' mission planning problem while allowing for more accurate estimation of the timeliness of deliveries subject to constraints imposed by appearance and range varying time windows.

Acknowledgements. This research was co-founded by the National Science Centre, Poland, grant number 2019/03/X/ST6/00429.

References

1. Crevier, B., Cordeau, J.-F., Laporte, G.: The multi-depot vehicle routing problem with inter-depot routes. Eur. J. Oper. Res. **176**(2), 756–773 (2007). https://doi.org/10.1016/j.ejor.2005.08.015
2. Dantzig, G., Ramser, J.: The truck dispatching problem. Manag. Sci. **6**(1), 80–91 (1959)
3. Dorling, K., Heinrichs, J., Messier, G.G., Magierowski, S.: Vehicle routing problems for drone delivery. IEEE Trans. Syst. Man Cybern. Syst. **47**, 70–85 (2017)
4. Gulczynski, D., Golden, B., Wasil, E.: The split delivery vehicle routing problem with minimum delivery amounts. Transp. Res. Part E: Logist. Transp. Rev. **46**(5), 612–626 (2010). ISSN 1366-5545 https://doi.org/10.1016/j.tre.2009.12.007
5. Molina, J.C., Salmeron, J.L., Eguia, I., Racero, J.: The heterogeneous vehicle routing problem with time windows and a limited number of resources. Eng. Appl. Artif. Intell. **94**, 103745 (2020). https://doi.org/10.1016/j.engappai.2020.103745
6. Patella, S.M., Grazieschi, G., Gatta, V., Marcucci, E., Carrese, S.: The adoption of green vehicles in last mile logistics: a systematic review. Sustainability, **13**(6) (2021)
7. Poikonen, S., Go, B.: Multi-visit drone routing problem MVVRP. Comput. Oper. Res. **113**, 104802 (2020). ISSN 0305-0548. https://doi.org/10.1016/j.cor.2019.104802
8. Radzki, G., Nielsen, I., Golińska-Dawson, P., Bocewicz, G., Banaszak, Z.: Reactive UAV fleet's mission planning in highly dynamic and unpredictable environments. Sustainability **13**, 5228 (2021). https://doi.org/10.3390/su13095228
9. Ralphs, T.K., Kopman, L., Pulleybank, W.R., Trotter, L.E.: On the capacitated vehicle routing problem. Math. Program. **94**, 343–359 (2003). https://doi.org/10.1007/s10107-002-0323-0
10. Sung, I., Nielsen, P.: Zoning a service area of unmanned aerial vehicles for package delivery services. J. Intell. Robot. Syst. **97**, 719–731 (2020). https://doi.org/10.1007/s10846-019-01045-7
11. Thibbotuwawa, A., Nielsen, P., Banaszak, Z., Bocewicz, G.: Energy consumption in unmanned aerial vehicles: a review of energy consumption models and their relation to the UAV routing. In: Świątek, J., Borzemski, L., Wilimowska, Z. (eds.) Information Systems Architecture and Technology. Advances in Intelligent Systems and Computing, vol. 853, pp. 173–184. Springer, Cham (2018). https://doi.org/10.1007/978-3-319-99996-8_16

12. Troudi, A., Addouche, S.-A., Dellagi, S., Mhamedi, A.E.: Sizing of the drone delivery fleet considering energy autonomy. Sustainability **10**, 3344 (2018)
13. Xiaobing, G., Yan, W., Shuhai, L., Ben, N.: Vehicle routing problem with time windows and simultaneous delivery and pick-up service based on MCPSO. In: Selected Papers from the 7th International Conference on Computational Intelligence and Security (CIS'2011) (2012). Article ID 104279. https://doi.org/10.1155/2012/104279
14. Chodorek, A., Chodorek, R.R., Sitek, P.: UAV-based and WebRTC-based open universal framework to monitor urban and industrial areas. Sensors **21**, 4061 (2021). https://doi.org/10.3390/s21124061

Early-Stage Faults Detection Using Harmony Search Algorithm and STFT-Based Spectral Kurtosis

Muhammad Ahsan(✉) and Dariusz Bismor

Silesian University of Technology, 44-100 Gliwice, Poland
ahsanmuhammad@aol.com, dariusz.bismor@polsl.pl

Abstract. The predictive maintenance of rotating machines helps to prevent economic loss and personal damage. Vibrational data of bearing is used to measure the early-stage faults but sometimes this vibrational data consists of many unwanted signals that need to be removed before applying the signal processing techniques. This paper proposes an algorithm to design a dynamic filter that can adjust its bandwidth and central frequency according to different faults that occurs in the rotating bearing. The harmony search algorithm is used to optimize the band-pass filter parameters according to the faults and the short-time Fourier transform-based Spectral Kurtosis helps measure the strength of fault frequencies. The filter parameters are selected that correspond to the maximum value of the short-time Fourier transform-based Spectral Kurtosis. The simulation results verify the performance of the proposed algorithm.

Keywords: Predictive maintenance · Early-stage faults · Signal processing · Harmony search algorithm · Short-time Fourier transform · Spectral Kurtosis

1 Introduction

Bearings are the backbone of every rotating machine, and the health monitoring of bearing is very important to prevent economic loss and personal damage. Aging is the natural phenomena, but early-stage faults diagnose can prevent future damage by executing the required maintenance actions, thus attempting to re-establish the health of a bearing. Many early-stage faults diagnostic methods and algorithms are available in the literature [1, 2], and these methods are used to acquire the vibrational data, to detect faults, and estimate the remaining life of the bearing. Different signal processing techniques used include frequency response analysis, resonance demodulation technique, envelope analysis [3–7]. A summary of the data acquisition, processing, diagnosis and prognostics algorithms for predictive maintenance is presented in [8].

Signal processing approaches applied to the vibration signals for identification of faults in rotating element bearing are studied in [9]. Simulated and experimental vibration signals of rolling element bearing are used to study the effect of different faults in [10]. Inner race defect, outer race defect, ball defect and a mixture of the three under different load and speed combinations are examined in [11]. Inner race and outer race faults

R. Szewczyk et al. (Eds.): AUTOMATION 2022, AISC 1427, pp. 75–84, 2022.
https://doi.org/10.1007/978-3-031-03502-9_8

are measured using the experimental benchmark for rolling element bearing in [12]. Applications of vibration analysis for rolling element bearing are provided in [5, 13, 14].

The vibration signal acquired from the rotating element bearing is composed of actual rotor frequency, other frequencies caused by the other parts of the machine, noise, and the fault frequencies. The fault frequencies could be for single or multiple faults. Different sorts of faults could arise in the bearing, such as inner race fault, outer race fault, rolling element fault, etc. A bandpass filter can be used to filter out the unwanted noise and then signal processing techniques can be applied to detect the faults. Different methods are proposed to improve the bandpass filter [15–18]. The bandpass filter should have dynamic central frequency and bandwidth because of the different faults having unknown bandwidth and central frequency.

This paper proposes an alternative algorithm to design the bandpass filter that have the dynamic bandwidth and central frequency. The bandwidth and central frequency are adjusted according to the fault spectrum. The harmony search (HS) algorithm is employed to optimize the bandwidth and central frequency of the bandpass filter and spectral kurtosis (SK) estimated based on the short time Fourier transform (STFT) is used to measure the strength of the faults. HS algorithm is an efficient algorithm that can solve the multi-objective optimization problem. The parameters of the bandpass filter are obtained such that the spectrum of filtered signal have the highest value of the STFT based SK. Finally, the proposed algorithm is verified by extracting the early-stage faults in the bearing.

2 Proposed Algorithm

The overall proposed algorithm is depicted in Fig. 1. The fitness function helps to construct the best dynamic filter and then vibration signal is filtered out. The fault frequency can be visualized from the filtered signal. In the following sections, representation of the vibration signal acquired from the faulty bearing is reviewed. Before going straight to the bandpass filter designing procedure, the building blocks such as the SK and the HS algorithm are reviewed.

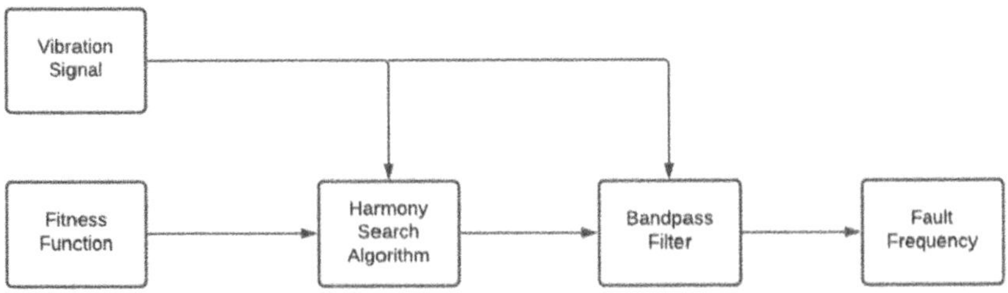

Fig. 1. Block diagram of the proposed algorithm.

2.1 Vibrational Signal

The vibrational signal measured from the rotary machine is composed of shaft rotating frequency x_v, fault frequency x_f, and noise signals x_n. Mathematically, it can be represented as follow:

$$x = x_v + x_f + x_n, \tag{1}$$

Noise is always present in the measured vibration signal, and it could be generated because of different reasons, such as subharmonic frequency, onefold or twofold frequencies, vibration of other parts of the machines or background noise etc. x_v is the rotary vibration signal of the shaft and it consist of fundamental rotating shaft frequency and its multiples as follows:

$$x_v = \alpha_1 \cos(2\pi f_1 t) + \alpha_2 \cos(2\pi f_2 t) + \alpha_3 \cos(2\pi f_3 t), \tag{2}$$

where f_1 is the shaft rotating frequency and f_2 and f_3 are the multiples of f_1. $\alpha_1, \alpha_2, \alpha_3$ are the magnitudes of the signals.

The fault frequency x_f is a periodic signal and it could be outer race fault, inner race fault, fundamental train fault, or ball spin fault. All these fault frequencies can be calculated using the bearing specification.

The health of the machine or bearing can be divided into three conditions, i.e., healthy, early-stage fault, damage. In the healthy condition, the bearing works well and there is no need of any maintenance. The second stage is early-stage fault, and in this stage, bearing may not cause much noise and rotatory machine may look healthy but this stage needs most attention because this stage has very short life and after that third stage arrives. In the third stage, the bearing is completely damaged, and the rotating machine cannot work. To prevent from damage, it is very important to predict early-stage fault.

2.2 Spectral Kurtosis

Kurtosis is a statical feature that uses probability density function (PDF) to determine the strength of transient impulses in signal. The value of the kurtosis is directly proportional to transient impulses. It can be defined as follows:

$$K = \frac{E\left(|\tilde{x}_i|^4\right)}{\left(E\left(|\tilde{x}_i|^2\right)\right)^2}, \tag{3}$$

where E(.) is the expectation operator and $\tilde{x}_i$ is the analytic signal that can be derived from the time-domain zero-mean filtered signal x_i as follows:

$$\tilde{x}_i = x_i + j.H(x_i), \tag{4}$$

where H(.) is the Hilbert Transform function.

When faults take place in the bearing, transient impulses generated periodically increase the value of the kurtosis. Using this phenomena, transient faults can be determined and bearing faults can be diagnosed. However, vibration signals of a bearing contain many other components – as described before – that can also increase the value of the kurtosis. Hence, kurtosis fails to distinguish between the noise and the fault impulses, and it is difficult to characterize the resonant band of the filtered signal.

To overcome this issue, a new term called the SK was introduced. The SK is computed over each frequency band and discovers the hidden transients and the frequency band in which these transients occur. The SK is calculated using the following equation:

$$K_X(f) = \frac{\lim_{N\to\infty} \frac{1}{N} \sum_N |H(n, f)|^4}{\left(\lim_{N\to\infty} \frac{1}{N} \sum_N |H(n, f)|^2\right)^2} - 2, \tag{5}$$

where H(n, f) is the complex envelope of x(n) and therefore 2 is subtracted. In the proposed algorithm, the complex envelope H(n, f) in Eq. (5) is replaced with the STFT. The STFT of the vibration signal x(n) is given by the following equation

$$STFT(n, f) = \sum_{k=1}^{N} x(k)w(n - k)e^{\frac{j2\pi fk}{N}}, \tag{6}$$

where w(n − k) is the short-time window with the shift k. Therefore, the fitness function used in the proposed algorithm is the STFT-based SK and the goal is to find a dynamic bandpass filter that maximize the value of fitness function. Consequently, the fault frequency spectrum could be determined from the filtered signal. The overall proposed STFT based SK is given by the following equation

$$K_X(f) = \frac{\lim_{N\to\infty} \frac{1}{N} \sum_N |STFT(n, f)|^4}{\left(\lim_{N\to\infty} \frac{1}{N} \sum_N |STFT(n, f)|^2\right)^2} - 2, \tag{7}$$

2.3 Harmany Search Algorithm

Harmony Search (HS) algorithm is a population-based metaheuristic optimization algorithm, and the object of the HS algorithm is to optimize the multi-objective fitness function. The fitness function is given by:

$$\begin{aligned} &\max F(s) \text{ or } \min F(s) \\ &\text{subjected to } g(s) \le 0 \text{ and } h(s) = 0, \end{aligned} \tag{8}$$

where s is the set of solution candidates and the goal is to find the best solution candidate that optimize Eq. (8).

Harmony memory size (HMS), harmony consideration rate (HMCR), and pitch adjustment rate (PAR) are the three main parameters that define the way the algorithm finds the optimization solution. The range of the values of these three components are frequently chosen as $10 \le HMS \le 50$, $0.7 \le HMCR \le 0.95$, and $0.2 \le PAR \le 0.5$.

The HS algorithm is explained in more detail in [19, 20]. The algorithm consists of five main steps. In the first step, fitness function is defined, and suitable values are assigned to the three main components (HMS, HMCR, PAR). In the second step, a random harmony memory (HM) of order HMS $\times$ N is generated as follows:

$$HM = \begin{bmatrix} s_1^1 & s_2^1 & \cdots & s_N^1 \\ s_1^2 & s_2^2 & \cdots & s_N^2 \\ \vdots & \vdots & \ddots & \vdots \\ s_1^{HMS} & s_2^{HMS} & \cdots & s_N^{HMS} \end{bmatrix}, \quad (9)$$

where $s_i^j, i \in HMS, j \in N$ is the element of the solution candidates $s^j = \left\{s_1^j, s_2^j, s_3^j \ldots s_N^j\right\}$, N is the number of elements in each solution candidate. Each element s_i^j of the solution candidate has an upper bound s_{ub} and a lower bound s_{lb}

$$s_{lb} \le s_i^j \le s_{ub}, \quad (10)$$

The fitness value is calculated for each solution candidate s^j and the worst fitness value is determined.

In the third step, a new harmony solution s^{new} is generated and its fitness value is calculated. In the literature, different techniques are reported to generate the new harmony solution s^{new}. In general:

$$s^{new} = \begin{cases} \text{Random selection from HM with probability HMCR} \\ \text{New solution candidate with probability } (1 - \text{HMCR}) \end{cases}, \quad (11)$$

And then pitch adjustment is applied to s^{new}.

$$s^{new} = \begin{cases} s^{new} \pm bw \text{ with probability PAR} \\ s^{new} \text{with probability } (1-\text{PAR}) \end{cases}, \quad (12)$$

where $bw = r \times (s_{ub} - s_{lb})$ and r is a random number between zero and one.

In the fourth step, the fitness values of the s^{new} are compared with the worst fitness value. If the fitness value of s^{new} is better than the worst fitness value, then the new solution is updated in the HM and the worst solution candidate is discarded, otherwise the new solution is discarded.

In the fifth step, a loop is applied to the steps three and four. Different criteria are used to terminate the loop, such as epochs reached its maximum value or fitness function attains a specific value. At the end, the best solution candidate is determined. This best solution candidate fulfills the optimization problem.

2.4 Design of Dynamic Bandpass Filter

For the proposed algorithm, the STFT-based SK is the fitness function. The goal is to find the optimized parameters for the dynamic bandpass filter that correspond the maximum value of the STFT based SK. It can be written as

$$\max K_x(f) = \max\left(\frac{\lim\limits_{N\to\infty}\frac{1}{N}\sum_N |STFT(n,f)|^4}{\left(\lim\limits_{N\to\infty}\frac{1}{N}\sum_N |STFT(n,f)|^2\right)^2} - 2\right) \tag{13}$$

The central frequency, the bandwidth, and the order of the bandpass filer is optimized using the HS algorithm described in the above section. This is a multi-objective optimization problem, and the HS algorithm is best fit for this sort of optimization problems. More than one new solutions are generated in each epoch to save the computational time. The overall HS algorithm is described in the Fig. 2. This modified HS algorithm is designed to optimize the bandpass filter according to the fault frequency.

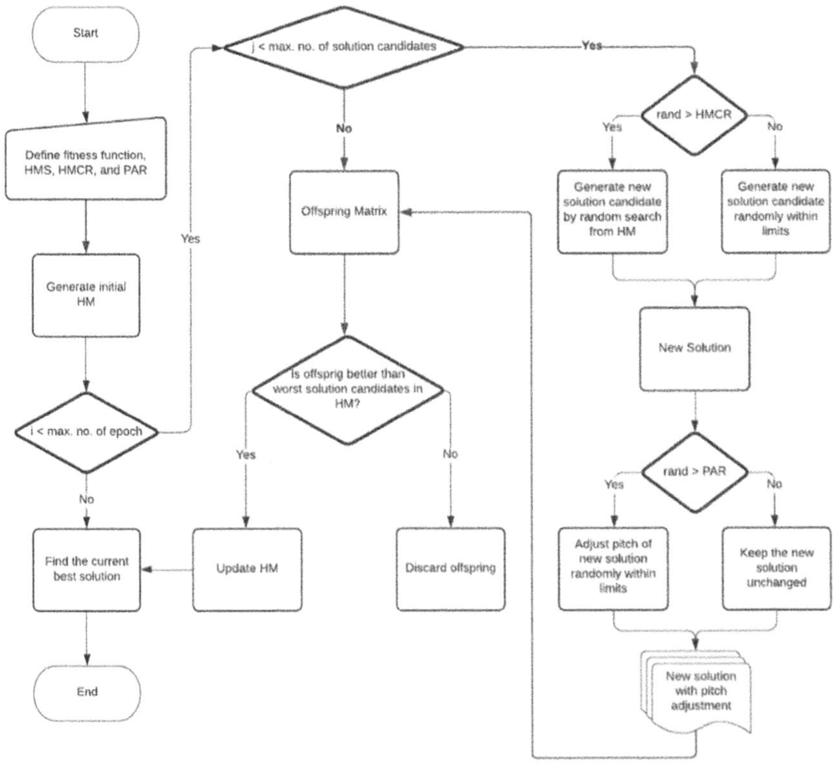

Fig. 2. Block diagram for the proposed HS algorithm to optimize the dynamic bandpass filter.

3 Simulation

A vibration signal was generated in the MATLAB using the Eq. (1) with shaft rotating frequency $f_1 = 22.5$ and its multiples $f_2 = \frac{13}{35}f_1$ and $f_3 = 13f_1$. The magnitudes were considered as $\alpha_1 = 0.4$, $\alpha_2 = 0.2$, and $\alpha_3 = 0.9$, respectively. The inner race fault was added to the vibration signal and was generated using the following formula:

$$BPFI = \frac{1}{2}nf_1\left(1 + \frac{d}{D}\cos\theta\right) \tag{14}$$

where $n = 8$ is the number of rolling element bearings, $d = 0.02$ is the diameter of the rolling element bearings, $D = 0.12$ is the pitch diameter of the bearing, and $\theta = 15$ is the bearing contact angle in degree. The proposed algorithm can detect the different sort of fault frequencies such as inner race fault frequency, outer race fault frequency, fundamental train fault frequency, and ball spin fault frequency.

The vibration signal with inner race fault was considered to check the validity of the proposed algorithm. The faulty signal is depicted in Fig. 3 and its spectral plot can be seen in Fig. 4. From Fig. 4, it is not possible to conclude about the fault.

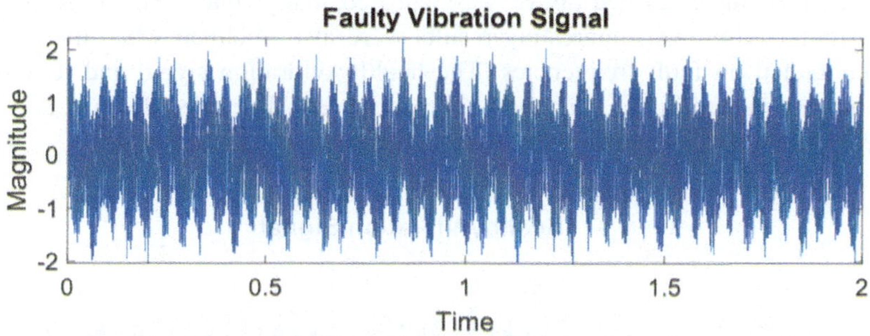

Fig. 3. Vibration signal with inner race fault.

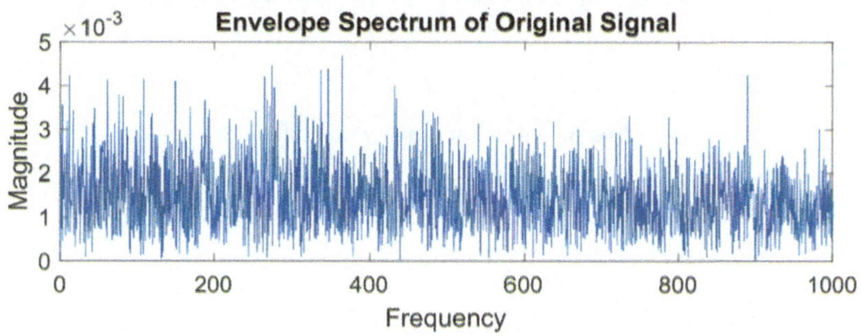

Fig. 4. Envelope spectrum of faulty signal shows no fault.

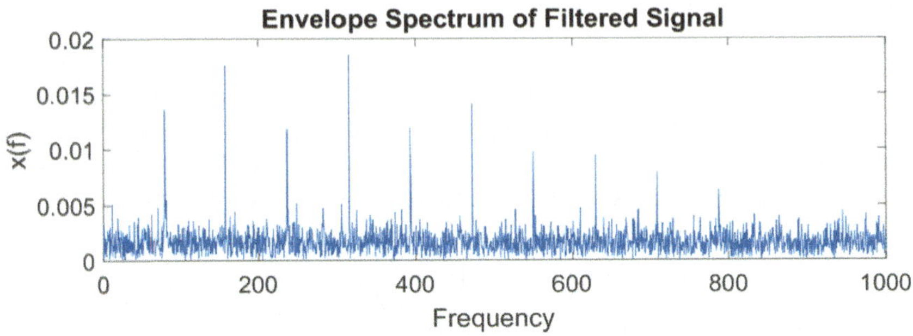

Fig. 5. Envelope spectrum of filtered signal.

The proposed HS algorithm was deployed with the following parameters: HMS = 20, HMCR = 0.82, PAR = 0.26. The algorithm performed 50 iterations, after which the fitness function attained the value 1.79 (the starting value was 1.12). As a result, the bandpass filter was designed, with the center frequency 4393.8 kHz and the width 6963 kHz. The faulty signal was filtered using the designed filter, and the resulting spectrum is showed in Fig. 5. The fault frequencies are now clearly visible.

The experiment described above was repeated many times and the results were always similar. Moreover, interesting results were also obtained when the simulated signal contained no fault frequencies. The healthy vibration signal and its envelope spectrum are depicted in the Figs. 6 and 7, respectively. The filtered signal also shows that there is no fault that can be seen in the envelope spectrum of Fig. 8.

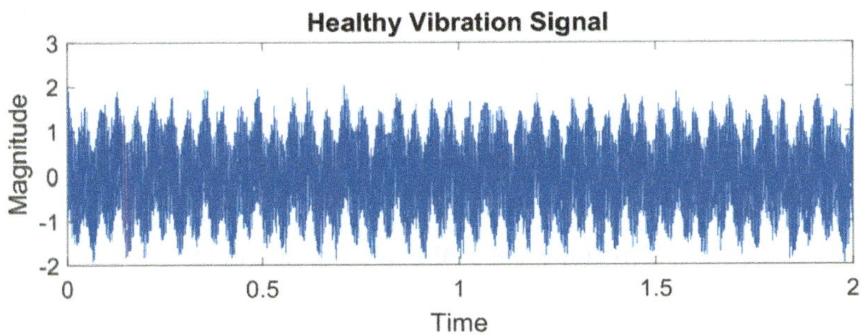

Fig. 6. Vibration signal with no fault.

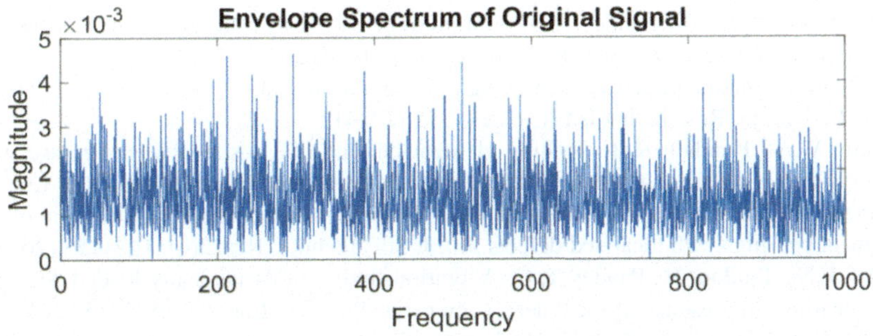

Fig. 7. Envelope spectrum of healthy signal.

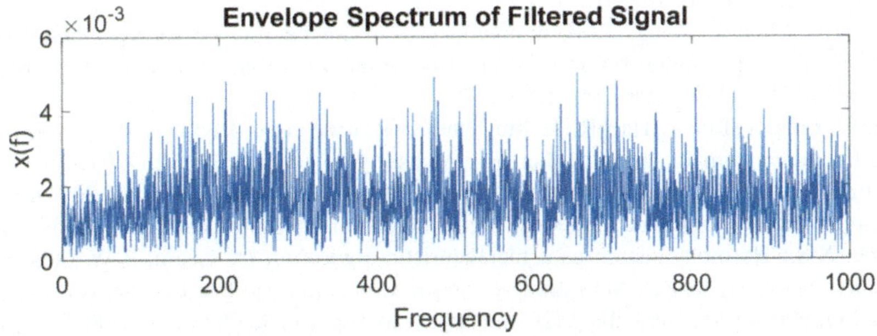

Fig. 8. Envelope spectrum of filtered signal.

References

1. Randall, R.B., Antoni, J.: Rolling element bearing diagnostics—A tutorial. Mech. Syst. Signal Process. **25**(2), 485–520 (2011)
2. Wang, W.: Early detection of gear tooth cracking using the resonance demodulation technique. Mech. Syst. Signal Process. **15**(5), 887–903 (2001)
3. Zoubek, H., Villwock, S., Pacas, M.: Frequency response analysis for rolling-bearing damage diagnosis. IEEE Trans. Industr. Electron. **55**(12), 4270–4276 (2008)
4. Rai, A., Upadhyay, S.H.: A review on signal processing techniques utilized in the fault diagnosis of rolling element bearings. Tribol. Int. **96**, 289–306 (2016)
5. Cong, F., Chen, J., Dong, G., Pecht, M.: Vibration model of rolling element bearings in a rotor-bearing system for fault diagnosis. J. Sound Vib. **332**(8), 2081–2097 (2013)
6. Omar, D.M., Rantatalo, M., Aidanpää, J.-O., Kumar, U.: Vibration signal analysis for gear fault diagnosis with various crack progression scenarios. Mech. Syst. Signal Process. **41**(2), 176–195 (2013)
7. Kan, M.S., Tan, A.C.C., Mathew, J.: A review on prognostic techniques for non-stationary and non-linear rotating systems. Mech. Syst. Signal Process. **63**, 1–20 (2015)
8. Jardine, A.K., Lin, D., Banjevic, D.: A review on machinery diagnostics and prognostics implementing condition-based maintenance. Mech. Syst. Signal Process. **20**(7), 1483–1510 (2006)
9. Tandon, N., Choudhury, A.: A review of vibration and acoustic measurement methods for the detection of defects in rolling element bearings. Tribol. Int. **32**(8), 469–480 (1999)

10. Patel, V.N., Tandon, N., Pandey, R.K.: Vibrations generated by rolling element bearings having multiple local defects on races. Procedia Technol. **14**, 312–319 (2014)
11. Saruhan, H., Sandemir, S., Çiçek, A., Uygur, I.: Vibration analysis of rolling element bearings defects. J. Appl. Res. Technol. **12**(3), 284–395 (2014)
12. Smith, W.A., Randall, R.B.: Rolling element bearing diagnostics using the case western reserve university data: a benchmark study. Mech. Syst. Signal Process. **65**, 100–131 (2015)
13. Dube, A.V., Dhamande, L.S., Kulkarni, P.G.: Vibration based condition assessment of rolling element bearings with localized defects. Int. J. Sci. Technol. Res. **2**(4), 149–155 (2013)
14. Patel, V.N., Tandon, N., Pandey, R.K.: Vibration studies of dynamically loaded deep groove ball bearings in presence of local defects on races. Procedia Eng. **64**, 1582–1591 (2013)
15. Chen, B.Q., Zhang, Z.S., Zi, Y.Y., He, Z., Sun, C.: Detecting of transient vibration signatures using an improved fast spatial–spectral ensemble kurtosis Kurtogram and its applications to mechanical signature analysis of short duration data from rotating machinery. Mech. Syst. Signal Process. **40**, 1–37 (2013)
16. Wang, H., Chen, J., Dong, G.: Fault diagnosis of rolling bearing's early weak fault based on minimum entropy de-convolution and fast Kurtogram algorithm. Proc. Inst. Mech. Eng. C J. Mech. Eng. Sci. **229**, 2890–2907 (2015)
17. Gu, X., Yang, S., Liu, Y., Hao, R.: Rolling element bearing faults diagnosis based on kurtogram and frequency domain correlated kurtosis. Measur. Sci. Technol. **27**, 125019 (2016)
18. Wang, Y., Tse, P.W., Tang, B., Qin, Y., Deng, L., Huang, T.: Kurtogram manifold learning and its application to rolling bearing weak signal detection. Measurement **127**, 533–545 (2018)
19. Yang, X.S.: Harmony search as a metaheuristic algorithm. In: Geem, Z.W. (eds) Music-Inspired Harmony Search Algorithm. Studies in Computational Intelligence, vol. 191, pp. 1–14. Springer, Berlin, Heidelberg (2009). https://doi.org/10.1007/978-3-642-00185-7_1
20. Zhang, T., Geem, Z.W.: Review of harmony search with respect to algorithm structure. Swarm Evol. Comput. **48**, 31–43 (2019)

Comparison of Nonlinear and Linear Models' Behaviour Regarding Rocket Stability

Dariusz Miedziński(✉) and Robert Głębocki

Institute of Aeronautics and Applied Mechanics, Warsaw University of Technology, Nowowiejska 24, 00-665 Warsaw, Poland
{dariusz.miedzinski2.dokt,robert.glebocki}@pw.edu.pl

Abstract. The article presents the comparison of nonlinear and linear models' responses regarding the determination of rocket static and dynamic stability. A full nonlinear model was shown along with the derivation of the linearized model utilizing the small perturbation theory. Criteria for determination of static and dynamic stability were also presented. A simulational model was created, and the results of its validation based on the flight data of the FOK rocket, as well as plots and tables containing the comparison of both models were provided. The results show that the linear simplification of the full nonlinear model is sufficient in order to determine rocket stability during all phases of the flight.

Keywords: Modelling · Simulation · Rocket · Linearization

1 Introduction

Modeling and simulation of flying vehicles is an important aspect of the design process [1,2]. Flights of rockets are usually more dynamic than flights of other vehicles. Rockets fly with higher velocities, experience higher values of longitudinal and lateral accelerations, and therefore usually require a highly nonlinear dynamic model to correctly predict their behaviour during the flight [3]. Rocket flight simulation is extensively used when designing guidance and control methods and identifying the accuracy of impact points [4,5]. However, nonlinear models are complicated and computationally complex, which results in the use of linearized models [6]. Many methods of linearization process exist, and some of them are presented in [7,8]. Linear models are preferable in use, due to the existence of many widely used methods for determining stability as well as methods for tuning various regulators using linear models. The classical approach to linearization, however, with a steady, rectilinear flight, as used to examine aircraft flight, is usually not accurate enough when applied to rockets. In this work, an

D. Miedziński and R. Głębocki—These authors contributed equally to this work.

R. Szewczyk et al. (Eds.): AUTOMATION 2022, AISC 1427, pp. 85–95, 2022.
https://doi.org/10.1007/978-3-031-03502-9_9

approach for the linearization of a nonlinear model using the small perturbation theory will be presented along with the criteria for static and dynamic stability calculations. The created simulation model, validated with the use of flight data will be presented, along with the results of the performed analyses.

2 Models

2.1 Nonlinear Model

The nonlinear rocket model was based on the linear and angular momentum change theorems given in (1) and (2) supplemented with the kinematic equations (3) and (4) [9].

$$
\begin{aligned}
m\dot{\mathbf{V}} + m\boldsymbol{\Omega} \times \mathbf{V} &= m\mathbf{C}_N^F\mathbf{g} + \mathbf{F_A} + \mathbf{F_T} && (1)\\
\mathbf{I}\dot{\boldsymbol{\Omega}} + \boldsymbol{\Omega} \times \mathbf{I}\boldsymbol{\Omega} &= \mathbf{M_A} + \mathbf{r_{ref}} \times \mathbf{F_A} && (2)\\
\dot{\mathbf{R}} &= (\mathbf{C}_N^F)^T\mathbf{V} && (3)\\
\begin{bmatrix}\dot{\Phi}\\ \dot{\Theta}\\ \dot{\Psi}\end{bmatrix} &= \begin{bmatrix}1 & \sin\Phi\tan\Theta & \cos\Phi\tan\Theta\\ 0 & \cos\Phi & -\sin\Phi\\ 0 & \frac{\sin\Phi}{\cos\Theta} & \frac{\cos\Phi}{\cos\Theta}\end{bmatrix}\omega && (4)
\end{aligned}
$$

The equations describe the movement of the rigid body with 6 degrees of freedom and are solved in body coordinate system (F), with its centre coincident with the centre of mass, with respect to navigation coordinate system (N), presented in Fig. 1, with the flat, non-rotating Earth assumptions. The state vector is comprised of linear velocity vector $\mathbf{V} = \begin{bmatrix}U\ V\ W\end{bmatrix}^T$, angular velocity vector $\boldsymbol{\Omega} = \begin{bmatrix}P\ Q\ R\end{bmatrix}^T$, position vector $\mathbf{R} = \begin{bmatrix}X\ Y\ Z\end{bmatrix}^T$ and orientation angles of roll Φ, pitch Θ and yaw Ψ. The transformation matrix from navigation to body coordinate system $\mathbf{C}_N^F$ is given in [9]. m is the rocket's mass, $\mathbf{I}$ is the moment of inertia matrix and $\mathbf{r_{ref}}$ vector describes the location of the aerodynamic moment reference point with respect to the instantaneous mass centre. Aerodynamic force and moment vectors are given by $\mathbf{F_A}$ and $\mathbf{M_A}$ respectively, $\mathbf{g}$ is the gravity vector and $\mathbf{F_T}$ is the thrust vector, collinear with the rocket's longitudinal axis. Aerodynamic force and moment vectors are given by (5) and (6).

$$
\mathbf{F}_A = qS_{ref}\mathbf{C}_A^F\begin{bmatrix}-C_D\\ 0\\ -C_L\end{bmatrix} \tag{5}
$$

$$
\mathbf{M}_A = qS_{ref}d_{ref}\begin{bmatrix}C_l\\ C_m\\ C_n\end{bmatrix} \tag{6}
$$

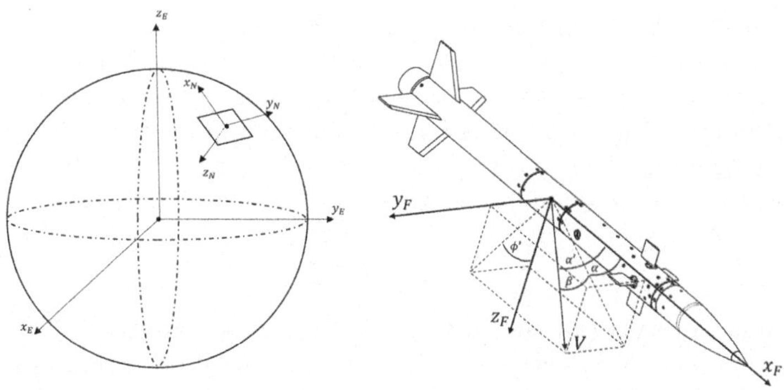

Fig. 1. Definition of navigation (left) and body (right) coordinate systems.

The transformation matrix from the aerodynamic coordinate system to the body coordinate system $\mathbf{C}_A^F$ is given in [10], where $q = \frac{1}{2}\rho|\mathbf{v}|^2$ is the dynamic pressure, ρ is the atmospheric density, S_{ref} and d_{ref} are reference area and reference length respectively. C_D and C_L are the coefficients of drag and lift, whereas C_l, C_m and C_n are the coefficients of rolling, pitching, and yawing moments. Aerodynamic force and moment coefficients are given by the semi-empirical equations comprised of static and dynamic derivatives presented in (7)–(12). The definitions of aerodynamic incidence angles α', ϕ', α and β are taken from [10], and are also presented in Fig. 1.

$$C_D = C_{D_0} + C_{D_{\alpha'2}} \sin^2 \alpha' \tag{7}$$

$$C_L = C_{L_0} + \frac{1}{2} C_{L_{\alpha'}} \sin 2\alpha' + C_{L_{q^*}} \frac{q^* d_{ref}}{2|\mathbf{v}|} \tag{8}$$

$$q^* = p \sin\alpha' \sin\phi' + q \cos\phi' + r \cos\alpha' \sin\phi' \tag{9}$$

$$C_l = C_{l_0} + C_{l_p} \frac{p d_{ref}}{2|\mathbf{v}|} \tag{10}$$

$$C_m = C_{m_0} + C_{m_\alpha} \sin\alpha + C_{m_{\alpha 3}} \sin^3 \alpha + C_{m_q} \frac{q d_{ref}}{2|\mathbf{v}|} \tag{11}$$

$$C_n = C_{n_0} + C_{n_\beta} \sin\beta + C_{n_{\beta 3}} \sin^3 \beta + C_{n_r} \frac{r d_{ref}}{2|\mathbf{v}|} \tag{12}$$

2.2 Linearization

Nonlinear state equations can be described in the general form as (13).

$$\dot{\mathbf{X}} = \mathbf{f}(\mathbf{X}, \mathbf{U}) \tag{13}$$

To get the linear state equations, the small perturbation theory was used [9,10]. This approach was then used to determine the static and dynamic stability of the rocket. According to the small perturbation theory, the nonlinear model from (1)–(4) can be linearized around a reference point obtaining the object's response to small perturbations. The state parameters of the rocket can be divided into reference state values and small perturbations as presented in (14).

$$X = X_{ref} + x \tag{14}$$

In this manner, every 12 of the state variables can be expressed. The reference state values were denoted with capital letters and small perturbations from this state in small letters. Using this approach, small perturbations were added to the nonlinear equations and the results are presented based on the linear momentum equation (15).

$$m(\dot{\mathbf{V}} + \dot{\mathbf{v}}) + m(\boldsymbol{\Omega} + \boldsymbol{\omega}) \times (\mathbf{V} + \mathbf{v}) = m\mathbf{C}_N^F(\Phi + \phi, \Theta + \theta, \Psi + \psi)\mathbf{g} + \mathbf{F} + \mathbf{f} \tag{15}$$

The transformation matrix according to [11], can be expressed as (16). Notation $[]_x$ describes the matrix representation of the vector used when transforming cross products to matrix multiplication and is presented in (17).

$$\mathbf{C}_N^F(\Phi + \phi, \Theta + \theta, \Psi + \psi) \approx \left(\mathbf{1} - \left[\begin{bmatrix} \phi \\ \theta \\ \psi \end{bmatrix}\right]_x\right) \mathbf{C}_N^F(\Phi, \Theta, \Psi) \tag{16}$$

$$\mathbf{A} = \begin{bmatrix} a_1 \; a_1 \; a_3 \end{bmatrix}^T \rightarrow [\mathbf{A}]_x = \begin{bmatrix} 0 & -a_3 & a_2 \\ a_3 & 0 & -a_1 \\ -a_2 & a_1 & 0 \end{bmatrix} \tag{17}$$

After the multiplication of all the terms and a few transformations, the Eq. (15) can be rewritten as (18). The terms underlined on the left side represent the original equation and therefore equal zero, and the term marked on the right side is the product of small perturbations which, according to the theory, can be neglected.

$$\begin{gathered} \underbrace{m\dot{\mathbf{V}} + m\boldsymbol{\Omega} \times \mathbf{V} - m\mathbf{C}_N^F(\Phi, \Theta, \Psi)\mathbf{g} - \mathbf{F}}_{=0} + m\dot{\mathbf{v}} + m\boldsymbol{\Omega} \times \mathbf{v} + m\boldsymbol{\omega} \times \mathbf{V} + \\ + \underbrace{m\boldsymbol{\omega} \times \mathbf{v}}_{=0} = -m \left[\begin{bmatrix} \phi \\ \theta \\ \psi \end{bmatrix}\right]_x \mathbf{C}_N^F(\Phi, \Theta, \Psi)\mathbf{g} + \mathbf{f} \end{gathered} \tag{18}$$

The next step was to change all the cross products to matrix multiplications and then the equation (18) could be expressed in a matrix form. The same approach was applied to every other nonlinear equation arriving at (19), where $\mathbf{C}_{Ang}^{Eul}$ is the transformation matrix from (4).

$$\begin{bmatrix} m\mathbf{1} & \mathbf{0} & \mathbf{0} & \mathbf{0} \\ \mathbf{0} & \mathbf{I} & \mathbf{0} & \mathbf{0} \\ \mathbf{0} & \mathbf{0} & \mathbf{1} & \mathbf{0} \\ \mathbf{0} & \mathbf{0} & \mathbf{0} & \mathbf{1} \end{bmatrix} \begin{bmatrix} \dot{\mathbf{v}} \\ \dot{\boldsymbol{\omega}} \\ \dot{\mathbf{r}} \\ \begin{bmatrix} \dot{\phi} \\ \dot{\theta} \\ \dot{\psi} \end{bmatrix} \end{bmatrix} =$$

$$= \begin{bmatrix} -m[\boldsymbol{\Omega}]_x & m[\mathbf{V}]_x & \mathbf{0} & m[\mathbf{C}_N^F(\Phi,\Theta,\Psi)\mathbf{g}]_x \\ \mathbf{0} & [\mathbf{I}\boldsymbol{\Omega}]_x - [\boldsymbol{\Omega}]_x\mathbf{I} & \mathbf{0} & \mathbf{0} \\ \left(\mathbf{C}_N^F(\Phi,\Theta,\Psi)\right)^T & \mathbf{0} & \mathbf{0} & -\left(\mathbf{C}_N^F(\Phi,\Theta,\Psi)\right)^T[\mathbf{V}]_x \\ \mathbf{0} & \mathbf{C}_{Ang}^{Eul} & \mathbf{0} & \mathbf{C}_{Ang}^{Eul}\mathbf{D} \end{bmatrix} \cdot$$

$$\cdot \begin{bmatrix} \mathbf{v} \\ \boldsymbol{\omega} \\ \mathbf{r} \\ \begin{bmatrix} \phi \\ \theta \\ \psi \end{bmatrix} \end{bmatrix} + \begin{bmatrix} \mathbf{f} \\ \mathbf{m} + [\mathbf{r_{ref}}]_\mathbf{x}\mathbf{f} \\ \mathbf{0} \\ \mathbf{0} \end{bmatrix} \tag{19}$$

$$\mathbf{D} = \begin{bmatrix} 0 & \cos\Theta\dot{\Psi} & 0 \\ \sin\Phi\dot{\Theta} - \cos\Phi\cos\Theta\dot{\Psi} & \sin\Phi\sin\Theta\dot{\Psi} & 0 \\ \cos\Phi\dot{\Theta} + \sin\Phi\cos\Theta\dot{\Psi} & \cos\Phi\sin\Theta\dot{\Psi} & 0 \end{bmatrix} \tag{20}$$

Small perturbations of aerodynamic forces and moments are presented based on the axial force f_X. The force was expanded into Taylor series giving (21).

$$f_X = \frac{\partial f_X}{\partial u}u + \frac{\partial f_X}{\partial w}w \tag{21}$$

As the first step, the partial derivatives from (21) were expanded obtaining partial derivatives of the force coefficients, example in (22), where $U_0 = |\mathbf{v}|$ and C_{X_1} is the value of axial force coefficient at the reference point.

$$\frac{\partial f_X}{\partial u}u = \left(\rho U U_0 S_{ref} C_{X_1} + q S_{ref}\frac{\partial C_X}{\partial \frac{u}{U_0}}\right)\frac{u}{U_0} = X_u u \tag{22}$$

The next step was to change the axial force coefficient into coefficients expressed in the aerodynamic coordinate system, drag and lift coefficients, using (23), where α_1' is the value of the total angle of attack at the reference point.

$$C_X = -\cos\alpha_1' C_D + \sin\alpha_1' C_L \tag{23}$$

After some transformations, the small perturbation of the axial force could be expressed in a linear form as (24).

$$f_X = X_u u + X_w w \tag{24}$$

The same approach was used in the derivation of every other small perturbation of the aerodynamic force. For the moments' equations, the step of changing the coordinate system was omitted, because the moments were already expressed in the body coordinate system (F).

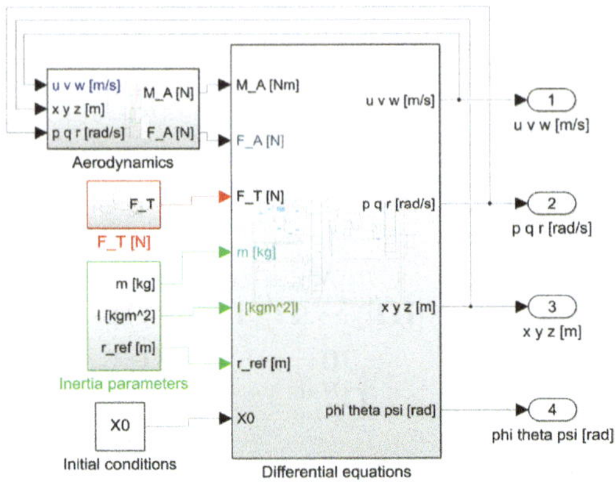

Fig. 2. Simulation architecture block diagram.

2.3 Simulation Model

A simulation model was created in MATLAB/Simulink 2020b environment. The top level architecture of the program is presented in Fig. 2. The geometrical, inertial, and aerodynamic parameters of the FOK rocket, developed by the Students' Space Association of Warsaw University of Technology [12], were used to generate the results. The model was validated based on the results of the FOK rocket's flight tests, and the results of this validation, comparing altitude, axial velocity, and axial acceleration are presented in Fig. 3. The blue line on the plots represents the flight data and the red line the simulation results. The rocket reached the apogee of around 1400 m m after approximately 15 s of flight. The maximum velocity was a little less than 200 m/s and was achieved after the motor burned out around 2.5 s after launch. Based on those results, it was assumed that the simulation model was correct.

3 Results and Discussion

3.1 Models Comparison

The linear model was first compared with the nonlinear model. The two models were excited by a small perturbation in the linear or angular velocity. The results comparing the propagation of vertical linear velocity and pitching angular velocity perturbations are presented in Fig. 4. The nonlinear model, the blue line, is characterised by greater damping and slightly higher natural frequency.

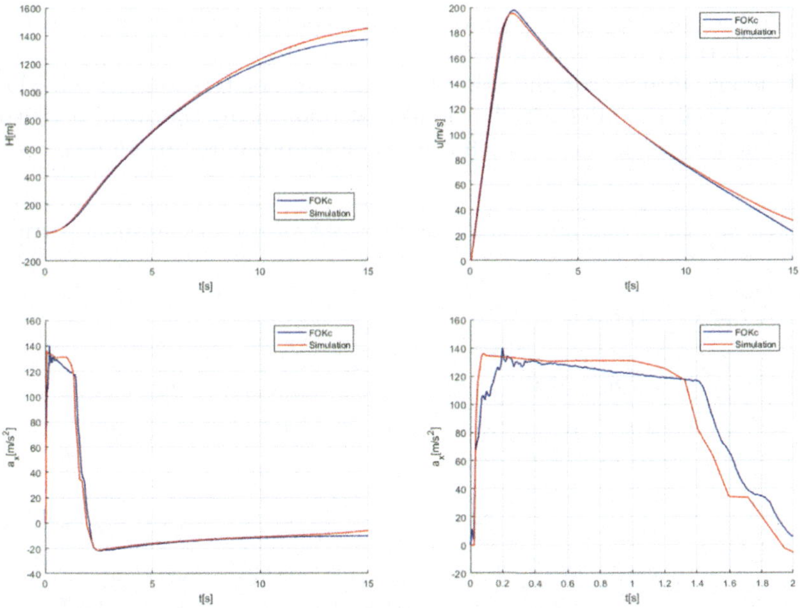

Fig. 3. Comparison of the simulation with the results from the FOK rocket's flight tests.

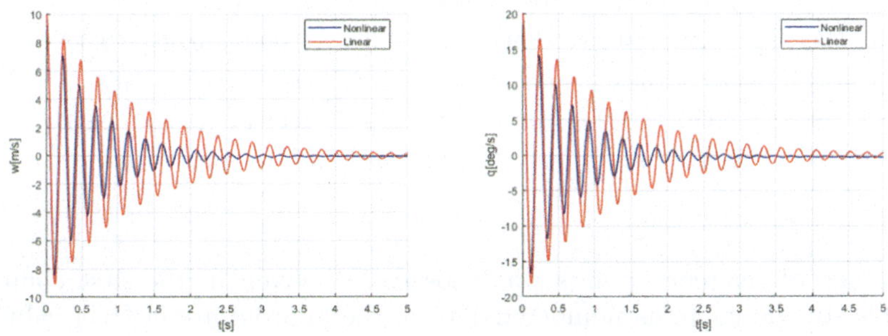

Fig. 4. Comparison of linear and nonlinear models response to small perturbations of vertical velocity and pitching angular velocity.

3.2 Static Stability

Static stability in aviation is an initial tendency of an object to return to the equilibrium after a distortion. The concept of static margin and static stability criteria from [9] were applied using nonlinear and small perturbation models to determine static stability. Criteria 1 to 3 describe the derivatives of aerodynamic force components with respect to the respective linear velocity components, whereas criteria 4 and 5 describe the derivatives of pitching and yawing moment with respect to the angle of attack and sideslip angle respectively. Criteria 6 to 8 describe the derivatives of aerodynamic moment components with respect to the angular velocity components. To obtain a reference point, the simulation with the use of the nonlinear model was stopped after t_{ref} time and the instantaneous state vector and other necessary parameters were saved. The nonlinear model was then linearized around that point using the small perturbation theory. The results of static stability criteria are presented in Table 1.

Table 1. The results of static stability criteria obtained by using small perturbation model.

$t_{ref}[s]$	Criterion							
	1	2	3	4	5	6	7	8
1	−0.36	−2.4	−4.9	−226	219	−0.02	−0.21	−0.18
	yes	*yes*	*yes*	*yes*	*yes*	*yes*	*yes*	*yes*
3	−0.49	−4.9	−4.9	−426	426	−0.03	−0.29	−0.29
	yes	*yes*	*yes*	*yes*	*yes*	*yes*	*yes*	*yes*
5	−0.37	−3.6	−3.9	−264	263	−0.02	−0.22	−0.22
	yes	*yes*	*yes*	*yes*	*yes*	*yes*	*yes*	*yes*
7	−0.29	−2.7	−3.1	−163	161	−0.02	−0.17	−0.17
	yes	*yes*	*yes*	*yes*	*yes*	*yes*	*yes*	*yes*
10	−0.19	−1.9	−1.9	−72	72	−0.01	−0.11	−0.11
	yes	*yes*	*yes*	*yes*	*yes*	*yes*	*yes*	*yes*
14	−0.09	−0.95	−0.96	−18	18	−0.006	−0.06	−0.06
	yes	*yes*	*yes*	*yes*	*yes*	*yes*	*yes*	*yes*
18	−0.07	−1.2	−0.21	−9	10	−0.004	−0.04	−0.03
	yes	*yes*	*yes*	*yes*	*yes*	*yes*	*yes*	*yes*

A range of reference points was chosen, t_{ref} given in the first column of the table. In the columns numbered 1 to 8, the appropriate criteria values are presented with the information whether they are met. All the criteria are always fulfilled, which means that the rocket is statically stable in all the flight regimes. In Table 2 static margin evaluations in two planes for different reference times and using both models are shown. It can be seen that the value of static margin for the pitch channel and higher velocities (lower reference time) is similar in both models and is always greater than 3, which means that the rocket is statically stable, and which is in correlation with the results from Table 1. In the yaw channel, for the linear model, the results are far more oscillatory, which might be caused by small differences in the Taylor series expansion of various aerodynamic forces and moments. For the reference time of 10 s, for the nonlinear yaw channel and linear pitch channel, a negative value appears, which indicates the instability of the rocket. The cause of this effect is yet unknown. The results indicate that the linear model in higher velocity regimes for the pitch channel might be used to determine the static margin.

Table 2. Static margin results for pitch and yaw channels obtained by nonlinear and linear models.

$t_{ref}[s]$	Nonlinear		Linear	
	$SM_{xz}[-]$	$SM_{xy}[-]$	$SM_{xz}[-]$	$SM_{xy}[-]$
1	3, 17	3, 17	3, 58	8, 58
3	3, 47	5, 57	3, 50	205, 32
5	3, 40	2, 39	3, 50	−37, 94
7	3, 54	4, 49	3, 58	142, 14
10	5, 48	−2, 94	−17, 85	6, 4
14	3, 28	3, 42	3, 12	22, 6
18	4, 77	4, 85	6, 10	6, 55

3.3 Dynamic Stability

The definition of stability says that the object's response to a finite excitement is also finite. Dynamic stability in aviation means then that the object returns to the equilibrium after a distortion. The natural frequency ω_n, damping coefficient ξ, shown in (25) and (26) from [13] as well as rolling time constant T_R given in (27) from [9] were chosen to determine the dynamic stability.

$$\omega_n = \sqrt{-M_\alpha} \tag{25}$$

$$\xi = \frac{Z_\alpha \omega_n}{2M_\alpha} \tag{26}$$

$$T_R = -\frac{4I_{xx}}{C_{l_p}\rho|\mathbf{v}|S_{ref}d_{ref}^2} \tag{27}$$

M_α and Z_α are the dimensional derivatives of the pitching moment and normal force respectively calculated from the linear model, I_{xx} is the longitudinal moment of inertia and C_{l_p} is the rolling moment dynamic derivative. To obtain these values from the simulation results, the logarithmic decrement method from [14] was used to determine the natural frequency and damping coefficient. The rolling time constant was determined from the definition as the time after which the response of a system will reach 63% of the steady state value. In Table 3 the results from the nonlinear and linear simulations along with the values obtained from equations (25)–(27) are presented. The results are in agreement with the graphs from Fig. 4, where the nonlinear model has a higher damping coefficient and a slightly higher natural frequency. The rolling time constant has the same value for both models. The natural frequency and rolling time constant are also in agreement with the values obtained from (25) and (27). The damping coefficient value from (26), however, is placed between the values from the nonlinear and linear models. With the increase of t_{ref}, so a decrease in the rocket's velocity, the natural frequency decreases, and the rolling time constant increases, which is a correct and expected behaviour.

Table 3. Natural frequency, damping coefficient and rolling time constant from non-linear and linear models simulations and literature equations values.

$t_{ref}[s]$	Nonlinear			Linear			Literature [9,13]		
	ω_n	ξ	T_R	ω_n	ξ	T_R	ω_n	ξ	T_R
1	19,01	0,0539	0,211	18,71	0,0238	0,21	18,84	0,0314	0,211
3	26,66	0,0516	0,156	25,76	0,027	0,156	26,21	0,0318	0,156
5	20,97	0,0511	0,201	20,34	0,0263	0,201	20,63	0,0319	0,202
7	16,47	0,0509	0,26	15,99	0,0252	0,259	16,18	0,0319	0,261
10	11,06	0,0493	0,394	10,63	0,0196	0,393	10,79	0,0303	0,395
14	5,42	0,0204	0,802	5,32	0,0274	0,806	5,38	0,0303	0,801
18	3,96	0,0199	1,096	3,87	0,0246	1,126	3,49	−0,0346	1,1

4 Conclusions

The article presented the derivation of a linearized rocket dynamic model with use of the small perturbation theory. The nonlinear model was validated based on the flight data of the FOK rocket. The static and dynamic stability criteria were formulated and tested on the linear model, and the results were compared with the nonlinear model. The results show that the linear model gives the correct values of the natural frequency and rolling time constant, but underestimates the damping coefficient. The static margin and static stability criteria show that the rocket is stable in all phases of the flight, which is in agreement with the flight data. From the results, it can be assumed that the linear simplification of the nonlinear model is sufficient to determine the static and dynamic stability of rockets. Future work may include using the linear model for the flight control parameters tuning.

Acknowledgments. This research was funded by National Centre for Research and Development grant number DOB-SZAFIR/03/B/002/01/2021.

References

1. Dat, N.T., Son, T.N., Tra, D.A.: Development of a flight dynamics model for fixed wing aircraft. In: Proceedings of the 10th International Conference on Computer Modeling and Simulation. ICCMS 2018, pp. 202–205. Association for Computing Machinery, New York (2018). https://doi.org/10.1145/3177457.3177470
2. Silveira, G., Carrara, V.: A six degrees-of-freedom flight dynamics simulation tool of launch vehicles. J. Aerosp. Technol. Manag. **7**, 231–239 (2015). https://doi.org/10.5028/jatm.v7i2.433
3. Trevisi, F., et al.: Simulation of a sounding rocket flight's dynamic, pp. 296–300 (2017). https://doi.org/10.1109/MetroAeroSpace.2017.7999584
4. Głębocki, R., Jacewicz, M.: Parametric study of guidance of a 160-mm projectile steered with lateral thrusters. Aerospace **7**(5) (2020). https://doi.org/10.3390/aerospace705006

5. Ożóg, R., Jacewicz, M., Głębocki, R.: Modified trajectory tracking guidance for artillery rocket. J. Theoret. Appl. Mech. **58**(3), 611–622 (2020). https://doi.org/10.15632/jtam-pl/121981
6. Nurhafid, A., Trilaksono, B., Sasongko, R., Riyadl, A., Irwanto, H.: Modelling and simulation of RKX200 rocket flight dynamics. In: Instrumentation, Communications, Information Technology, and Biomedical Engineering, pp. 230–236 (2013). https://doi.org/10.1109/ICICI-BME.2013.6698498
7. Myszkowski, P.: Classification criteria of linearization methods. Int. J. Electron. Telecommun. **55**(4), 659–668 (2009)
8. Guzzella, L., Isidori, A.: On approximate linearization of nonlinear control systems. Int. J. Robust Nonlinear Control **3**(3), 261–276 (1993). https://doi.org/10.1002/rnc.459003030
9. Napolitano, M.R.: Aircraft Dynamics: From Modeling to Simulation. Wiley, Hoboken (2012)
10. Zipfel, P.H.: Modeling and Simulation of Aerospace Vehicle Dynamics. AIAA education series. American Institute of Aeronautics and Astronautics, Reston (2000)
11. Barfoot, T., Forbes, J., Furgale, P.: Pose estimation using linearized rotations and quaternion algebra. Acta Astronaut. **68**, 101–112 (2011). https://doi.org/10.1016/j.actaastro.2010.06.049
12. Miedziński, D., Sochacki, M., Bresler, K., Małecki, S., Mochol, A., Wojciechowski, K.: Low cost rocket guidance and control development platform. In: 71st International Astronautical Congress - the CyberSpace Edition, Paris, France (2020)
13. Zarchan, P.: Tactical and Strategic Missile Guidance. AIAA Tactical Missile Series. American Institute of Aeronautics and Astronautics, Washington, DC (2007)
14. Inman, D.J.: Engineering Vibration, 3rd edn. Pearson/Prentice Hall, Upper Saddle River (2008)

Multithreading Errors in Data Reading Automation

Wojszczyk Rafał(✉) and Giebas Damian

Koszalin University of Technology, 75-453 Koszalin, Sniadeckich 2, Poland
rafal.wojszczyk@tu.koszalin.pl, damian.giebas@s.tu.koszalin.pl

Abstract. With the development of the Internet of Things (IoT) domain, it is now possible to build IT systems fulfilling the role of control and measurement systems by non-professionals. Such systems created following the idea of "Do It Yourself" (DIY) are successfully implemented not only in homes, but also in small enterprises that cannot afford to hire a professional team specialising in building such systems. These systems are usually built by ignoring many principles of software engineering, e.g. testing, which causes that these systems, for example, do not meet safety standards or operate unstably. In this study, an attempt was made to locate errors in the server application of an IT system playing the role of alarm and some elements of measurement.

Keywords: Multithreaded applications · Static analysis · Automation of data reading

1 Introduction

Thanks to the development of the IoT, many electronic devices that allow you to monitor and control many aspects of our lives can be bought for little money or built by yourself. It is not difficult to find on the Internet many projects of automation systems or alarm systems developed by passionate, which are often built by amateurs and then implemented in homes and small businesses. This is possible thanks to the idea of WEB 2.0, which is based mainly on the community, the community also has a large share in the development of such systems, building their own DIY solutions from scratch. Similar phenomena can also be observed in the industry (especially small, family businesses), where on the basis of experience of DIY community dedicated solutions are built. These solutions are not as highly professional as commercial equivalents, but very often in terms of basic control and measurement functions they are sufficient for small companies. The unquestionable advantage of such solutions is much lower price of tailoring the functions to specific requirements of the company. On the other hand, such systems may be insufficiently tested (tested only by the programmer) or may not use solutions that guarantee reliability. Simplicity of the solution and ease of implementation are more important in DIY than compliance with good practices and know-how. The C language is often used to learn programming,

R. Szewczyk et al. (Eds.): AUTOMATION 2022, AISC 1427, pp. 96–105, 2022.
https://doi.org/10.1007/978-3-031-03502-9_10

which is why many people know it. Many people will choose an easier DIY project and will not be aware of the disadvantages of such a solution.

Development of application is very expensive so it is common to reduce the costs. One of the most common factors to reduce the costs of software development is using open source projects. Many of the commonly used tools verify the source code using analysis techniques. Analysis techniques are divided into three types: static analysis, dynamic analysis, and mixed analysis. Every analysis types have pros and cons. The most popular methods belong to the dynamic analysis type which is more precise, and the results have reduced the number of false-positives. The cons of the dynamic analysis type are time-consuming. Also, they observe how programs work, and they report bug only when it occurs during work. Static analysis methods do not require observing the program, instead, these methods require the source code of the program. Thanks to that, this type of method is quicker, but precision is not as high as in dynamic methods. Another pro of the static analysis method is an easy way to speed up the whole analysis process. To do that there is only a need to run a process on a faster computer, because processors are faster and faster these types of methods are becoming increasingly cost-effective. Mixed analysis methods merge some pros and cons of previous methods, and it is very dependent on the method.

Authors attempt to diagnose a dedicated alarm system with some elements of measurement and repair occurring errors. This type of system can be implemented in seasonal hotels or other temporary abandoned buildings. The system described was built by a person who is not an automation or computer science specialist, which used software and hardware solutions used from the DIY community. Described system after the correction was tested again to be sure corrections do not provide new errors. The tests were carried out using a tool called rdao detector implementing static analysis method. The research carried out was intended to demonstrate the usefulness of the Multithreaded Application Source Code Model which is used in implemented in the tool. The use of the tool will reduce the cost of software production despite the lack of experience in multi-threaded applications.

The next Sect. 2 describes the context of the problem, i.e., the idea of DIY and the specifics of the implementation of the discussed alarm system. The Sect. 3 briefly describes the types of errors in multithreaded applications and methods how to detect them. The Sect. 4 presents the use of MASCM method. The Sect. 5 is a summary of the work.

2 Background to the Problem

2.1 Alarm Systems Manufactured at Home

The DIY idea is very well known in electronics and computer science. There are many examples of people trying to build their clones of commercial products. The biggest communities are concentrated around hardware platforms like Raspberry Pi, Arduino and their clones [14]. Using one of mentioned platforms, it is easy to build an alarm system, weather station, and this is enough to use it as a control and measurement system in home applications (or similar).

This paper describes the case study of a small company that has old seasonable hotel. The hotel is not open from autumn to spring, while the weather conditions in this location are not conducive to leaving the building unattended. From autumn to spring it is required that the system allows:

- detection of sudden movement and opening of the window in order to detect burglary,
- temperature reading to detect the aeration of the radiator,
- humidity reading to detect window leaks.

If motion is detected, the window is opened, or the temperature and humidity below the set value, the system should send information about the event to the employee. After receiving the information, the employee takes action. Manual checking of these things will take more time. Devices should be in every room.

Building intelligent systems in homes and companies is driven by cheap electronic components. This makes it possible to build cheap systems to read the temperature and humidity in small companies or in home, instead buying systems from the vendors. Creating own solutions than buying ones gives more possibility of expansion, but on most case they are not professional like this one on the market created by professional companies. A compromise solution in this case turned out to be the use of microcontrollers from the ESP8266/ESP32 family, popular temperature sensors (e.g. DS18B20, DHT11/22, BME/BMP), motion sensors and communication via Wi-Fi network [15]. The total cost of hardware components turned out to be no higher than in the case of other mentioned solutions, while the implementation of the program code was done by one of the employees (who isn't a programmer), basing on the tutorials available on the Internet. The solution developed in this way automated data reading, and the total cost was within the planned budget and allows for further expansion.

2.2 System Architecture

The architecture of the system is shown in Fig. 1. The application called server was written in C with pthread library and connects all components of the systems and manage the collecting data. The clients are ESP8266 modules with connected sensors, but in more detail, each ESP module contains firmware with a micro HTTP server. The server application periodically sends HTTP requests to the ESP modules, which responses in JSON format are saved in files and then can be viewed using basic GUI. The main thread of the application is responsible for displaying the current information, while other threads are used to read data from the ESP modules. The ESP modules have been programmed in the Arduino environment and the main loop contains code waiting indefinitely for requests, after receiving the request the reading from the sensors is performed. Communication between the client modules and the server application is via Wi-Fi (built into the ESP8266), using TCP/IP. The equipment used for reading is respectively: DHT11 sensor, HC-SR501, reed switch.

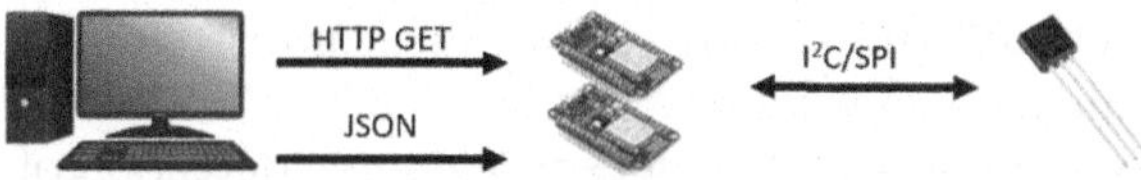

Fig. 1. The architecture of the deployed system, there are many modules ESP8266.

3 Errors in Multithreaded Applications

3.1 Problem Diagnosis

After several months of the implementation solution's author tried to resolve the problems with the server application not responding (going down) or unexpected shutdown. The result of the occurrence of the problems was the lack of up-to-date sensor data. The ad-hoc solution invented by the author to the problems was to restart the server application. Because of the low level of author knowledge about software engineering he could not find the cause of the problem, which was the occurrence of data races or deadlock in the source code. Deadlock caused the server application to go down, while unexpected application shutdown could be the result of data races i.e., errors belonging to the classes of race condition, atomicity violation, and order violation.

With the basic knowledge gained from tutorials, the author was not able to diagnose the cause on his own. Using additional tools to analyze Wi-Fi network traffic did not bring any results. Only after extending the application with an event log system was it possible to conclude that the problem does not lie in the communication process, nor the status of data, but in the process of handling them, which takes place in many threads. Gain this knowledge by the author, the analysis of the program did not bring the expected results, due to the size of the application and its architecture.

3.2 Ways of Detecting Multithreaded Errors

Detecting errors requires a good knowledge of software engineering and software development processes. The first process used to detect errors in software is the code review, which is part of the software development process. The assumptions of the code review in software engineering assume that this process is performed by members of the same team. However, the fact is that many times in the software development, the team consists of a single programmer, which makes it impossible to perform such a process correctly. Additionally, the probability of locating race condition and deadlock in the code review decreases as the amount of program code and its complexity increases.

The second process to detect errors in software is through software testing performed during the testing process. This is a multi-stage process that consists of tests of different purpose and scope. Depending on the stage, the testing process involves a programmer, a tester or tests are conducted jointly by both of them. Tests in the testing process are created in accordance with the test pyramid [1], which consists of unit tests, integration tests, system tests and acceptance tests in the following order.

The testing process at any level does not assume detection of multithreading errors. Some of the multithreading errors with low probability can be detected fastest in integration tests, and latest in acceptance tests. The higher the level of testing, the less information about the errors that occur can be obtained and there is less chance of its reproduction. As a result, it is true to say that the testing process may or may not reveal multithreading errors.

In addition to the code review and the testing process, the programmer to detect data races and deadlock can use specialized tools developed for this purpose. One of the best developed tools is ThreadSanitizer [2] developed for the C++ language, and it is included in the clang compilers starting from version 3.2 and gcc from version 4.8 for this language. This tool is limited to detecting race condition class errors, Unix family systems, and the C++ language itself. Due to the limitations and cons of the tool, it cannot be used to detect deadlocks occurring in the server program. For the same reasons, the RacerX [3] and RELAY [4] tools, which have similar limitations and cons, should also be rejected, and due to their low effectiveness, work on them was abandoned. In addition to the tools already mentioned, LOCKSMITH [5], SVD [6], ERIGONE [7] and Bauhaus [8] tools have also been abandoned. Currently, only two methods are being developed whose applications make it possible to locate errors in programs written using the C language. The first tool is DeepRace [9], which uses techniques from the field of deep learning. This tool is in the early stages of development and only allows the detection of race condition. The second tool is rdao detector [10], which is developed based on the Multithreaded Application Source Code Model. Rdao detector makes it possible to locate data races and deadlock errors in the process of static source code analysis of programs written in C using the phtread library.

Available methods and tools for detecting errors in multithreaded applications revealed a significant gap in the area of error detection in multithreaded programs written in C using the phtread library. Most of the tools have been abandoned, due to low effectiveness implemented methods. The remaining tools requirements exclude it's used in most cases i.e. detecting a multiple class of errors, mature stage of development or language support. A comparison of multiple tools can be found in [10]. All this leads to a situation where the development of multi-threaded applications becomes highly risky, as there are no tools to support their production.

4 Use of MASCM Method

4.1 MASCM

The case of the control and measurement system presented in this paper, served as a case study to verify the MASCM method and rdao detector tool described in detail in [10]. Multithreaded application source code model allows for the mapping of a multithreaded computer program in the form of a tuple:

$$C_P = (T_P, U_P, R_P, O_P, Q_P, F_P, B_P) \tag{1}$$

where:

1. P is the program index,
2. T is the set of all threads,
3. U is a sequence of sets containing threads working in the same time frame,
4. R is a family of shared resources,
5. O is a set of all atomic operations of the program,
6. Q is the set of mutexes available in the program,
7. F is the set of edges,
8. B is sequence of sets; where every set contains pairs of operations related by forward relationship or backward relationship or symmetrical relation.

The method uses static source code analysis, so that error detection in the server application could be performed off-site without the need to move ESP modules. The method works by transforming the C language source code into a formal representation that highlights the features of multithreading and facilitates the detection of errors of the Race condition, Deadlock, Atomicity violation, and Order violation classes. Part of model instance which part of source code can be found on listings 1 and 2 for server application is as follows:

$T_{SRV} = \{..., t_4, ..., t_{11}, t_{12}, ...\}$
$U_{SRV} = (..., u_3 = \{t_4, ...\}, ..., u_6 = \{t_{11}, t_{12}, ...\}, ...)$
$R_{SRV} = \{..., r_{11} = \{data\_handler\}, ...\}$
$O_{SRV} = \{..., o_{4,34}, o_{4,35}, ..., o_{4,63}, o_{4,64}, ..., o_{4,78}, ..., o_{11,12}, ..., o_{11,54}, ..., o_{12,7}, ...\}$
$Q_{SRV} = \{..., q_7 = (data\_lock, PMN), ...\}$
$F_{SRV} = \{..., (q_7, o_{4,34}), (o_{4,34}, o_{4,35}), ..., (q_7, o_{4,63}), (o_{4,64}, o_{4,65}), ..., (o_{4,78}, q_7), ..., (o_{11,12}, r_{11}), ..., (o_{11,54}, r_{11}), ..., (o_{12,7}, r_{11}), ...\}$
$B_{SRV}^{FWD} = \{..., (o_{11,12}, o_{12,7}), ..., (o_{11,54}, o_{12,7}), ...\}$

4.2 Occurrence of a Race Condition Class Error

Incorrect use of mutex in server application source code provides incorrect operation of a server program. In the case of data race errors, the cause of their occurrence is the failure to create a critical section, which should include operations on shared resources and operations in order relations with them [10,11]. Deadlock, on the other hand, is caused by mutually exclusive pairs of mutexes, a retry to establish an already established mutex, or a missed mutex release operation [10,12]. The problems with the server application described in Sect 3.1 allow us to assume the occurrence of at least one deadlock class error resulting in the server application going down and at least one data race error resulting in the application unexpected shutdown. Source code analysis with the rdao detector tool allowed to obtain information about errors in the source code confirming previous presumptions. For this paper only fragments of the program are presented in the form of formal notation and their graphical representation. Analysis of the source code revealed the presence of one deadlock class error and

```
void * decode_base64_data(const char *encoded_data) {
  ...
  pthread_mutex_lock(data_lock);
  void *data = malloc(sizeof(char)*DECODED_DATA_MAX_SIZE);
  ...
}

bool validate_data_schema(const char *data, const char * path_to_schema) {
  ...
  pthread_mutex_lock(data_lock);
  char * schema = load_schema(path_to_schema);
  ...
  pthread_mutex_unlock(data_lock);
}
```

Listing 1: Part of source code from thread T_4 of the server application.

```
char * allocate_memory() {
  ...
  data_handler = (char*)malloc(sizeof(char)*DATA_HANDLER_SIZE);
  ...
    data_handler = (char*)realloc(data_handler, sizeof(char)*DATA_HANDLER_MAX_SIZE);
  ...
}

void clean_data_buffer(char * data_handler, const long buffer_length) {
  ...
  memset(data_handler, 0, buffer_length);
}
```

Listing 2: Part of source code from threads T_{11} and T_{12} of the server application.

two order violation class errors. The deadlock error located in the source code of the server application is the result of an attempt to re-establish an already established mutex.

A graphical representation of the error can be found in Fig. 2, and a fragment of the formal notation and parts of the source code on the listings 1 and 2 can be found in previous subsection. In the depicted fragment of the graphical representation, it is easy to locate a path in the graph in which there is a second operation of establishing a mutex without first releasing it. In other words, the condition that causes the deadlock error [10] is met. For order violation class errors, the situation is much more complicated. A fragment of the model instance representing the error can be found in previous subsection, and its graphical representation is in Fig. 3. As noted in the paper [10], the order violation is caused by a change in the execution order of at least a pair of operations that access two (or more) memory areas. Rdao detector reported two errors with the same source of the problem.

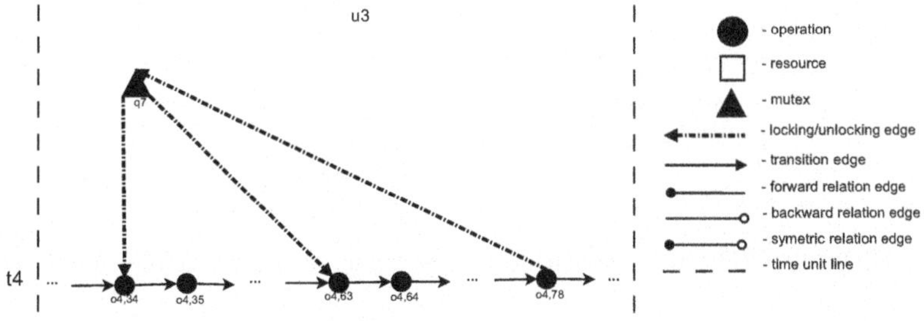

Fig. 2. Graphical representation of deadlock error from server application.

Namely, the cause of the order violation errors is the execution of an operation to free memory occupied by a resource in a thread T_{12}, while another thread T_{11} tries to perform operations on the said resource. The effect of the T_{11} operation thread is an unexpected shutdown of the server program. The solution to the problem is to use a condition variable, which will allow the T_{12} thread to wait for the completion of the operation of the T_{11} thread, and after receiving a signal to release the memory occupied by the shared resource.

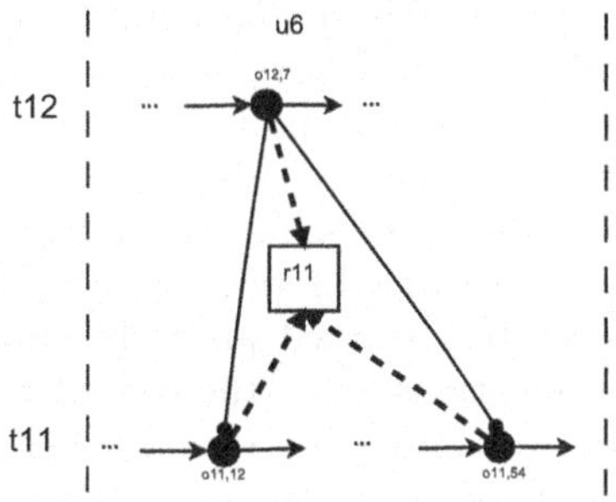

Fig. 3. Graphical representation of order violation error from server application.

With the help of the method based on the Multithreaded Application Source Code Model implemented in the rdao detector tool, it was possible to locate the causes of errors in the server application. The cause of the application crashing was an error of deadlock class, while the cause of unexpected application switching off was an error of order violation class. The analysis of the source code allowed us to introduce appropriate modifications, which eliminated the errors.

Good practices in software development involved running job during CICD process. Part of the CICD for multithreaded application should be static analysis of source code made by rdao detector. Example of that process is shown on the Fig. 4. Automation of the analysis process has few pros, when some CICD job fails then deploy/delivery job cannot be performed and team is alarmed about that. If programmer forget run analysis locally CICD process do it anyway.

To the cons should be calculated cost of the CICD process. Not all of the public repos systems support CICD. A lot of them allow run job's pipelines as a part of premium account. Even that which allow run job's pipelines for free introduce limits e.g. number of minutes per job or numbers of parallel jobs.

4.3 Verification of System Operation After Modifications

After correcting the errors described in the previous section, the system was rebooted in the target environment. With the new deployment, only the server

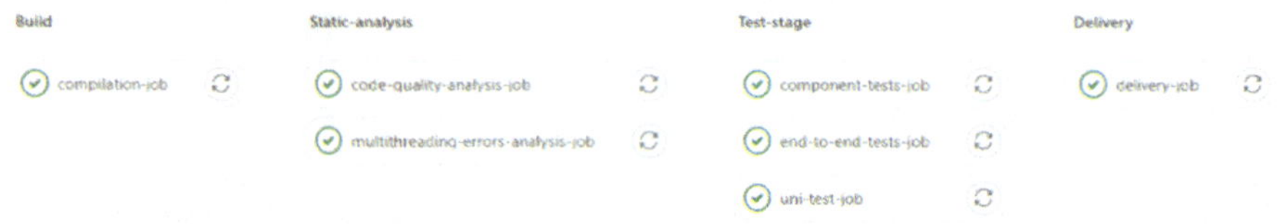

Fig. 4. GitLab's CICD process with static analysis of multithreaded application.

application code was modified, the other hardware and software components were not altered. In terms of expected performance, the program did not show any changes, while incorrect performance (no response) was no longer present.

It is worth emphasizing that the consequences of the errors that occurred earlier could be very costly for such a company. An employee, seeing outdated data (and not being aware that the data is false), would not make any reaction. Lack of reaction leads, among other things, to the appearance of a fungus on the walls, or in the event of a break-in financial loss. The malfunctioning of the system occurred intermittently, posing an additional risk. It was estimated that the system worked properly about 70–75% of the time. After performing the modifications using the rdao detector tool, no abnormal operation was noticed. Thus, it should be considered that with the MASCM method, the reliability of the operation of the control and measurement system described in this paper was increased to 100%.

5 Summary

This paper presents a case study in which an hobbyist developed an alarm system that was not tested. To resolve the problems with the stability of the systems forum, tutorials and any other non-professional sources of knowledge are not sufficient. Particularly in cases where errors come from multithreading. To resolve this type of problem from specific domain expertise is required.

The presented server application was analyzed with rdao detector tool, which is a verification of MASCM method. As a result, specific errors were located that would be very difficult to detect in a typical testing process. Example described in this paper shown how important is building tools and methods to locating multithreading errors.

The choice of the rdao detector tool was practical, because this tool can locate as many as four classes of errors [10–13] in the process of static source code analysis. There is little research interest in this area and there are not many new tools for static analysis of multithreaded programs. For this reason, there are few competing tools and most of them have not been developed for more than ten years. In other words, programmers and hobbyists do not have tools to support writing code of multithreaded application.

Further development of the MASCM method includes increasing the accuracy of the analysis and reducing the number of false positives. In the case of rdao detector tool, the work is expected to be connected with making it easier to use.

References

1. Certified Tester, Foundation Level Extension Syllabus – Agile Tester International Software Testing Qualifications Board Version (2014)
2. Serebryany, K., Iskhodzhanov, T.: ThreadSanitizer: data race detection in practice. In: Proceedings of the Workshop on Binary Instrumentation and Applications, pp. 62–71. ACM (2009). https://doi.org/10.1145/1791194.1791203
3. Engler, D., Ashcraft, K.: RacerX: effective, static detection of race conditions and deadlocks. ACM SIGOPS Oper. Syst. Rev. **37**(5), 237–252 (2003). https://doi.org/10.1145/1165389.945468
4. Voung, J.W., Jhala, R., Lerner, S.: RELAY: static race detection on millions of lines of code. In: Proceedings of the the 6th Joint Meeting of the European Software Engineering Conference and the ACM SIGSOFT Symposium on The Foundations of Software Engineering, pp. 205–214 (2008)
5. Pratikakis, P., Foster, J.S., Hicks, M.: LOCKSMITH: practical static race detection for C. ACM Trans. Program. Lang. Syst. (TOPLAS) **33**(1), 1–55 (2011). https://doi.org/10.1145/1889997.1890000
6. Xu, M., Bodík, R., Hill, M.D.: A serializability violation detector for shared-memory server programs. SIGPLAN Not. **40**(6), 1–14 (2005). https://doi.org/10.1145/1064978.1065013
7. Ben-Ari, M.M.: Tool presentation: teaching concurrency and model checking. In: Păsăreanu, C.S. (ed.) SPIN 2009. LNCS, vol. 5578, pp. 6–11. Springer, Heidelberg (2009). https://doi.org/10.1007/978-3-642-02652-2_5
8. Koutsopoulos, N., Northover, M., Felden, T., Wittiger, M.: Advancing data race investigation and classification through visualization. In: 2015 IEEE 3rd Working Conference on Software Visualization (VISSOFT), pp. 200–204. IEEE (2015). https://doi.org/10.1109/VISSOFT.2015.7332437
9. Tehrani, A., Khaleel, M., Akbari, R., Jannesari, A.: DeepRace: finding data race bugs via deep learning. arXiv preprint arXiv:1907.07110 (2019)
10. Giebas, D., Wojszczyk, R.: Detection of concurrency errors in multithreaded applications based on static source code analysis. IEEE Access (2021). https://doi.org/10.1109/ACCESS.2021.3073859
11. Giebas, D., Wojszczyk, R.: Atomicity violation in multithreaded applications and its detection in static code analysis process. Appl. Sci. **10**(22), 8005 (2020). https://doi.org/10.3390/app10228005
12. Giebas, D., Wojszczyk, R.: Deadlocks detection in multithreaded applications based on source code analysis. Appl. Sci. **10**(2) (2020). https://doi.org/10.3390/app10020532
13. Giebas, D., Wojszczyk, R.: Order violation in multithreaded applications and its detection in static code analysis process. Appl. Comput. Sci. 16(4), 103–117 (2020). https://doi.org/10.23743/acs-2020-33
14. Al-Tabatabaie, K.F., Abdulameer, S.D.: Applying arduino for controlling car parking system. Appl. Comput. Sci. **15**(2), 53–62 (2019). https://doi.org/10.23743/acs-2019-13
15. Shun, W.G., Muda, W.M.W., Hassan, W.H.W., Annuar, A.Z.: Wireless sensor network for temperature and humidity monitoring systems based on NodeMCU ESP8266. In: Anbar, M., Abdullah, N., Manickam, S. (eds.) ACeS 2019. CCIS, vol. 1132, pp. 262–273. Springer, Singapore (2020). https://doi.org/10.1007/978-981-15-2693-0_19

The Multivariable Control for Dynamic Partially Observable Objects

Viacheslav Ivashchuk[1] and Igor Korobiichuk[2(✉)]

[1] National University of Food Technologies, Volodimirska str. 68, Kyiv 01601, Ukraine
ivaschuk@nuft.edu.ua

[2] Łukasiewicz Research Network – Industrial Research Institute for Automation and Measurements PIAP, Warsaw, Poland
igor.korobiichuk@piap.lukasiewicz.gov.pl

Abstract. The article is devoted to the development of multivariable control algorithm as for functionally complex objects class, which has a nonlinear feedback. In most cases, only part of the system is measured by output that demands the linearity of some input. Obviously, the observable nonlinear systems are not diffeomorphic to linear systems. So that the output actions and inputs of those systems does not distinguish some set of different initial states. That condition changes order of complexity of the singular analysis of inputs, although a complete theory, which allows the design of an observer, for these unobservable systems does not exist. This paper presents the results of hybrid control algorithm that involved by response of inverse dynamic. The control of drying process, where the filling of chamber is increased, has been taken as a base for comparison of results of parametrical stability. We made this for the spray drying process to minimize the quality losses of product by adjustments of the input flows of dryer according to internal disturbances and process constraints.

Keywords: Hybrid algorithm · Partially observable · Multivariable control · Lyapunov functional · Industrial control systems

1 Introduction

The technological systems, those are used in production, changes own purpose for the modern requirements of product customers as for to establish a new assortment [1]. The task of required rate of change for parameters of the environment state, which has the strict regulations, requires the sources, which are unattainable by the technological characteristics of equipment [2]. The indicated features are explained by the economic calculation of equipment. Therefore, the rate of flows for acceleration of technological modes affects the increase of gradient of influence by the technological variables and attaches the product loss, as a result. The many parametric changes of control object produce a new task for automation that typically has a compact structure for the limited purpose of parametric variances, when the control action is produced by single variable [3, 4]. This problem is relevant in accordance with dimension of space of technological rules. So that, the dynamic control error cause the product damage and the

R. Szewczyk et al. (Eds.): AUTOMATION 2022, AISC 1427, pp. 106–115, 2022.
https://doi.org/10.1007/978-3-031-03502-9_11

failure occurrence of technological equipment. The complex physical processes (change the aggregation state, making the field of medium by atomizing liquids and direct heat transfer) are often irreversible, and the gradient of energy distribution in technological medium complicates the observing task of object for effective control [5]. All these above-mentioned characteristics may be allocated as a special class of functionally complex objects (FCO) for such processes as spray drying of fluid fraction. Consequently, the change of regulatory demands of technological products accelerates the transitional modes, because it cannot be powered by sparse control models for a single technological variable (Fig. 1).

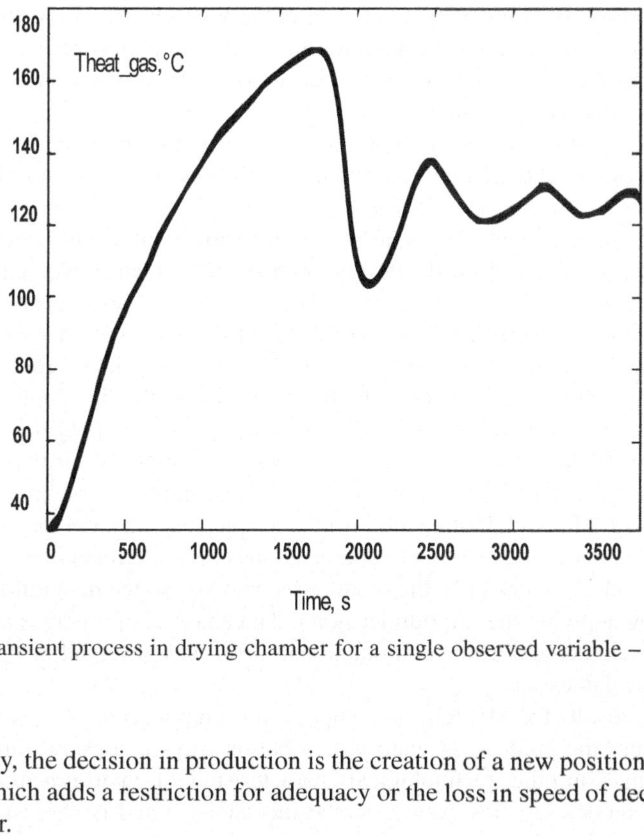

Fig. 1. The transient process in drying chamber for a single observed variable – temperature of heating gases

As usually, the decision in production is the creation of a new position – operator of apparatus, which adds a restriction for adequacy or the loss in speed of decision making in usual order.

The real-time tuning of regulator is a complex task, because the coefficients should be modified with taking into account the subjective people ware feedback, without the estimation of meaningful physical parameters. The use of decision maker in control loop transforms the fast processes in multidimensional control into the not searchable task.

The typical problem of monitoring the state of control object is the limitation of technological complexes for local estimation. This can be explained by: the turbulence of moving medium around of sensitive elements that causes the accumulation of errors; the ability of particles of medium forms the contaminants around of sensitive elements that reduce sensor activity; heterogeneity of technical medium by parameters in closed space of equipment that explains the loss of estimations adequacy; the aggressiveness of technological medium that limits the appliance of instrumental measurements. These features determine the partial observation of control object, so the dimension of available variables, which influence on process, is greater than the number of observable parameters of state.

2 Related Works

In case of weak observability, as was pointed in work [6], if the influence of some input is small in a suitable sense, then exact observability is preserved. The operator of action for that input could be omitted from control chain without loss of generality. But, if we are not making any changes of observability, then one way expect the observation to converge into the limited observable subspace of system, which does not suite for identification of dissipative systems state.

In work [7], the group of researchers gets aim by asymptotic state feedback stabilizability by the much weaker assumption than uniform observability, for poorly observable object.

For some classes of nonlinear systems, unobservable control may informs about of sufficient conditions those allow the design of observers, which converge irrespectively of the inputs. In work [8, 9], the observers had been synthesized as a separately high gain structure. But one can easily come to logical conclusion that any error of the observer (especially for high gain observers) will get to seriously impairs of control quality by generating large transitional errors [10]. In other way is the developing in order of relaxation of system structure as a partial non-linear observer [11], by use a similar canonical form. In this case, the single input does not distinguish of two initial states, which are issued respectively two trajectories. So those unobservable modes are stable, as for linear case. Although that method is in principle very rich, but may be difficult to apply in a very general case as observer-based control, as underlined in the introduction, as author conclude. In work [12], the model was involved so the heat influence is taken into account by applying the multidimensional data analysis of processes in apparatus. However, fast degradation in the computational effort was observed as the problem of dimension of variables set.

Significant results for MIMO (multi input multi output) control were achieved for processes, where the gradient of energy distribution stays a constant direction as the spread of changes in other technological characteristics of environment. In that case, the change of modes satisfies event-oriented mechanisms and is at least linear to the change of vector of state parameters that stay in the dynamically stable state of medium components (the mixing of stable chemical products; the processing of solid synthetic products that does not have influence from the reversible states and others physical transformations).

3 Formalization

The idea to use the discontinuous control may is applied to nonlinear functions of system variables those can be represented by high polynomial orders, i.e. in the presence of stiff models. Thus, there is a problem of iterative process that implements the steps for multi-dimensional control by vector of variables of object $y_i \in \overline{y}$ that could be characterized by a difference in time constants

$$\mathrm{T}_{\mathrm{y_i}\_U} >> n^{-1} \sum_{i=1}^{n} T_{y_i\_U} \quad (1)$$

where the descent will cause a difference of dynamic shifts $|\Delta y_i| >> |\Delta y_{i+1}|$. Because the chosen object is characterized as partially observable $\dim A < \dim C$, then the use of linearized mathematical models for control will lead to large errors $\Delta y_i > \Delta y_{i\_zad}$ and to loss of stability $\Delta y_i \underset{i \to n}{\longrightarrow} \infty$, as a result.

The selected characteristics of behavior of FCO require the execution of logical conditions for to alternate the continuous dynamic, which require the creation of a hybrid modeling of system. For that system the part of continuous dynamic is an approximation of physical principles and the discrete part is the primary source of logical conditions, which set switches depending from internal variations of system variables.

In particular, the use of hybrid systems can be defined for: control problems with differentiated scales of variation, such as parallelism and sparseness in the object model; to find solutions with a reserve of safety; the ability to process the conditions of event that has a priori unknown time; the ability to handle situations that have the number of variables or equations, which are changed relative to a particular state; the ability to process high-order functionals.

The making of hybrid control will be performed by the generalizing of complex description of sparse dynamic system that will allow applying the research methods, which typically are applied for continuous systems.

The making of order of alternative dynamic descriptions creates a set of trajectories of hybrid system for each of the alternatives is similar to changes of the right part of differential equations system. That order would return the result of solutions that creates a set of switching gs of state S_i. The function of initialization for a new state of automata will be established according to each switching s $s \in Sk$.

The states of hybrid system are responsible to differential behavior of the phase vector, from which will change other alternatives. The hybrid automata model is a set of rules, which determines the differential behavior that will have been integrated on defined time. The errors of determining points of switching between states are accepted as the conditions of poles for a change of alternatives, which produces qualitatively different trajectories.

The use of a discontinuous control system [13], which involves the dynamic parameters of object, will be stay symmetrical respect to the switching surface. Thus, the stability of hybrid behavior will be ensured by the stability of dynamic behavior that is executed only for a single pole (condition of necessity) or provided by a single source of propagation of influence (condition of sufficiency). In addition, the sliding control involves the presence of a stable drift around the switching surface, which is not always the case for technological objects (in particular, irreversible processes) and is characterized as excessive use of source power. If the solution of differential equation with discontinuous coefficients belongs to the discontinuity surface, then the projection getting the nil, so further motion is impossible, because the right part is not defined on the discontinuity surface [14]. So the right part will require additional definition. For that, the condition of sufficiency of the asymptotic approximation will have been confirmed. The derivative of functional along the direction of solution for only the one from many alternatives must remains as invariant for to continue of solution in the following moments of time. This property of functionals was formalized in Kim's work [15], where the concepts of invariant derivative and invariant differentiation had been introduced.

For to define the "variable modes" of functional-differential equations with the alternative (discontinuous) right-hand side, there is the same difficulties that could be found in the theory of stability that use Lyapunov functionals, which are associated with the calculation of functionals derivatives that are along the solution direction [16]. This determines the effective use of "sliding modes" or description in form a set of separate right parts from functional space or in form of localized state spaces.

4 Proposed Algorithm

For following by methodology of sliding mode, we will determine the surfaces where the sliding motion will show the properties of invariance to behavior or will be limited by additional conditions. The limitations, which will be determined by estimating the dynamics of trend that desposes the presented method of order reducing. So the dimension of control algorithm corresponds to the number of switching surfaces. Therefore, if the control model of the output coordinate of object is presented as Cauchys' form:

$$\begin{cases} \dot{x} = Ax + Bu \\ y = Cx \end{cases} \tag{2}$$

where the vector of state $x \in R, \dim R = n$, the matrix that describe the influence of system variables $A \in R^{\mathrm{n}}, \dim A = [n \times n]$ and $B \in R, \dim B = [n \times m]$ – the matrix of gain coefficients for multidimentional control $u \in R, \dim u = n$, which allows to resolve of control task by result of observation of outputs, by matrix of coefficients $C \in R^{\mathrm{m}}, \dim C = [m \times m]$, then the input actions, which is caused by control channels and responsible for the correction of control trajectory of the output coordinate of state:

$$B = \begin{pmatrix} B_1 \\ B_2 \end{pmatrix} \tag{3}$$

where $B_1 \in R, \dim B_1 = [(m - l) \times m]$ and $B_2 \in R, \dim B_2 = [m \times m]$, for condition $\det B_1 \neq 0, \det B_2 \neq 0$, l – the dimension of elements those are a part of control implementation, which is responsible for the descent by technologically efficient variables, those are maximally correlate with the target variables of process, which are selected as:

$$l : \dim B \underset{\|\overline{y}\| \geq \delta_y}{\longrightarrow} m \tag{4}$$

The construct of sliding control will consist from two stages that will carry out the conditioned sparse the order of sliding elements that maintains the dimension of problem in accordance with the control vector $\overline{\mathrm{U}}$. First of all, the order of changes $\Delta\overline{x}$ by selected variable of control has been determined a priori. Secondly, there is a distribution of control along the trajectory of changes $\Delta U : y_i \underset{\tau_{\Delta y}}{\longrightarrow} y_{i+1}$.

The hybrid controller will apply the system model and the current state of system to observe the violation of constrains for future state over a finite prediction horizon $\sum_{i=1}^{Ny_2} \left[y^*(k+i) - y_{zad}(k+i)\right]$. So the optimal adaptive action is derived over a control horizon $\Delta u(k+i-1)$. Then only the one (each next, if task has not been changed) control input is applied to the real system and same process will be repeated for the next time step. Therefore, the hybrid input can be regarded as a nonlinear state feedback control input that is obtained by repeatedly minimizing of objective function that adheres by the system dynamics, which bounded by given constraints at every time step.

The model of observation of outputs may be changed to form in matrix interpretation:

$$\begin{bmatrix} \dot{x}_1 \\ \dot{x}_2 \end{bmatrix} = \begin{bmatrix} A_{11} \; A_{12} \\ A_{21} \; A_{22} \end{bmatrix} \cdot x \tag{5}$$

The representation in form of state space system is effective for calculations when it is necessary to localize the empirical settings. The impulse response for matrix of control will be found as

$$G(t) = \begin{cases} 0 & , t < 0 \\ Ce^{At} B_2 B_1^T & , t \geq 0 \end{cases} \tag{6}$$

where e^{At} forms a resolvent matrix that will be taken into account by observation matrix I, which will be used to estimate the offset and the increment of control error by the increment of prediction error along the trajectory

$$I = diag(\sum_{i=1}^{Ny_2} \left[y^*(k+i) - y_{zad}(k+i)\right]^2 + \lambda_k \sum_{i=1}^{N_{u1}} [\Delta u(k+i-1)]^2) \tag{7}$$

where λ_k – linear estimation that is founded by $\lambda_{k\_i} = K_i^u / K_i^y$, that forms a gain matrix of control channel, $y = Gu$. Then, the model of the controlled object with observer (7) takes the form

$$\begin{cases} \dot{x} = (sI - A)^{-1} x + B_2 B_1^T u \\ y = C((sI - A)^{-1} x + B_2 B_1^T u) \end{cases}$$

and the matrix of coefficients of transfer function G in static appliance, taking into account (6), will have got the form:

$$G = C(sI - A)^{-1} B_2 B_1^T \tag{9}$$

or

$$G = \frac{1}{\det(sI - A)} \left[CB_2 B_1^T\right] \tag{10}$$

It needs remember that any system represented in state space form can be involved to create a transfer function, but the inverse operation is often incorrect. As a necessity condition for apply the sliding control algorithm, even with minimal alternatives (as for the conditions of partial observation), is the number of variations by the way to reach the target value with accuracy $\|\overline{y}\| < \delta_y$. That number of steps must be finite.

For implement this control, we switch in the surface area S

$$S : B_2 u = B_1 u + Ax \tag{11}$$

The control for the sliding alternative will be found as.

$$\Delta u = -k_g \cdot sign(S(y)), \tag{12}$$

where k_g – selected as the amplitude of control signal, $S(y)$ – the estimation of object response. The exist error Δy_i of deviation of sliding mode in system predicates the conditions of necessity for to ensure the asymptotic stability of control system $y \cdot y'' < 0$.

The problem of "chattering" will be solved by using the sliding function in form of a saturation link instead of the typical relay element $sign(\bullet)$. The most significant is gain k_g of control link, which provides the necessary accuracy of descent of control by estimating of approximations for the vector of target values $\overline{y}$ and condition that initiates redirect. In this way, during the simulation of the sliding control mode, the additional limitation is involved by variables of related processes of the technological flow, for to ensure the phase margin of stability of control system.

The stability of dynamic part is may investigated by checking the condition of Routh-Hurwitz stability criterion for the roots of characteristic equation, which describes the continuous control behavior and eigenvalues of the state space for the characteristic equation.

The founded control influence can be represented by

$$k_{gB2} = [A_{22} + CA_{12}]\left[C\left(sI - CB_2^{-1}\right)B_1^T\right]^T \tag{13}$$

$$k_{gB1} = [A_{21} - A_{22}C + CA_{11} - CA_{12}C]\left[sI - B_1^T B_2\right]^T \tag{14}$$

or as the generalized control

$$u = -diag\left(k_{gB1}\, K_{gB2}\right)_m sign[S(y)] \tag{15}$$

with an estimation of surface of the state change.

$$S(y) = \left(C \;\; sI\right)\begin{bmatrix}\dot{x}_1\\ \dot{x}_2\end{bmatrix} \tag{16}$$

where a change of control alternative occurs by rule.

$$sign(y) = \begin{cases} -1, y < +\frac{\delta}{2} \\ 1, y > -\frac{\delta}{2} \end{cases} \tag{17}$$

If use the variable x_2 as for indirect state control of x_1, for to reduce the order of system, the problem of matrix determine C is such that $A_{11} - A_{12}C$ satisfies the Hurwitz criterion, and the solution will be reduced to definition of linear feedback, which stabilizes $x_2 = -Cx_1$ for a typical solution of linear matrix equations $\dot{x}_1 = A_{11}x_1 + A_{12}x_2$.

5 Experimental Evaluation

We demonstrate the application of the proposed hybrid algorithm for control of industrial A1-OR2CH spray dryer that is produced by Kalinivskij machine making factory, which produces approximately 1020 kg/hr of skim milk powder. That object is one of the largest energy consumers in food industry, had been represented as example. The control was applied to nonlinear dynamic model for heat processes in drying chamber as for partially observable object. Result had been compared to the conventional PID controller. Our first results show that the hybrid improves the profit by lack of colored particle and general appearance of product, the product rate by 108 kg/hr and the energy efficiency by 7 °C of temperature error at comparable ambient air humidity conditions (Fig. 2).

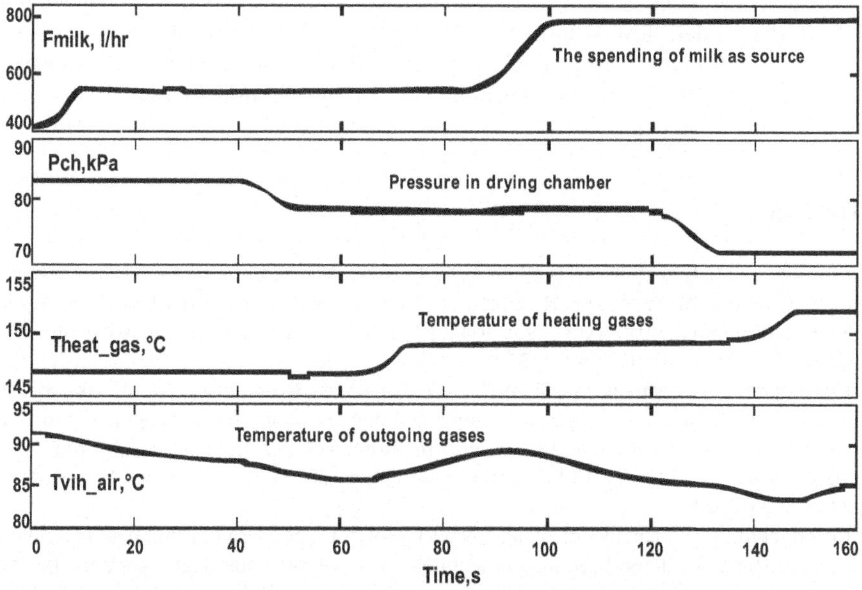

Fig. 2. The increase of fulfill of drying chamber that is being driven by hybrid algorithm

The conventional PID control strategy is simple, but insufficient for providing the optimal operation. In case to change of source flow for another fat content in raw, at start of approximation to technological norm of drying and adds a loss of heat along of finishing of drying operation also. We made this for the spray drying process so that minimizes the quality loss of product by adjustments of the dryer inputs, according to the present internal disturbances and process constraints. The hybrid system in the software application of industrial controller had been successfully implemented by use of Petri automata in form of Graphset language.

6 Conclusion

This paper takes aim to propose a hybrid control algorithm incorporating the inverse dynamic response of estimation and the nonlinear dynamic model for heat processes of the partially observable object. The gain change scheme had been made within the mixing model. The action had been decoupled into separate states for incorporates constraints on the rate of variation without affecting the production of driven flows, for apply the actual dynamic characteristics. The constraints had been imposed on the workspace. In the heat influence model, the actual gain had been calculated by use the inversion chain of nonlinear dynamics for approximation of the heat transfer in object model and had been estimated as states within the estimator workspace.

However, a more advanced comparison by important characteristics of controllers' performance is lacking, such as: the robustness of the controllers under changes in the heat fulfilling model; flow of influence on dynamic behavior; the use of atomizing of fluid flow, and ultimately, irreversible physical influence on product. In this research, we have compared the result of controller by some of the aimed parameters from the working drying chamber.

References

1. Korobiichuk, I., Ladaniuk, A., Ivashchuk, V.: Features of control for multi-assortment technological process. In: Szewczyk, R., Krejsa, J., Nowicki, M., Ostaszewska-Liżewska, A. (eds.) MECHATRONICS 2019. AISC, vol. 1044, pp. 214–221. Springer, Cham (2020). https://doi.org/10.1007/978-3-030-29993-4_27
2. Korobiichuk, I., Ladanyuk, A., Shumyhai, D., Boyko, R., Reshetiuk, V., Kamiński, M.: How to Increase efficiency of automatic control of complex plants by development and implementation of coordination control system. In: Szewczyk, R., Kaliczyńska, M. (eds.) SCIT 2016. AISC, vol. 543, pp. 189–195. Springer, Cham (2017). https://doi.org/10.1007/978-3-319-48923-0_23
3. Korobiichuk, I., Ladanyuk, A., Zaiets, N., Vlasenko, L.: Modern development technologies and investigation of food production technological complex automated systems. In: Paper Presented at the ACM International Conference Proceeding Series, pp. 52–56 (2018). https://doi.org/10.1145/3185066.3185075
4. Kharlamenko, V., Ruban, S., Korobiichuk, I., Petruk, O.: Adaptive control of dynamic load in blooming mill with online estimation of process parameters based on the modified kaczmarz algorithm. In: Szewczyk, R., Kaliczyńska, M. (eds.) SCIT 2016. AISC, vol. 543, pp. 227–233. Springer, Cham (2017). https://doi.org/10.1007/978-3-319-48923-0_28
5. Korobiichuk, I., et al.: Synthesis of optimal robust regulator for food processing facilities. In: Szewczyk, R., Zieliński, C., Kaliczyńska, M. (eds.) ICA 2017. AISC, vol. 550, pp. 58–66. Springer, Cham (2017). https://doi.org/10.1007/978-3-319-54042-9_5
6. Tucsnak, M., Weiss, G.: Observation and Control for Operator Semigroups. Birkhäuser Advanced Texts. Basler Lehrbücher, Birkhäuser Verlag (2009)
7. Sacchelli, L., Brivadis, L., Andrieu, V., Serres, U., Gauthier, J.-P.: Dynamic output feedback stabilization of non-uniformly observable dissipative systems. In: 21st IFAC World Congress (Virtual), Berlin, Germany, pp. 4997–5002 (2020)
8. Besancon, G., Hammouri, H.: On observer design for interconnected systems. J. Math. Syst. Estim. Control **8**(3), 1–25 (1998)

9. Chan, J.C.L., Lee, T.H., Tan, C.P.: Observer-based fault-tolerant control for non-infinitely observable descriptor systems. In: Park, J.H. (ed.) Recent Advances in Control Problems of Dynamical Systems and Networks. SSDC, vol. 301, pp. 123–145. Springer, Cham (2021). https://doi.org/10.1007/978-3-030-49123-9_6
10. Długosz, M., Baranowski, J.: Observer design for estimation of nonobservable states in buildings. Math. Probl. Eng. **2020**, 1–8 (2020). https://doi.org/10.1155/2020/3404951
11. Jo, N., Seo, J.: Observer design for non-linear systems that are not uniformly observable. Int. J. Control **75**(5), 369–380 (2002). https://doi.org/10.1080/00207170110112287
12. Jover, C., Alastruey, C.F.: Multivariable control for an industrial rotary dryer. J. Food Control **17**(8), 653–659 (2006)
13. Utkin, V.: Sliding modes in control and optimization. Springer, Heidelberg (1992). https://doi.org/10.1007/978-3-642-84379-2
14. Kolesov, Y., Senichenkov, Y.: Mathematical Modeling of Complex Dynamical Systems. Polytech-Press, St. Petersburg (2019)
15. Kim, A.V.: I-Smooth Analysis. Theory and Applications: monograph. Institute of Mathematics and Mechanics, Ural Branch of Russian Academy of Sciences (2015)
16. Pukdeboon, C.: Lyapunov optimizing sliding mode control for attitude tracking of spacecraft. Int. J. Pure Appl. Math. **80**(1), 25–40 (2012)

The Optimization Model of the Production Process for Corporate Architecture

Zbigniew Juzoń(✉), Jarosław Wikarek, and Paweł Sitek

Kielce University of Technology, Al. Tysiąclecia P.P. 7, 25-314 Kielce, Poland
{zjuzon,j.wikarek,sitek}@tu.kielce.pl

Abstract. The article presents and discusses the issues related to the area of production process management in terms of enterprise corporate architecture to increase the volume and efficiency of production processes. The enterprise architecture also defines the parameters of the technological process and also enables the management of a complex set of auxiliary activities. At the stage of implementing automation and robotization of production processes, the aim of which is to replace (or at least significantly reduce) the involvement of people in the production process, modern technologies are usually applied (mobile robots, AVG, PLC, IoT, RFID, cloud computing, etc.). As a result, there is a problem with the proper allocation and configuration of production resources. All this has a strong influence on the company's position in the market. There is a contradiction between planning and value orientation in production. The ability to manage production plans is the key to market advantage. The article proposes a model for the optimization of the production process for corporate architecture in the form of a mathematical programming problem.

Keywords: Process management · Enterprise architecture · Automation · Robotization · TOGAF · Mathematical programming

1 Introduction

The framework for Corporate Architecture is the standard Open Group Architecture Framework (TOGAF) [1], which can be freely used by any organization. Although the TOGAF standard works as a whole, it is advisable for an organization planning to implement the TOGAF meta-model, also at the level of technological layer management, to use only those elements that will have an impact on maintaining the architectural order also at the level of the manufacturing process. A modern factory, in accordance with the idea of "Industry 4.0" [2], is a combination of intelligent machines and systems that are introduced at the level of the manufacturing process in order to increase production efficiency and maintain the highest production standards. It should be emphasized that Industry 4.0 concerns not only the technology itself but also a new, structured and defined approach to the way work is performed and the role of people in the industry. The introduction of corporate architecture at the organizational level as a result of a business decision makes it possible to apply the automation and robotization of these

R. Szewczyk et al. (Eds.): AUTOMATION 2022, AISC 1427, pp. 116–125, 2022.
https://doi.org/10.1007/978-3-031-03502-9_12

production processes, where repetitive activities are identified and it is possible to plan the sequence of work at the stages of the production line. For modeling enterprise architecture, ArchiMate [3] open and independent enterprise architecture modeling language can be used, which supports description, analysis and visualization within the four architectural domains (*business, data, application and technology*) defined under the TOGAF standard. ArchiMate is based on two paradigms:

- layering *(it introduces the business, data and application and technical layers),*
- service *(business, application and infrastructure services are identified).*

The Business Process Model and Notation (BPMN 2.0) [5] standard described using the ISO/IEC 19510 standard is ideally suited to describe business services. Currently, there are many IT tools available on the market that support the process of modeling corporate architecture. An example of such a tool is the commercial Enterprise Architect (EA) software from Sparx Systems.

2 Problem Description

After the implementation of the corporate architecture at the production company level, the next step for the selected production processes will be the use of automation and robotization to replace or at least significantly reduce the involvement of people. The typing of technological processes to be automated and robotized is done by using in particular the Process Mining or Task Mining concepts. Dedicated data collecting and processing systems are responsible for automation and optimization, and the ability to manage production plans is the key to market advantage. It should be emphasized that modern technologies in the field of automation, robotics and IT. Of course, implementation of the process approach at the process level production significantly facilitates the process of implementing automation and robotization but also reveals the problem of allocation and configuration of these production resources. However, in the case of industrial robots, it is necessary to take into account the necessary auxiliary activities, such as configuration, kinematics and dynamics, and the use of classical machine learning algorithms (*e.g. decision trees*). However, in the case of a software robot, the problem comes down to the construction, testing and implementation of dedicated simulation and decision-making software in the absence of a complete data set based solely on numerical algorithms built on the effects of the Artificial Intelligence engineering process [4]. Therefore, in this context, there are, among others, key questions:

(Q1): *How to plan the management of the production process to maintain the company's high position on the market?*

(Q2): *How to optimize the production process?*

Below is the answer to the question Q1.

To answer for Q1 a process approach should be implemented at the level of the production process. Therefore in the first step, it is necessary to introduce management standards, which should define the following elements according to the Open Group Architecture Framework (TOGAF) architectural framework at the enterprise level:

- business architecture that describes the business strategy, ways of managing the organizational structure and key business processes,
- a data architecture that describes the main types and sources of data necessary for the functioning of the corporation,
- application architecture that describes individual software systems, their location, mutual interaction and relations between these systems and the main business processes of the corporation,
- technological architecture that describes the technological layer.

The organization should develop, inter alia, the Corporate Architecture Management process, which defines the principles of building and maintaining the architectural order at the level of the entire organization to the extent adapted to the level of architectural maturity of the organization (Fig. 1).

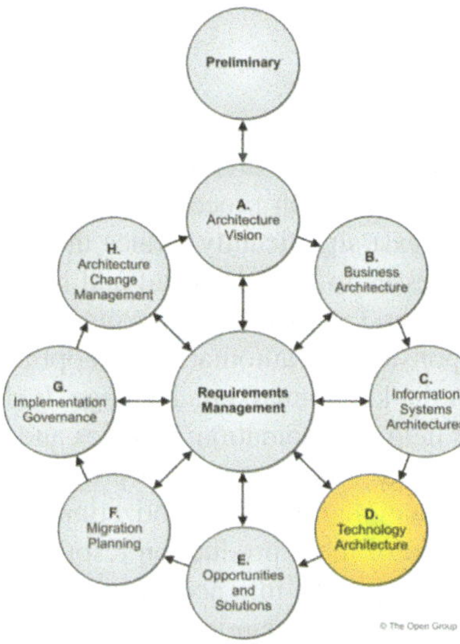

Fig. 1. Architecture development cycle *(source:* https://pubs.opengroup.org/architecture/togaf9-doc/arch/Figures/adm.png*)*

The basis for maintaining the architectural order is the implementation of appropriate corporate architecture management mechanisms. Implementation of the corporate architecture management process allows for effective business decision-making in the area of a manufacturing enterprise while maintaining a controlled and rapid introduction of changes at the level of the production process. An important step at the stage of implementing the process approach at the organization level is the modeling of business processes. Process modeling aims to determine how an organization operates and can be used to define the target course of action. The process approach allows for the automation and robotization of production processes, delivery processes or the maintenance of appropriate stock levels. Skillful management of corporate architecture elements at the organizational level will translate into ensuring a high position of the company in the changing economic environment.

3 Illustrative Example

The factory produces 2 products that can be marked with the symbols "A" and "B". These products are made of 4 types of semi-finished products: a, b, c, d. A single product "A" is made of semi-finished products used in the following amounts: 2a, b, 3c, 2d.

A single product "B" is made of semi-finished products used in the following amounts: 2a, 2b, 2c, 3d. In turn, the unit of intermediate "c" consists of: 2a, b. Whereas the unit of intermediate "d" consists of: 2a, 2b. The production capacities of individual departments in the production plant where these semi-finished products are made are as follows. Production department "D1" can produce at most 1200 "A" intermediates. Production department "D2" can produce at most 1500 semi-finished products "B".

Production department "D3" can produce at most 500 "c" intermediates. Production department "D4" can produce at most 600 "d" intermediates. The assembly department "D5" has a production capacity of no more than 200 units of products "A" and "B". The presented example shows the problem of the "lack of production capacity" of individual departments.

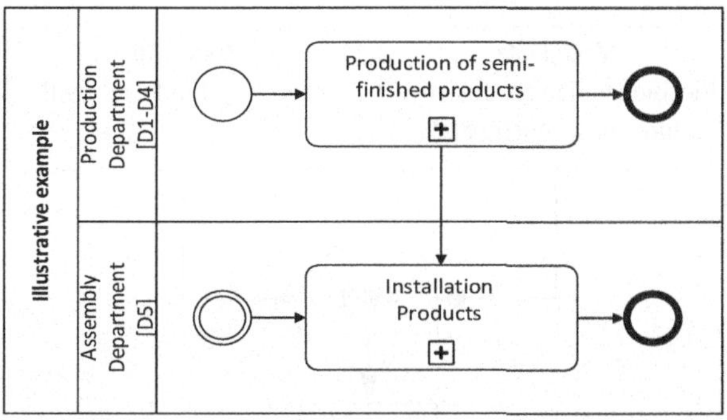

Fig. 2. BPMN diagram for illustrative example production process

The price for product "A" is at level 1 USD and the price for product "B" is 2 USD. BPMN (Business Process Model and Notation) diagram for illustrative example production process has been shown in Fig. 2.

4 Mathematical Model of Production Process Optimization

At the stage of designing corporate architecture, and in particular at the level of the technological layer, all auxiliary activities related to the production resources used at the level of the manufacturing process should be taken into account. All these above-mentioned elements have one thing in common, they are changeable over time and are configurable. Finding the answer to this question is the key to the optimal implementation of production tasks, because thanks to the appropriate configuration of production re-sources, we will be able to meet the production schedule, regardless of the lack of availability of the resource or its functionality. Proper configuration can ensure the performance of specific tasks in accordance with the agreed schedule when re-sources are unavailable. Therefore, in order to answer the question (*Q2: How to optimize the production process?*), it is necessary to develop an appropriate mathematical model and on its basis select parameters for the configuration of production resources. With the use of the AMPL system [6], it is possible to solve the problem of allocation of production resources and their configuration on the basis of the developed mathematical model, and then select the appropriate configuration for the entire production system. It is important that also on the basis of such a mathematical model, change management can also be built by modifying the parameters of the configuration of production resources, introducing a change at the level of parameters of indicators and defining new constraints. By introducing appropriate modifications on the mathematical model, you can easily and quickly manage the area of production costs while maintaining the continuity of the production process. The AMPL system consists of several basic elements, such as: algebraic modelling language, solvers and an interactive command environment. The method of functioning of the AMPL system is presented in Fig. 3.

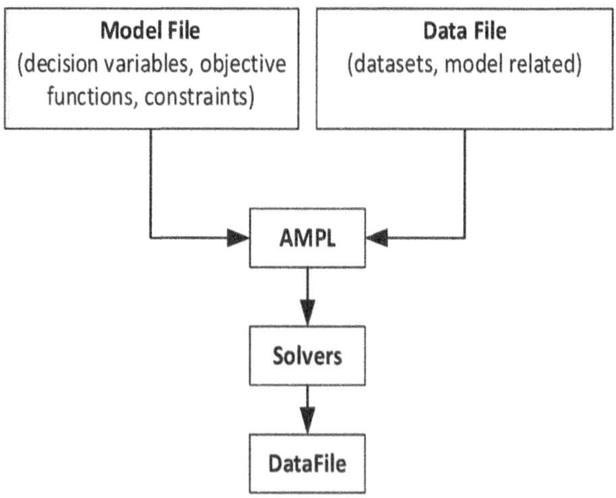

Fig. 3. Structure of AMPL (developed on the basis of [5])

In this program, the model file can be run without referring to other files, or it is possible to separate data from the model. This is beneficial, especially when creating and solving very large and complex optimization problems. One of the many MP (Mathematical Programming) or CP (Constraint Programming) solvers such as Gurobi, CPLEX, Baron, LINGO, etc. [10] can be used to solve the AMPL model.

Formalization of Mathematical Model

The company can produce products from components. We mark any product or component with $\boldsymbol{i}$ where **I** means a collection of all components/products ($\boldsymbol{i} \in \boldsymbol{I}$). The any product $\boldsymbol{i}$ it can be produced and it can be a component. With the help of the coefficient $\boldsymbol{a_{i,k}}$ we will determine how much component $\boldsymbol{k}$ need to manufacture any product $\boldsymbol{i}$ ($\boldsymbol{i} \in \boldsymbol{I}$, $\boldsymbol{k} \in \boldsymbol{I}$). All products are made on individual departments in this production plant, therefore enter the next designations ($\boldsymbol{d}$ – a factories index; $\boldsymbol{D}$ – a collection of all factories $\boldsymbol{d} \in \boldsymbol{D}$). The coefficient of $\boldsymbol{b_{d,i}}$ means how much of a given product $\boldsymbol{i}$ can produce factories $\boldsymbol{d}(\boldsymbol{i} \in \boldsymbol{I}, \boldsymbol{d} \in \boldsymbol{D})$. In the considered case, we assume that the production capacity of individual departments in this production plant will be denoted by the coefficient $\boldsymbol{b_{d,i}}$ while using the coefficient $\boldsymbol{c_i}$ let us denote the profit from the manufactured product $\boldsymbol{i}(\boldsymbol{i} \in \boldsymbol{I})$ (Table 1).

Table 1. Defining elements of the mathematical model

Symbol	Components of the model	Description
$\boldsymbol{I}$	**Sets**	Set of products
$\boldsymbol{D}$		Set of factories
$\boldsymbol{i}, \boldsymbol{k}$	**Indexes**	Product index $\boldsymbol{i} \in \boldsymbol{I}$, $\boldsymbol{k} \in \boldsymbol{I}$
$\boldsymbol{d}$		Factories index $\boldsymbol{d} \in \boldsymbol{D}$
$\mathbf{a_{i,k}}$	**Parameters**	How much component $\boldsymbol{k}$ needs to manufacture the product $\boldsymbol{i}(\boldsymbol{i} \in \boldsymbol{I}, \boldsymbol{k} \in \boldsymbol{I})$
$\mathbf{b_{d,i}}$		How much product $\boldsymbol{i}$ can produce factories $\boldsymbol{d}(\boldsymbol{i} \in \boldsymbol{I}, \boldsymbol{d} \in \boldsymbol{D})$
$\mathbf{c_i}$		Profit from the production of the product $\boldsymbol{i}(\boldsymbol{i} \in \boldsymbol{I})$
$\boldsymbol{X_i}$	**Decision variables**	Production volume $\boldsymbol{i}(\boldsymbol{i} \in \boldsymbol{I})\boldsymbol{X_i} \geq 0, \boldsymbol{X_i} \in \boldsymbol{N}$
Z	**Determined values**	Profit on sales
(1)	**Constraints**	Profit on sales (goal function)
(2)		Production capacity of departments
(3)		Product structure (components into the product)
(4)		Integrity of variables
(5)		Non-negative

$$Z = \sum_{i \in I} c_i \cdot X_i \tag{1}$$

$$\sum_{i \in I \forall b_{d,i} \neq 0} \frac{X_i}{b_{d,i}} \leq 1 \forall d \in D \tag{2}$$

$$\mathbf{X_k} = \sum_{\mathbf{i} \in \mathbf{I}} \mathbf{a_{i,k}} \cdot \mathbf{X_i} \forall \mathbf{k} \in \mathbf{I} \wedge \sum_{\mathbf{i} \in \mathbf{I}} \mathbf{a_{i,k}} \neq 0 \quad (3)$$

$$X_i \in C \, \forall i \in I \quad (4)$$

$$X_i \geq 0 \, \forall i \in I \quad (5)$$

5 Numerical Experiments

In order to select the optimal production process and determine the maximum production value of products "A" and "B" for an illustrative example, the mathematical model (1)..(5) was used. This is because solving this problem manually, even for small data instances, is time-consuming and error-prone. The experiments were conducted for different data instances (Table 2). All experiments were made on a computer with the following specification:

- Processor: Intel(R) Core(TM) i5-7300U CPU @ 2.60GHz 2.70 GHz
- RAM: 8 GB
- OS: Windows 10 Pro

Table 2. Details of the conducted experiments (*sets, profit of sales, calculation time*)

Experiment number	Set of products [SI]	Set of factories [SD]	Profit on sales [Z]	calculation time
1	4	5	**200**	**0.27 s**
2	4	10	**196**	**0.29 s**
3	4	15	**196**	**0.25 s**
4	8	5	**200**	**0.31 s**
5	8	10	**196**	**0.35 s**
6	8	15	**196**	**0.30 s**
7	12	5	**200**	**0.33 s**
8	12	10	**196**	**0.29 s**
9	12	15	**196**	**0.33 s**

The data, the implementation model in AMPL and the runtime script for experiment 1 are presented in Fig. 6, Fig. 4 and Fig. 5, respectively. The obtained detailed results for experiment number 1/values of all decision variables/are given in Fig. 7 of Appendix A. The Gurobi environment [12] was used to solve the modeled problem.

6 Conclusion

Although the TOGAF standard works as a whole, it is advisable that the organization for the development process already at the stage of implementing the TOGAF meta-model, use only those elements that will affect all layers of the organization's corpo-rate architecture. The application of corporate architecture at the level of the production system enables the grouping of technological processes and the classification of various technological solutions. The advantage of such an action is the ordering of business processes and, consequently, the identification of tasks that can be automated. Unfortunately, the very process of creating an architectural model cannot be automated. Therefore, in the first place, it is necessary to build a set of good practices for modeling enterprise architecture that has worked well in various projects. Typically, the stages of working on architectural models come down to determining the scope of information based on which the future model will be based. However, an important feature of this activity related to the use of corporate architecture elements is the ability to model business processes, which facilitates the implementation of process management and, consequently, building a mathematical model. The above may be the basis for supporting decisions in the production system, in particular in the selection and configuration of resources, optimization of production costs, acceptance or rejection of a new project, etc. Implementation of a mathematical model using specialized software, e.g. AMPL enables planning of production processes using mathematical programming and constraint programming solvers, hybrid approach [7, 8] as well as artificial intelligence to solve larger problems related to the configuration and allocation of production re-sources. In future works, it is planned to apply the proposed model or its extended version to specific production systems organized in the form of smart production lines [9], production cells, and distribution processes with UAV [11].

Appendix A. Scripts for the Illustrative Example

```
set SI;
 set SD;
 param  a{SI,SI};
 param  b{SD,SI};
 param  c{SI};
 var    X{SI} >=0, integer;
 var    Z;
 subject to O1: sum{i in SI} c[i]*X[i]=Z;
 subject  to  O2  {d  in  SD}  :  sum{i  in  SI:
b[d,i]<>0} X[i]/b[d,i] <=1;
 subject to O3 {k in SI: sum{i in SI} a[i,k]
<> 0}:
   X[k]= sum{i in SI} a[i,k]*X[i];
 maximize fc: Z;
```

Fig. 4. Model in AMPL

```
model z.mod;
data  z.dat;
option solver gurobi;
solve;
display {j in 1.._nvars: _var[j] > 0}
(_varname[j], _var[j])>a1.txt;
```

Fig. 5. The source code for the runtime script

```
data;
set SI :=A B a  b c d;
set SD :=D1 D2 D3 D4 D5;
param c:= A 1 B 2 a 0 b 0 c 0 d 0;
param a: A B a b c d :=
  A      0 0 2 1 3 2
  B      0 0 2 2 2 3
  a      0 0 0 0 0 0
  b      0 0 0 0 0 0
  c      0 0 2 1 0 0
  d      0 0 2 2 0 0;
param b: A        B   a b c d:=
  D1     100    100   0 0 0 0
  D2     187.5 150    0 0 0 0
  D3     166.6 250    0 0 0 0
  D4     300    200   0 0 0 0
  D5     200    200   0 0 0 0;
```

Fig. 6. Input data

```
:  _varname[j] _var[j]     :=
2    "X['B']"       100
3    "X['a']"      1200
4    "X['b']"      1000
5    "X['c']"       200
6    "X['d']"       300
7    z              200
;
```

Fig. 7. Results

References

1. The Open Group Architecture Framework - TOGAF Standard. https://pubs.opengroup.org/architecture/togaf9-doc/arch/index.html, Accessed 2018
2. Bartodziej, Ch.J.: The Concept Industry 4.0: An Empirical Analysis of Technologies and Applications in Production Logistics. Springer Fachmedien GmbH, Wiesbaden (2016)
3. The Open Group. ArchiMate 3.0 Specification. Van Haren (2016)
4. Artificial Intelligence – AI. https://robonomika.pl, Accessed 2021
5. Bartoszewicz, G.: The use of BPM process management methods for the of purpose increasing the efficiency of complex logistics 4.0 processes in ERP 2.0 systems. In: Cyplik, P., Adamczak, M. (eds.) Selected Problems of Contemporary Logistics in the Light Research and Business Practice, pp. 12–37 (2017)
6. Holms, D.: AMPL at the University of Michigan. https://ampl.com, Accessed 1995
7. Sitek, P., Wikarek, J.: A multi-level approach to ubiquitous modeling and solving constraints in combinatorial optimization problems in production and distribution. Appl. Intell. **48**(5), 1344–1367 (2017). https://doi.org/10.1007/s10489-017-1107-9
8. Wikarek, J., Sitek, P.: A data-driven approach to constraint optimization. In: AUTOMATION 2019. Advances in Intelligent Systems and Computing, vol 920. Springer, Cham (2020). https://doi.org/10.1007/978-3-030-13273-6_14

9. Rahman, H., Janardhanan, M., Nielsen, P.: An integrated approach for line balancing and AGV scheduling towards smart assembly systems. Assemb. Autom. **40**(2), 219–234 (2020). https://doi.org/10.1108/AA-03-2019-0057
10. AMPL. https://ampl.com, Accessed 2021
11. Thibbotuwawa A., Bocewicz G., Radzki G., Nielsen P., Banaszak Z.: UAV mission planning resistant to weather uncertainty. Sensors (Basel) **20**(2), 515 (2020). https://doi.org/10.3390/s20020515
12. Gurobi. https://www.gurobi.com, Accessed 2021

Hardware and Software Design of Onboard Computer Controlling the Flight Stabilisation System

Maciej Filipowicz(✉) and Cezary Szczepański

Avionics Division, Łukasiewicz Research Network – Institute of Aviation, Warsaw, Poland
{maciej.filipowicz,cezary.szczepanski}@ilot.lukasiewicz.gov.pl

Abstract. The paper presents the design and performance of the flight stabilisation system onboard computer developed in Łukasiewicz Research Network – Institute of Aviation. The presented unit is an electronic block performing automatic flight stabilisation of an aeroplane using trimmers. It was designed based on the launch platform PZL-130 Orlik aircraft requirements. The article contains the unit's hardware and software design description, unit performance, tests results and conclusions with improvements possibilities.

Keywords: Flight trimming · Automatic flight stabilisation computer · Avionics · Onboard computer · Trim tab

1 Introduction

A typical aeroplane trimming system consists of a set of trim tabs placed at the aileron, elevator and rudder or any combination of these as a part of the control system in the pitch, yaw, and roll channels. The purpose of the trimmers is to compensate for various undesirable moments of forces that can be caused by improper weight distribution or non-symmetrical moments caused by a single high-power turbo propeller engine operation [1]. In addition, using trimmers to balance the aircraft reduces the manual effort of the pilot.

Presented Onboard Computer is part of the flight stabilisation system designed for a turboprop aeroplane PZL-130 Orlik aircraft with a complete three-channel trimming system. The trim tabs are controlled from the cockpit by wire (set of buttons and connected servo actuators) in this aircraft, and primary control surfaces use conventional manual flight controls. Such configuration is typical. Therefore, the developed system can be easily adapted from the hardware point of view on other aircraft types.

The automatic stabilisation system integrates the original functionality of manual trim tabs control with automatic features, which can be used in two modes of operation: full three-channel and rudder channel only. The second control mode is dedicated to single turboprops. In such an aeroplane, the propeller effect causes non-symmetrical thrust force and spiralling slipstream generated by the running propeller, which should be compensated by proper rudder trim tab position.

R. Szewczyk et al. (Eds.): AUTOMATION 2022, AISC 1427, pp. 126–132, 2022.
https://doi.org/10.1007/978-3-031-03502-9_13

2 Onboard Computer as a Part of Trimmer Stabilisation System

The purpose of designing a new stabilisation system was to improve the flying qualities, which increases the safety and comfort of the pilot. It was planned to be used on a specific platform – the PZL-130 Orlik aeroplane. Demonstration platform requirements shaped the architecture of the developed system, where the main design assumptions were:

- using the existing trim control buttons and indicators,
- allowing control from the two cabins,
- using available flight data,
- currently mounted trimmers servomechanisms should be replaced by a newly designed,
- electronics should be a standalone unit mechanically matched to the existing mount.

These assumptions defined the system architecture (presented in Fig. 1) as an electronic unit with integrated servomechanisms drivers and three dedicated motors electrically connected with the control unit.

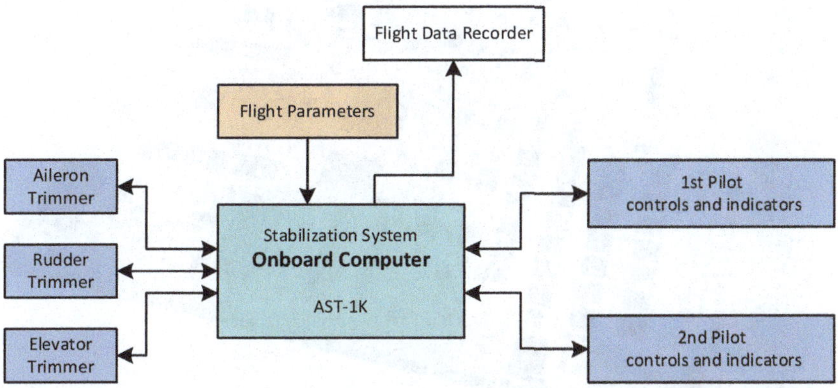

Fig. 1. Trimmer Stabilisation System architecture

The stabilisation system's central computing unit is the Onboard Computer. It is responsible for data signals exchange with a software controller algorithm. The developed system uses the chosen flight parameters to calculate desired positions of control trimming surfaces controlled by servomechanisms.

3 Onboard Computer Design

3.1 Hardware

The onboard computer hardware unit was designed as a specialised, low-power, low-cost electronic device. Therefore, we can specify two main hardware modules inside the unit: main PCB and servomotors drivers PCB. These two modules are interconnected by interface PCB, which additionally contains unit I/O connectors.

The main PCB module is presented in Fig. 2. It integrates CPU section, power-management circuit and input-output circuits. The CPU section contains a microcontroller (MCU), flash memory and specialised Commercial Off The Shelf (COTS) integrated circuits like transceivers and digital inputs interfaces. Used MCU is a relatively low-cost, high-performance Arm-based unit. Due to the research nature of the project and fast prototyping possibilities, COTS integrated circuits are not directly intended for aerospace usage; however, in restricted applications, they can be easily qualified.

Furthermore, high performance embedded MCU and large flash memory (8 Gbit capacity) significantly reduced the time needed to obtain a fully functional model of the final device. Specific advantages worth mentioning are the following: long, precise and frequent system state logging needed for analysis after laboratory tests, integration and flight tests, and fast USB communication. These additional maintenance functionalities do not slow down or interrupt the main control algorithm.

Fig. 2. The Main PCB module

The power management section, located on Main PCB, is responsible for digital and analogue circuits' protection and voltage conversion. The power module was designed and tested to meet DO-160G standard specification for lightning protection, ESD protection and digital circuits power sustained during short time failure. It protects the device from high voltage transients and stops surges from the power source. On the other hand, it also includes a timed, current-limited circuit breaker to keep the unit safe after internal failure. All input and output signals are protected according to the DO-160G specification.

Drivers PCB includes three circuits responsible for controlling motors as servomechanisms. Integration of the three independent actuators drivers inside the Onboard Computer allowed avoiding electronics exposure to environmental factors. It is enough to say that the angular position signal generated by MCU controls the drivers. Motors drivers are fully analogue. They include the regulator and use voltage feedback position signals from the actuators.

The Flight Stabilisation System Onboard Computer model (named AST-1K), containing PCB modules, is presented in Fig. 3. It is hard mounted without anti-vibration supports due to laboratory tests that have proven resistance to vibrations and shock impacts according to the DO-160G standard.

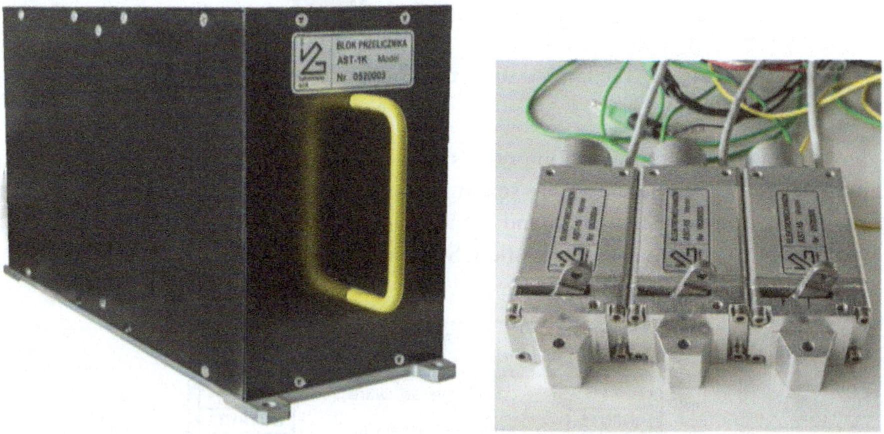

Fig. 3. Onboard Computer unit (left), Trim tabs servomechanisms (right)

3.2 Software

Onboard Computer software was implemented in a very perspective design method that combines hand-writing code and automatic code generation from the developed model. From a functional point of view, it can be well described by dividing software into two layers (Fig. 4): low level software typed by a programmer and high level generated automatically by a MATLAB/Simulink package. The software was implemented without operating system.

Low level software (LLS) code has been written in C programming language by a programmer. In this layer, all peripherals are handled. It is in particular prepared for porting high level software automatically generated code from the Simulink model containing system state machine and controller model. It can be understood that LLS is responsible for data flow to and from the control algorithm, but it also has other functionalities (Fig. 5). The LLS is responsible for failure detection and reporting it to the pilot. As mentioned in a hardware design description, the USB diagnostic interface is also entirely handled in LLS. This interface allows, among other things, detailed diagnostics, hardware configuration and controller model tuning (also during the test

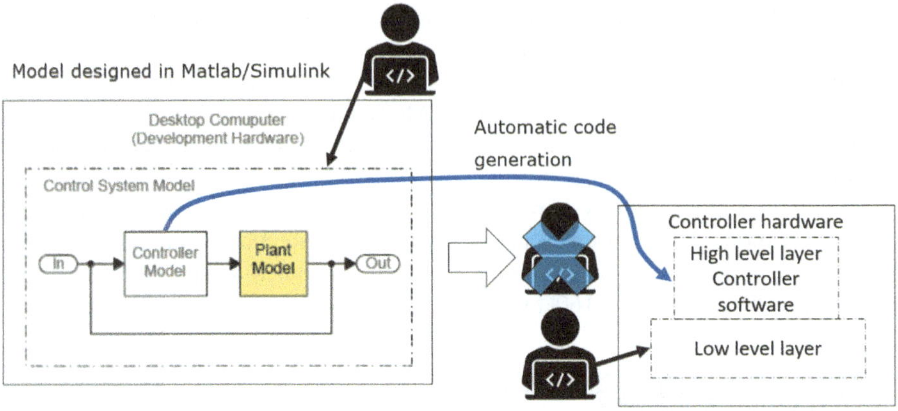

Fig. 4. The layered structure of software design and test

flights). In hardware configuration options also can be listed: servomechanisms ranges scaling, indicators and inputs ranges scaling – these parameters are stored in non-volatile EEPROM memory. System state and data are logged into internal flash memory, and they can be easily read via a diagnostic USB interface.

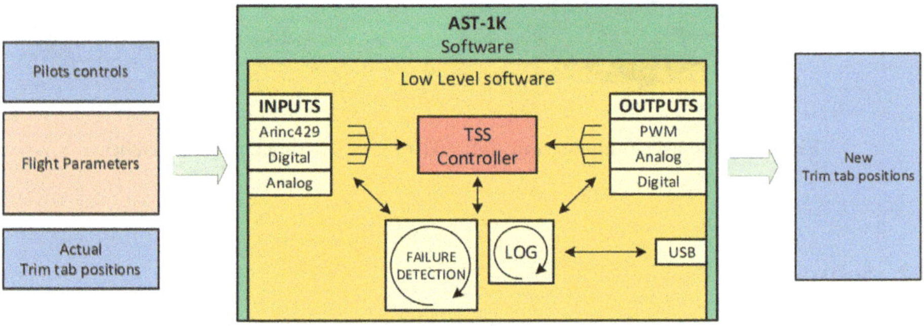

Fig. 5. Software architecture with inputs and outputs

High-level software (HLS) can be simplified to the system state machine and controller model translated into C code as a set of structures and functions. Advantages of such an approach are: traceability from code to model, the graphical software definition and feasibility of testing in simulations, including Aircraft Model and Pilot. Moreover, automatic code generation from the model is free from man-made programming mistakes.

The whole system, including software and hardware, was precisely tested during the extensive Hardware in the Loop (HIL) simulation. The laboratory HIL tests of the presented stabilisation system were documented in a separate article [2].

4 Unit Configuration Specification and Adaptation Possibilities

Stabilisation System integration, ground tests, hardware and system-level verification and validation, were performed on the PZL-130 Orlik MPT aeroplane at the Airbus Poland facility in Warsaw. As a result, the flight stabilisation system Onboard Computer complied with the environmental requirements according to the DO-160G standard and passed all functional ground tests previously confirmed in a laboratory. In addition, tests strictly connected with the onboard computer included electrical connections, data integrity and functional behaviour.

As was mentioned, Onboard Computer and the whole system were designed with PZL-130 Orlik aeroplane requirements. Design for a specified platform requires specific cockpit controls and onboard sensors, but it can be easily adapted for the other platforms. For a proper system operation, several onboard signals must be connected:

- airspeed,
- barometric height,
- pitch, roll and yaw angles,
- pitch and roll rates,
- landing gear position and touchdown information,
- flaps position information,
- engine torque (only for a single-engine turboprop).

Flight data can be classified as critical to performing stabilisation of the aeroplane, but information about landing gear position is also required for safety reasons. In addition, engine torque should be provided for application which needs compensation of propeller effect. All input data can be provided via reconfigurable interfaces, including Arinc 429 (four channels available), analogue inputs (seven channels with range 0–5 V with programmable linearisation available) and several logical inputs.

The applied power supply is typical for this class of aircraft – nominal input voltage is 28 V DC, but unit and servomechanisms can be supplied from 18 V to 32.2 V DC range without any restrictions. Therefore, power supply safety was classified as DO-160G cat. B. The operating temperature range of the Onboard Computer is –35–70 °C. Therefore, temperature and altitude meet the category C4 requirements of DO-160G.

5 Conclusions

Presented Flight Stabilisation System Onboard Computer is a fully functional model of that device. All design assumptions were achieved. After several testing cycles in a laboratory and inside the aircraft, the requirements were fully verified.

- Presented Onboard Computer is an electronic unit designed as a specialised, low-cost, low power consumption, highly integrated electronic device.
- Built-in servomechanisms drivers and integrated cabling interface help to decrease needed space and weight.
- The modular structure of electronics simplifies repairs and modifications.

- The system's structure and parameters allow for simplified implementation on other aircraft types:
  - porting the automatically generated code from the model as the main control algorithm,
  - multiple input interfaces,
  - standard power supply range input.
- Onboard Computer passed all the required ground tests according to the DO-160G standard and confirmed functionalities tested in HIL simulations.
- Built-in logging functionality can help with the diagnosis in case of system failures.

Acknowledgements. Researches described in that paper have been funded under the EU co-financed project number POIR.04.01.02–00-0006/17–00, titled "Innovative system of flight stabilisation with use of trimmers" – ISSLOT.

References

1. Krawczyk, M., Zajdel, A.: Trimming surfaces in the automatic plane control process. J. KONES **24**(1), 223–230 (2017). https://doi.org/10.5604/01.3001.0010.2818
2. Krawczyk, M., Zajdel, A., Szczepański, C.: Simulation and testing of flight stabilisation system using trimmers. In: Szewczyk, R., Zieliński, C., Kaliczyńska, M. (eds.) AUTOMATION 2021. AISC, vol. 1390, pp. 185–196. Springer, Cham (2021). https://doi.org/10.1007/978-3-030-74893-7_18
3. Szczepański, C., Krawczyk, M., Zajdel, A.: The airplane trim system – new functionalities. Aircraft Eng. Aerosp. Technol. **92**(9), 1401–1406 (2020). https://doi.org/10.1108/AEAT-12-2019-0241
4. Krawczyk, M., Rogalski, T.: General overview of experimental control system developed on I-23 manager aircraft. Trans. Inst. Aviat. **4**(206), 93–102 (2010)
5. European Aviation Safety Agency, CS-23-Certification Specifications for Normal, Utility, Aerobatic, and Commuter Category Aeroplanes (2012)
6. SAE International, Guidelines for Development of Civil Aircraft and Systems ARP4754A (2010)
7. Radio Technical Commission for Aeronautics, Environmental Conditions and Test Procedures for Airborne Equipment DO-160G (2010)
8. Eddleman, D.: High voltage surge stoppers ease MIL-STD-1275D compliance by replacing bulky passive components. LT J. Analog Innov. (2014)

Implementation of the Ship's Autopilot in the CPDev Environment

Zbigniew Świder(✉)

The Faculty of Electrical and Computer Engineering, Rzeszów University of Technology, Rzeszów, Poland
swiderzb@prz.edu.pl

Abstract. Ship autopilots can be divided into conventional, only capable of maintaining a given course (heading), and advanced, which can additionally keep the ship on the "path" (track) connecting the given navigation points along the route. The article presents the structure of a prototype autopilot of a ship implemented in the CPDev environment and the formulas allowing to determine the settings of the course controller (PID) and the track controller (PI) in the cascade control. For each of them, single design parameters were adopted to define the dynamics of the closed control loop. These rules were applied in the software of the autopilot prototype, created in cooperation with a Dutch company designing control and visualization systems for ships.

Keywords: Ship's autopilot · PID controller · Tuning rules · Engineering environment

1 Introduction

The ship's course is usually controlled by autopilots, which automatically correct the course using the built-in steering devices. Such an adaptive autopilot adjusts the course controller settings depending on the ship's speed, sea state and other factors, and in addition to maintaining a fixed course set by the operator, it is able to keep the ship on the "path" connecting subsequent navigation points, despite interference from e.g. sea currents or the influence of wind. Such a system requires the installation of an area navigation module that calculates any deviation from the programmed route and then automatically makes the necessary corrections. Companies such as Sperry Marine, Raytheon, Kongsberg and many others offer both conventional and highly advanced autopilots, including those with adaptive functions. Most of the literature on autopilot for ships concerns complex algorithms based on the model [1, 11]. The second group includes, for example, genetic algorithms [9], fuzzy regulators [12] or neural networks [13].

The author would like to thank the co-authors of the autopilot software and the entire CPDev community, in particular Leszek Trybus, Bartosz Trybus, Jan Sadolewski, Dariusz Rzonca, Andrzej Stec and Marcin Jamro.

R. Szewczyk et al. (Eds.): AUTOMATION 2022, AISC 1427, pp. 133–142, 2022.
https://doi.org/10.1007/978-3-031-03502-9_14

Work on a modern and universal engineering environment, compliant with the IEC 61131-3 standard, started over 10 years ago [7]. The environment developed at that time was called CPDev (*Control Program Developer*) and from the very beginning it supported the possibility of independent implementation on the target equipment, both by engineers in industry and students at universities. Then, extensions for 32- and 64-bit processors were added to the package, supplemented with LD, FBD, SFC graphic language translators, a HMI design tool and a multi-project runtime environment. The CPDev environment has been implemented and is still used by several major manufacturers of automation devices in the country and abroad.

Rzeszów University of Technology has been cooperating with Praxis Automation Technology B.V. in the Netherlands for many years, with the implementation of the CPDev [6] package for the development of ship monitoring and navigation systems. Recently, joint work on the autopilot prototype [8] has started, with Rzeszów responsible for its software implementation. One of the significant problems was the rules of tuning of individual PID blocks of the autopilot, because the classic manuals [3–5] are not detailed enough in this respect. For practical reasons, rules were sought which, apart from the requirements of no oscillation and overshoot, would be able to determine the rate dynamics using design parameters understandable to the operator. This dynamics strongly depends on the structure of the ship and the state of the sea.

The article presents the engineering environment CPDev, in which the autopilot module was implemented, and the basics of calculating the set-points for the course controller (classic PID controller block) and the "path" controller (PI block or state space controller). The autopilot prototype was tested on the full (non-linear) model of the 4-DOF container ship [14], also implemented in the CPDev package, and selected results from the simulation tests are presented below.

2 CPDev Engineering Environment

An engineering environment such as the CPDev package presented here can be considered: (1) universal, if it allows generating executable code for various hardware platforms, (2) open in terms of software and hardware, if tools for creating custom blocks are available for the end user functional and specifications of input and output interfaces, (3) flexible, if it is possible to freely select the appropriate data types for a given application.

The first version of CPDev, initially supporting only the ST language, was extended in the following years to support other languages of the IEC 61131-3 standard, graphic design of HMI interfaces, automatic documentation generator and additional modules for research applications. Figure 1 shows the stages of processing control programs written in different languages of the standard. Text languages (ST, IL) are here directly compiled into mnemonic codes developed for the needs of the VMASM specialized language environment. LD, FBD and SFC graphical charts are first converted to ST language and then compiled into VMASM. The ST language is thus the base language of the CPDev environment. The executable binary code of the program is generated from VMASM mnemonics and loaded into the virtual machine. Virtual machine sources are written entirely in C, which greatly facilitates portability to various hardware platforms.

The compiler uses a set of ST language rules, functional block libraries, and a list of elementary instructions executed by the virtual machine. In addition to the executable

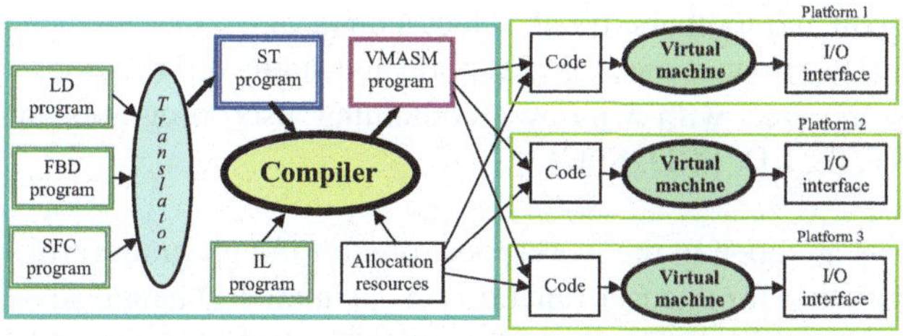

Fig. 1. Stages of software development for various hardware platforms

code, information is generated for tracking, running, and simulating the program. The hardware resource configuration covers memory, I/O interface, and communication interface, and the specification covers memory types and areas, I/O numbers and types, and communication channels.

The virtual machine acts as an interpreter by executing consecutive lines of program code. The interface for the target platform consists of low-level hardware and operating system dependent functions such as updating inputs and outputs or cycle time management.

In addition to the virtual machine performing the main control tasks, the runtime environment may also contain HMI modules handling individual graphical interfaces for communicating with the operator. The virtual machine and the HMI module then operate in parallel and independently, exchanging the necessary data via the communication channel via global variables.

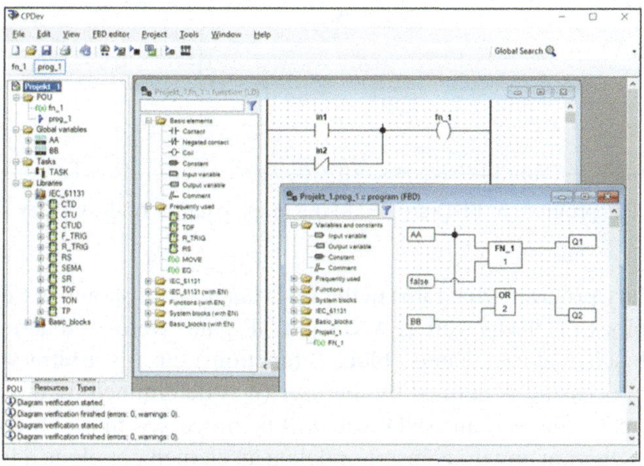

Fig. 2. The main program window and the user interface in the CPDev system

Figure 2 shows the main window of the CPDev interface, containing:

- (on the left) tree of the project structure, global variables, performed tasks, used libraries and program resources,

- (top) main menu and toolbars (quick select icons),
- (at the bottom) window of messages, reports and status information,
- (centrally) work area with windows containing user programs in IEC 61131-3 languages (ST, IL, LD, FBD, SFC).

Two windows are open in the workspace: one for the program (prog_1) in the FBD language and the other for the user function (fn_1) in the LD language – this function is used in the program prog_1. The interfaces of both graphic editors (FBD and LD) are identical, which significantly facilitates their operation by the end user creating control programs. Individual programs or functions (written in any language of the standard) can also exchange data through global variables. There are also ready-made libraries of IEC 61131-3-compliant functions and function blocks with and without EN input at your disposal. Own libraries are also added on an ongoing basis.

The common feature of the FBD and LD diagrams editors implemented in the CPDev package is the mechanism of automatic determination of connections between elements. As a result, schematic elements, such as blocks, functions, contacts or coils, do not have a tightly tied position in the working area (e.g. horizontally in the LD ladder) and can be positioned freely in the diagram. This significantly affects its transparency and allows you to easily modify it and add new elements as needed.

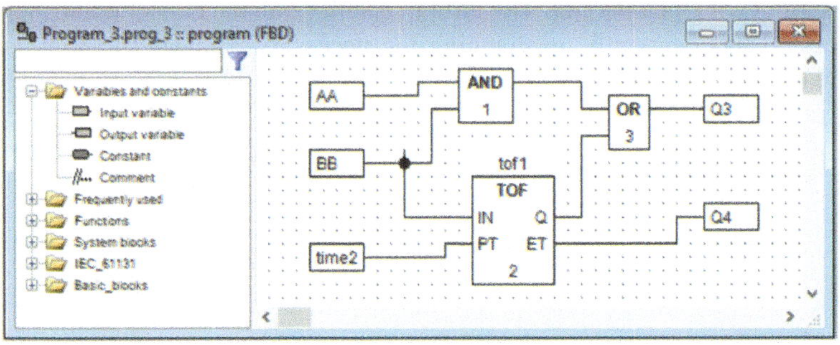

Fig. 3. The program in FBD created in the graphic editor of the CPDev package

An exemplary program (diagram) in the FBD language is shown in Fig. 3. Instances of the library block (TOF) as well as AND and OR gates were used here. Note that the position of the schematic elements (blocks, functions, etc.) is arbitrary here, and the order of their processing is defined by the user (it is determined by the number at the bottom of the rectangle, e.g. an AND gate will be processed first). Of course, the user can change this order at any time, insert or delete an element, or clone a fragment of the diagram (the order of elements will then be automatically renumbered).

The schemas created in this way are transformed into source programs in the ST language, where they can be freely modified. Similarly, functional blocks or functions, written by the user in other languages of the standard, are automatically depicted in the graphical editor and can be used there to create further schematics.

3 Course-Keeping Autopilot Structure

Ship Model. The full (non-linear) model of a 4-DOF ship is described by equations in the state space [14], but it is too complex to define analytical expressions for the settings of individual controllers. Therefore, for such a design, a simplified ship model is used, called the Nomoto model [3–5], in the form of a connection of the integrating element with the inertial element (1), which shows how the rudder deflection δ affects the angle ψ (course) under which the ship is sailing.

The gain k depends directly proportional and the time constant T is inversely proportional to the ship speed V, i.e.

$$\frac{k}{s(Ts+1)} = \frac{\psi(s)}{\delta(s)}, \qquad \text{where } k = k_0\frac{V}{V_0}, \qquad T = T_0\frac{V_0}{V} \tag{1}$$

The reference values $\{k_0, T_0, V_0\}$ in formulas (1) are assumed to be known from previous maneuvers during open sea trials or immediately after the ship left port. Here, the values also include the dynamics of the steering gear and the rudder itself (see Fig. 4).

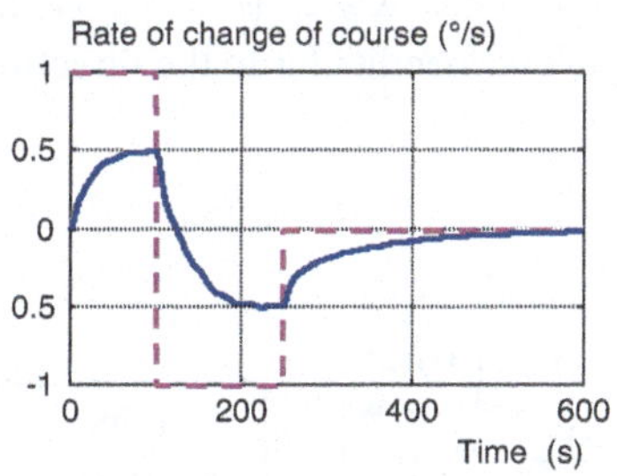

V_0 [m/s]	k_0 [1/s]	T_0 [s]
2.85	0.020	41
6.63	0.048	19
10.2	0.082	12

Fig. 4. Exemplary values of k_0, T_0 for different speeds V_0

Course-Keeping Controller. The structure of the prototype ship course-keeping controller, which is a modification of the controllers presented in [3, 4], is shown in Fig. 5. The values ψ_{ref}, ψ are the set point value and the actual course of the vessel, respectively, while δ denotes the rudder deflection angle. The quantities ψ_L, ψ_H are low and high frequency components, respectively, the latter being mainly induced by sea waves. The state observer module generates two estimates $\hat{\psi}_L$ and $\hat{\psi}_H$, of which only $\hat{\psi}_L$ (low frequency) is used for heading control. The elimination of high-frequency components significantly reduces the rudder response to the influence of sea waves. The state observer can be implemented as a Kalman filter or as a Luenberger observer. The additional ARMAX block estimates an unknown wave frequency ω_n, which is one of the coefficients in the observer's equations. Please note that this frequency depends on the state of the sea and the wind direction (related to the ship's course). Additionally, the slope rate limiter eliminates possible overshoots after a step change ψ_{ref}. The PID controller settings are automatically adjusted to the ship's speed V.

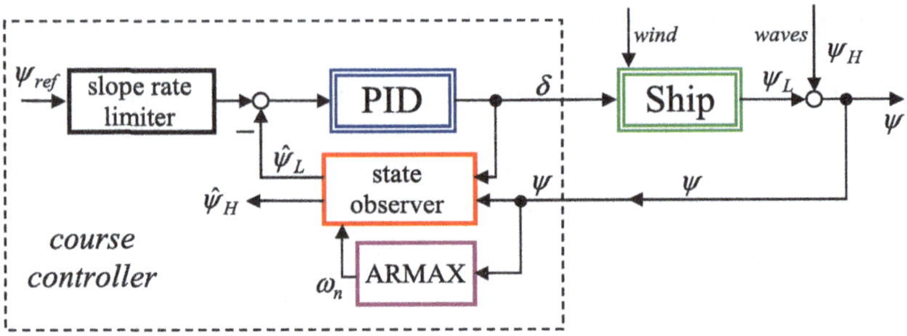

Fig. 5. Structure of course-keeping controller in autopilot

Track-Keeping Controller. The structure of the autopilot part for the cascade control of the track ("path") is shown in Fig. 6. If a deviation y of the ship's position from the set "path" is detected, then the track controller (main) corrects the set course value ψ_{ref} by $\Delta\psi_{ref}$, so the sum $\psi_{ref} + \Delta\psi_{ref}$ becomes a new value set for the rate regulator (auxiliary). The track controller (main) can be implemented as a state space controller or a classic PI controller. Kinematic equations including the ship's current position (e.g., from a GPS signal), speed V, and heading ψ (or yaw $\Delta\psi = \psi - \psi_{ref}$) determine both components of the motion (forward x and yaw perpendicular to the direction of motion y).

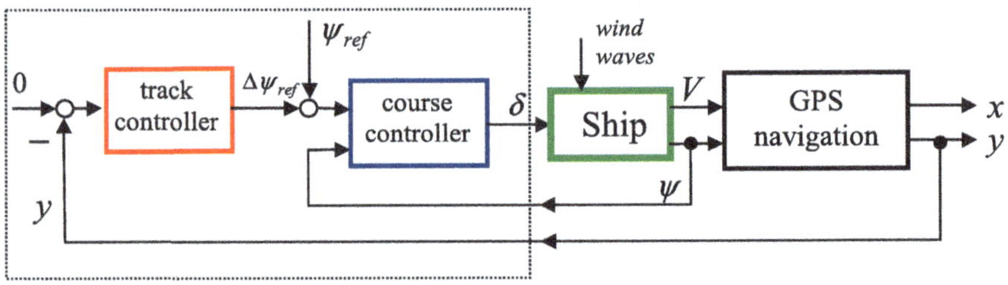

Fig. 6. Structure of the track-keeping controller in autopilot

4 Selection of Settings for the Course Controller

The course (heading) is controlled by a classic PID controller (see Fig. 5).

$$k_p(1 + \tfrac{1}{T_i s} + T_d s) = k_p \tfrac{(T_1 s+1)(T_2 s+1)}{T_i s} \tag{2a}$$

$$T_i = T_1 + T_2, \qquad T_d = \tfrac{T_1\, T_2}{T_1+T_2} \tag{2b}$$

Design Assumptions. The following assumptions were made for calculating the controller settings:

(a) no oscillation in the transient responses
(b) the closed loop time constant is defined as $T_{cl} = T/r, \quad r > 1$
(c) no overshoot for step changes of the set point value ψ_{ref}.

Condition (b) requires that the time constant T_{cl} was r times smaller than the time constant T in the ship model (1). The factor r is a design parameter here. By increasing r you can make the loop "faster". To determine the settings k_p, T_i, T_d, let's assume $T_2 = T$. Then the transfer functions of the open and closed systems, and the gain K, take the form:

$$G_{open}(s) = K\frac{T_1 s + 1}{s^2}, \qquad G_{loop}(s) = \frac{K(T_1 s + 1)}{s^2 + K\,T_1 s + K}, \qquad K = \frac{k_p k}{T_i} \tag{3}$$

Assumption (a) will be satisfied if the denominator of the transfer function G_{loop} is zero, which gives the condition for K and determines the values of the double pole $s_{1,2}$, while assumption (b) requires that the time constant $T_{cl} = 1/|s_{1,2}| = T_1/2$ be equal T/r, so T_1 it will be

$$K = \frac{4}{T_1^2}, \qquad s_{1,2} = -\frac{K\,T_1}{2} = -\frac{2}{T_1}, \qquad T_1 = \frac{2T}{r} \tag{4}$$

Finally, the PID controller settings obtained from the above equations are

$$k_p = \frac{1}{kT} r(r+2), \qquad T_i = T\frac{r+2}{r}, \qquad T_d = T\frac{2}{r+2} \tag{5}$$

Note that due to Eq. (1), the settings of the controller (5) depend on the speed of the vessel V. Sensitivity tests have shown that the settings should be modified if the speed change exceeds 20%. The step responses for the PID controller at settings (6) for different speeds V are shown in Fig. 7.

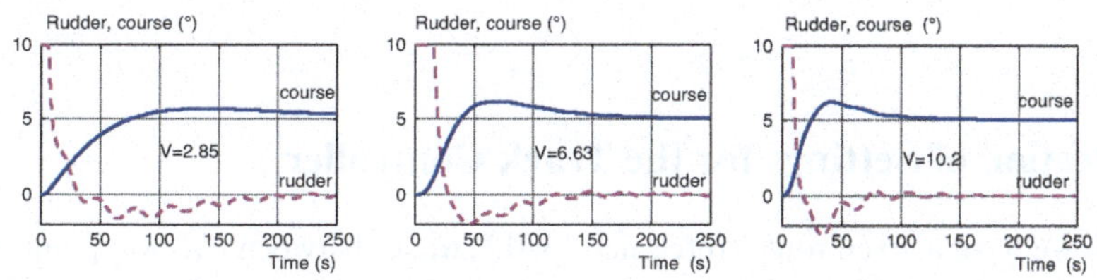

Fig. 7. Step responses for PID controller at different speeds V

Pre-filter. A system with only a classic PID controller will show some overshoot due to the term $T_1 s + 1$ in the transfer function G_{loop}. To reduce or eliminate this, a pre-filter can be used for the set-point as shown in Fig. 8. The transfer function of the closed system (without the slew rate limiter) will then be

$$G_{cl}(s) = \frac{1}{(T_{cl}s + 1)^2} = \frac{\psi(s)}{\psi'_{ref}(s)} \tag{6}$$

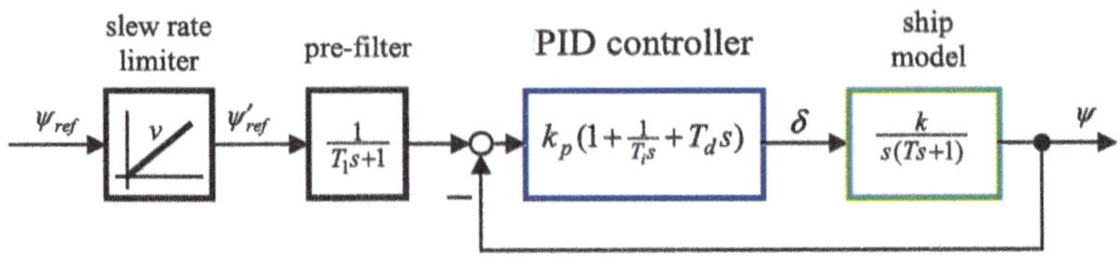

Fig. 8. Course control system for controller design

Slew Rate Limiter. Since in practice the rudder deflection δ cannot exceed a certain maximum limit δ_M (when $|\delta| < \delta_M$), so usually the pre-filter $1/(T_1 s + 1)$ is not sufficient to avoid overshoot with step changes ψ'_{ref} (typically max $\delta_M = 35°$). Additionally, the value δ_M may depend on the speed of the vessel V. The autopilot is thus equipped with a set-point slew rate limiter, which limits the rate of change ψ'_{ref} to a predetermined value v. For $\psi'_{ref} = v \cdot t$ the expression determining the rudder deflection angle will take the form

$$\delta(s) = \frac{1}{k}\frac{s(Ts+1)}{(T_{cl}s+1)}\frac{v}{s^2} = \frac{1}{k}\frac{Ts+1}{(T_{cl}s+1)^2}\frac{v}{s}, \tag{7}$$

which time form has a maximum for $T_{cl} < T$ $(r > 1)$. The course controller step responses in the presence of sea waves are shown in Figure 9.

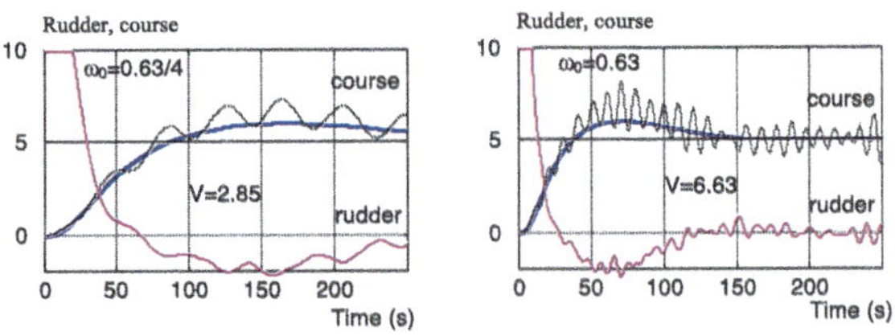

Fig. 9. Step responses of course controller in the presence of sea waves

5 Selection of Settings for the Track Controller

Suppose we have a given route (reference "path", track) between two waypoints for the ship to follow. Let it ψ_{ref} denote the assumed course on this path. The area navigation system is able to determine the perpendicular y-component of the deviation from the set path (see Fig. 4). As shown in [3], for small deviations of the course, this deviation is a weighted integral $\Delta\psi$, i.e.

$$y(s) = \frac{V}{s}\Delta\psi(s) \tag{8}$$

($\Delta\psi$ in radians). Using the standard PI controller (as the main controller in cascade control), consider the circuit shown in Fig. 10.

Specifications here are as follows:

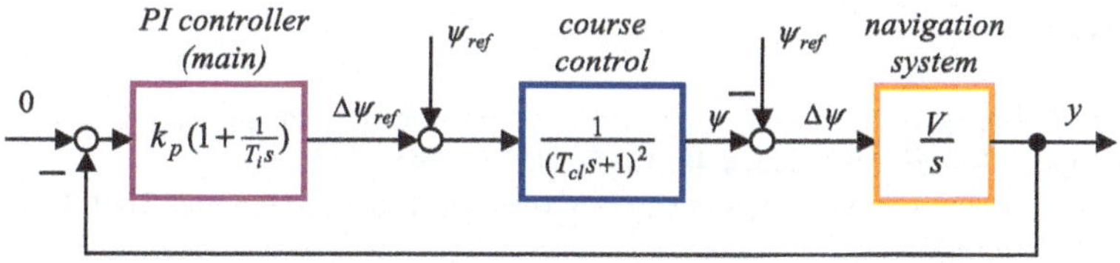

Fig. 10. Track-keeping system with standard PI controller

(a) no oscillation in the transients for the step response
(b) a short rise time of the step response (if possible).

The problem of selecting the PI controller settings can be solved by designing them using the square root method [2, 10]. These two requirements will be met if we find a controller zero $1/T_i$ for which a closed loop circuit has a true triple pole. Then the controller gain, integration constant and the rise time of the step response are defined by the formulas

$$T_i = 15\,T_{cl}, \qquad k_p = 0.2\frac{1}{T_{cl}V}, \qquad t_{tr} \cong \frac{4}{|s_{tr}|} = 25\,T_{cl} \tag{9}$$

Note that this settling time is more than 6 times longer than the settling time for the course controller (4 T_{cl}). Besides, for typical data, the gain k_p turns out to be small and the integration time constant T_i is large. So as expected from [2], we have a PI regulator with relatively low gain.

6 Summary

The article presents the structure of the ship's autopilot and the principles of analytical selection of set-points for the course (PID) and track (PI) controller implemented in the CPDev environment. For each of them, only single design parameters are required to determine closed-loop dynamics, which greatly simplifies the autopilot tuning process. In addition, the behavior of the autopilot can be easily adapted to the different structure of the ship and the state of the sea (wind, waves). The settings are automatically adjusted to the ship's speed, and the built-in pre-filter and the set-point rate limiter additionally eliminate any possible overshoots.

The autopilot software will be expanded with adaptive tuning of the pond (using a fuzzy controller or neural network) and new algorithms for identifying disturbances originating from wind and sea waves.

Acknowledgments. This project is financed by the Minister of Education and Science of the Republic of Poland within the "Regional Initiative of Excellence" program for years 2019–2022. Project number 027/RID/2018/19, amount granted 11 999 900 PLN.

References

1. Cimen, T., Banks, S.P.: Nonlinear optimal tracking control with application to super-tankers for autopilot design. Automatica **40**, 1845–1863 (2004)
2. Dorf, R.C., Bishop, R.M.: Modern Control Systems, 11th edn. Prentice Hall, Upper Saddle River, NY (2008)
3. Fossen, T.I.: Guidance and Control of Ocean Vehicles, 4th edn. Wiley, Chichester (1999)
4. Fossen, T.I.: Marine Control Systems. Marine Cybernetics, Trondheim (2002)
5. Lisowski, J.: Ship as Automatic Control Plant (in Polish). Wyd. Morskie, Gdańsk (1981)
6. Jamro, M., et al.: CPDev engineering environment for modeling, implementation, testing, and visualization of control software. In: Szewczyk, R., Zieliński, C., Kaliczyńska, M. (eds.) Recent Advances in Automation, Robotics and Measuring Techniques. AISC, vol. 267, pp. 81–90. Springer, Cham (2014). https://doi.org/10.1007/978-3-319-05353-0_9
7. Rzońca, D., Sadolewski, J., Stec, A., Świder, Z., Trybus, B., Trybus, L.: Mini-DCS system programming in IEC 61131-3 structured Text. J. Autom. Mob. Rob. Intell. Syst. **2**(3), 48–54 (2008)
8. Rzońca, D., Sadolewski, J., Stec, A., Świder, Z., Trybus, B., Trybus, L.: Ship autopilot software – a case study. In: Bartoszewicz, A., Kabziński, J., Kacprzyk, J. (eds.) Advanced, Contemporary Control. AISC, vol. 1196, pp. 1499–1506. Springer, Cham (2020). https://doi.org/10.1007/978-3-030-50936-1_124
9. McCookin, E.W., Murray-Smith, D.J., Li, Y., Fossen, T.L.: Ship steering control system optimization using genetic algorithms. Control. Eng. Pract. **8**, 429–443 (2000)
10. Trybus, L.: Control Theory (in Polish). Ofic. Wyd. PRz., Rzeszów (2007)
11. Tzeng, C.Y., Lu, G.H.: An Internal Model Control- based Neural Network ship auto-pilot design. J. Soc. Naval Arch. Marine Eng. ROC **22**(1), 13–23 (2003)
12. Vanec, T.W.: Fuzzy guidance controller for an autonomous boat. IEEE Control. Syst. **17**(2), 43–51 (1997)
13. Zhang, Y., Hearn, G.E., Sen, P.: A multivariable neural controller for automatic ship berthing. IEEE Control Syst. Mag. **17**(4), 31–45 (1997)
14. http://www.marinecontrol.org/Tutorial.html

Research of Attractive Behaviour and Chaos Characteristics of a Control Object at Its Computer Modeling

Yaroslav Smityuh(✉), Vasii Kyshenko, and Alexander Romashchuk

National University of Food Technologies, Kiev, Ukraine
smityuh1@gmail.com

Abstract. The analysis of time series of variables of the juice purification process at the sugar factory was carried out by methods of nonlinear dynamics, which allowed to determine the values of chaos and randomness based on the depth of prediction of object behavior. Reconstruction of the object attractors was carried out and their features were determined. The fractal dimension of time series of the main technological parameters was determined, which allowed to make a deep prediction of the object's behavior. The intensity of the influence of the value of the technological parameter on the process to increase the efficiency of management of the technological complex of the sugar plant is determined.

Keywords: Monitoring · Management · Sugar production · Nonlinear dynamics · Attractor

1 Introduction

Traditional models of process control and traditional analytical methods of analysis of performance indicators and forecasting are increasingly encountering problems that do not have an effective solution within the known solutions. Traditional approaches, which have become classical, have been developed to describe stable and non-radically variable processes – which do not deviate much from the state of equilibrium. In essence, these methods and approaches were not designed to describe and model rapid changes, unforeseen leaps and complex interactions of individual components of the technological process [1–3]. The changes taking place at the defecosaturation station are so variable and their quality indicators are so unexpected that for the analysis and prediction of initial parameters, i.e. the quality of diffusion juice purification, the synthesis of new analytical and computational approaches originating in various fields of human knowledge, has become an urgent practical necessity. It is necessary to study the dynamic processes occurring in irreversible multicomponent interactive adaptive systems, to consider the causes and mechanisms of new regimes and structures, to study the characteristic scales and speeds of transient and permanent processes, to anticipate probable changes in the system to manage unexpected cardinal changes arising in complex systems. The technological process of defecosaturation can be attributed to such systems. Recent

R. Szewczyk et al. (Eds.): AUTOMATION 2022, AISC 1427, pp. 143–150, 2022.
https://doi.org/10.1007/978-3-031-03502-9_15

years have shown increased interest in finding nonlinear models that could adequately reproduce complex patterns of dynamic processes, as it has become clear that a linear approach to the analysis of initial parameters does not simulate the highly irregular behavior of most technological systems [4]. There are several competing approaches that use the idea of nonlinearity. Traditional models are stochastic. However, the limitations used in the construction of the model in order to make it suitable for practical use, in fact, destroy the internal "complexity" inherent in the dynamic process [5].

In this regard, an alternative approach to the analysis of nonlinearities has been intensively developed recently, namely the approach based on the theory of deterministic chaos, which offers an explanation for irregular behavior and anomalies in systems that are not stochastic in nature. Chaos theory offers completely new concepts and algorithms for the analysis of time series, which can lead to a deeper and fuller understanding of the processes they reflect. The sugar factory's juicing department is a complex object of management. One of the characteristic manifestations of this complexity is, as studies have shown, the presence of intermittency: deterministic behavior (channel) changes to stochastic (jokers) with the formation of dissipative spatio-temporal structures (attractors). Successful monitoring and management of the juicing process requires effective prediction of attractive behavior, taking into account the phenomena of self-organization and evolution of structures. Solving this problem will allow for resource-efficient management of juicing processes by organizing low-power, but sensitive and topologically coordinated management actions.

2 Materials and Methods of Research

Modern methods of analysis are based on the search for models of nonlinear behavior of time series. This is due to the fact that nonlinear models can capture very complex patterns in time series [6]. The crucial task is the reconstruction of attractors according to the time series of technological variables of the control object, resulting in a predictive predictor function. To identify attractors and their characteristics, we use the Tuckens method [7]. The method is that for all "smooth" dynamical systems on the existing time implementation of one observed dynamic variable, you can construct a new attractor, the basic properties of which will be the same as the original. The main idea of applying the methods of chaotic dynamics to the analysis of time series is that the basic structure of a chaotic system containing all the information about the system, namely the attractor of a dynamic system (subset of phase space that attracts trajectories to infinite time) restored through the measurement of only one observed characteristic of this dynamic system, fixed as a time series. According to the method of Grasberger and Procacci, the procedure of reconstruction of the phase space and restoration of the chaotic attractor of the system in the dynamic analysis of the time series is reduced to the construction of the so-called phase space.

The following software products were used for dynamic data analysis: DATAN's Dataplore, TISEAN's modular suite of programs developed at the Institute of Physics and Theoretical Chemistry of the University of Frankfurt by Rainer Hegger, Holger Kantz and Thomas Schreiber. Chaos Data Analyzer (CDA), written by J. Sprott and J. Rowlands (Sprott J., Rowlands G.), as well as the Fractan program, developed at the

Institute of Mathematical Problems of Biology of the Russian Academy of Sciences V. Sychev. For fractal analysis, the paper used a specialized set of extensions (toolbox) of the MATLAB program – FracLab, created in the French laboratory Fractals team.

The main idea of applying the methods of chaotic dynamics to the analysis of time series is as follows. It turns out [4], the basic structure of a chaotic system, which contains all the information about the system, namely, its attractor (a subset of phase space that attracts trajectories to infinite time), can be restored by measuring only one observed parameter of this dynamic system, fixed as a time series.

Deterministic dynamical systems describe the evolution of a system over time in some phase space.

Time series can be considered as a sequence of observations $\{S_n = S(x_n)\}$. Because the sequence (usually scalar) $\{S_n\}$ does not in itself generate a multidimensional phase space of a dynamic system, it is necessary to use some technique to reveal the multidimensional structure using only available data.

According to the method of Grassberger and Procacci [7], the procedure of reconstruction of phase space and restoration of chaotic attractor of the system in dynamic time series analysis is reduced to the construction of so-called lag or restored space using the method of delays. Vectors in the new space, attachment space, formed from the values of the time series of scalar measurements with time delay:

$$\overline{S}_n = (S_{n-(m-1)r}, S_{n-(m-2)r}, \ldots, S_n) \tag{1}$$

The number of elements m is called the dimension of the attachment, the time τ is usually called the delay or lag. The theorems of Tuckens [6] and Sawyer [7] show that if the sequence $\{S_n\}$ actually consists of scalar measurements of the structure of a dynamical system, then, under certain assumptions, such a reconstruction of the phase portrait is an exact picture of the real set $\{x\}$, if m quite large. Or, in other words, the real attractor of a dynamic system and the "attractor" restored in the lag space in time series according to the above rule (pseudo-attractor), with adequate selection of the dimension of the attachment m, are topologically equivalent.

Restoration of phase space using the method of Grasberger and Procacci is as follows. For each point Si of the time series, its nearest neighbor S_j in m-dimensional space is sought. The distance between them is calculated:

$$||\overline{S_i} - \overline{S_j}|| \tag{2}$$

Next we calculate:

$$R_i = \frac{|\overline{S}_{i+1} - \overline{S}_{j+1}|}{||\overline{S}_i - \overline{S}_j||} \tag{3}$$

If R_i exceeds this heuristic threshold R_t, then the point is considered to have an erroneous nearest neighbor.

One of the main indicators by which the dimensionality of the attractor can be estimated is the Reni dimension [4]. For research by such method it is necessary to divide area into segments of some constant size ε. Suppose that the probability that some point of the attractor is on the *i*-th segment ε is equal to pi (ε). Then we get the Reni dimension, which is determined by:

$$D_q = \frac{1}{q-1} \lim_{\varepsilon \to 0} \frac{\ln \sum_{i=1}^{N_\varepsilon} p_i^q(\varepsilon)}{\ln \varepsilon} \tag{4}$$

where q is the order of dimension; p_i^q – the probability that a pair of points of the attractor belongs to the *i*-th segment; N_ε – the number of self-similar structures of linear size ε.

For a more complete assessment of the dimension, partial cases of the Reni dimension are used. Is most often used in studies and is limited to determining the correlation dimension D_2.

Correlation dimension D_2 – shows the probability that two arbitrarily selected points of the attractor can lie in the same zone ε (Fig. 1),

$$D_2 = \lim_{\varepsilon \to 0} \frac{\ln \sum_{i=1}^{\infty} p_i^2(\varepsilon)}{\ln \varepsilon} \tag{5}$$

that is, determines the dynamic inhomogeneity of the attractor.

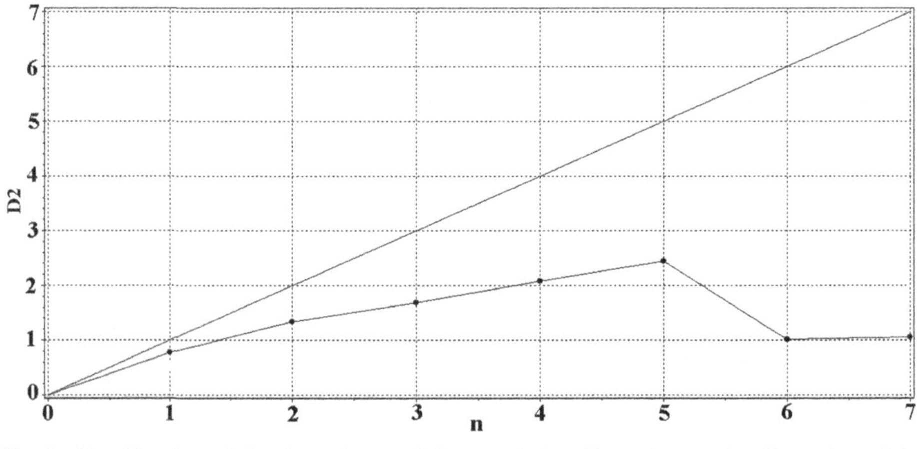

Fig. 1. Graphic view of the dependence of the correlation dimension on the dimension of the phase space *n*

The method of time delay of coordinates was used to restore the attractor. That is, *n* – dimensional phase space is a sequence of points of view:

$$x_k = \left(s_k, s_{k+\tau}, \ldots, s_{k+(n-1)\tau}\right)$$
$$k = 0, \ldots, m-1, m = M-(n-1)\tau \tag{6}$$

here τ – time delay, n – the dimension of the investment.

The method is as follows. If $M \to \infty$, then the set of points $x_k \in R^n$ specifies the embedding of the initial attractor for almost any choice of the observed variable, if n not less than twice the dimension of the initial attractor. The delay is chosen equal to the time of the first intersection of zero by the autocorrelation function.

3 The Results of Research

Time series of juice consumption, juice pH, saturation gas pressure, juice temperature, CO_2 content in saturation gas, juice density were studied. Reconstruction of the attractor is carried out in the MATLAB software environment (Fig. 2).

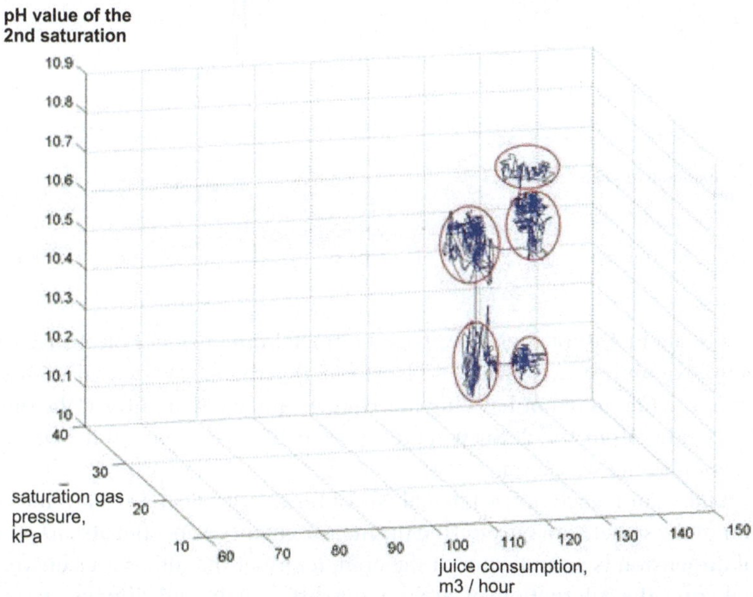

Fig. 2. Reconstructed phase portrait with areas of attraction

In the reconstructed phase space, the traction areas, the attractors of the dynamic system, which are shifted to the planes of pH and saturation gas pressure, are clearly visible.

The reconstructed attractors of time series of the pH value of the diffusion juice at the defecosaturation station are shown in Fig. 3.

The fractal dimension D was determined to study the time series. The method based on change of length of a curve depending on scale is applied. If the curve is close to fractal, then with decreasing scale, the length of the curve will increase incrementally. Fractal geometry begins with the Mandelbrot formula, which connects the length of an arbitrary curve L with the measurement scale ε:

$$L = C \cdot \varepsilon^{1-D}, \tag{7}$$

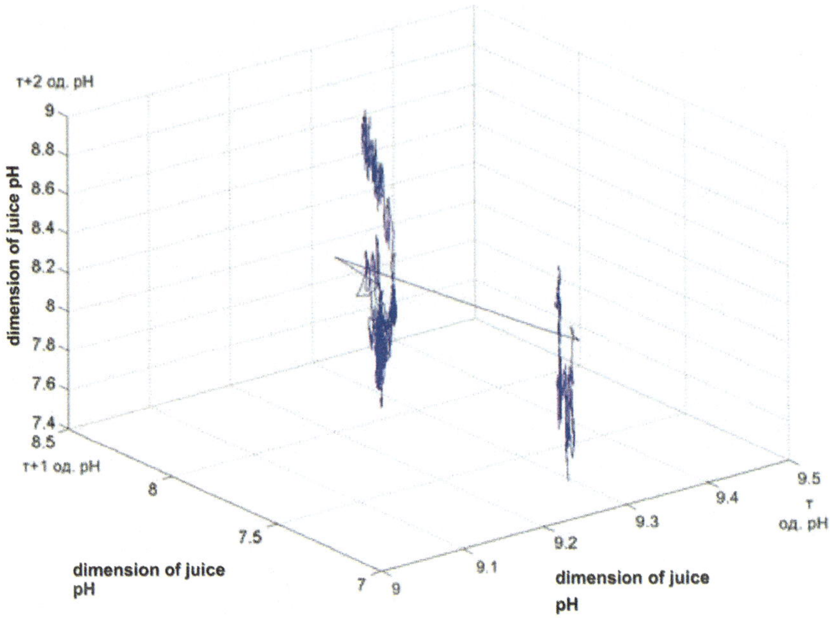

Fig. 3. Reduced time series attractors of pH value of diffusion juice of the second saturation

where C is a scale factor for each line; D is the fractal dimension. For ordinary smooth lines, which have almost everywhere at least one derivative, D is equal to one. In the General case, the line may not have any derivatives anywhere and for it the dimension D may have a value from 1 to 2. Such curves are self-like – any area of the curve has the same dimension as the whole line.

The special significance of fractal analysis of time series is that it takes into account the behavior of the system not only during the measurement period, but also its prehistory.

Fractal dimension is an indicator of the complexity of the time series curve.

By analyzing the alternation of areas with different fractal dimensions, you can learn to predict the behavior of the system. And, most importantly, diagnose and predict unstable conditions.

An essential point of this approach is the presence of a critical value of the fractal dimension of the time curve, when approaching which the system loses stability and becomes unstable, the parameters quickly increase or decrease, depending on the current trend.

The dynamics of pH change before the second saturation (Fig. 4) and pH of the second saturation (Fig. 5) are analyzed below.

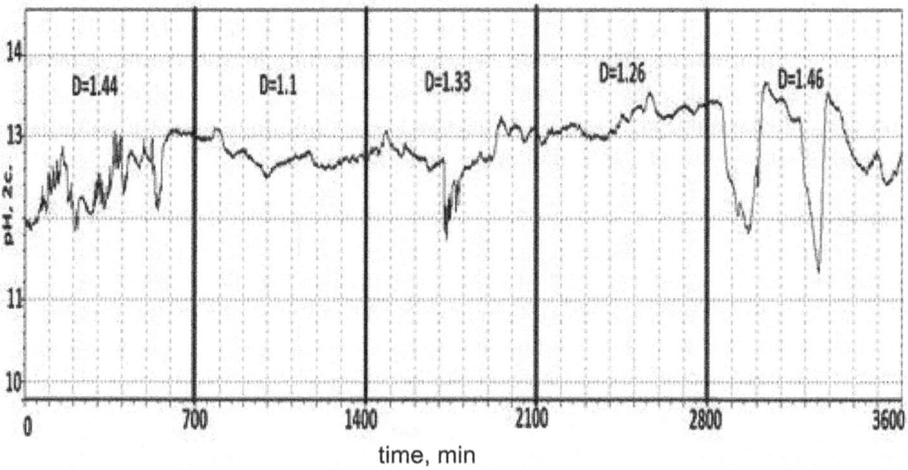

Fig. 4. Fractal dimension for a number of pH values before the second saturation

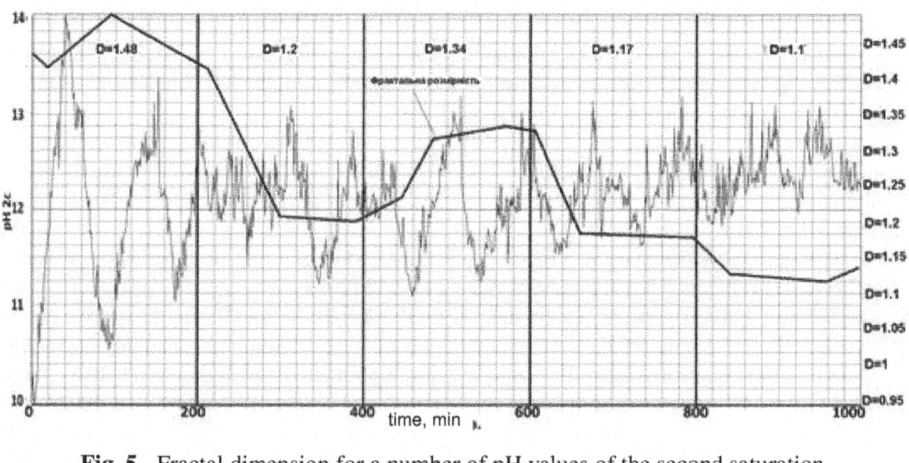

Fig. 5. Fractal dimension for a number of pH values of the second saturation

The fractal dimension of a certain quantity can be used as an "indicator" of a catastrophe. Analysis of experimental data shows that the trend line for the time series is well described by the equation:

$$\overline{y}(t) = \overline{y}(t_0) + \frac{K_f(t_0)(t - t_0)}{(D - D_0)^{\beta}} \quad (8)$$

where – the average value for the period preceding the forecast; K_f and β coefficients; t_0 is the period of time preceding the forecast; t – time at which the forecast is made; D_0

– fractal dimension in the period preceding the forecast. Also, the value of the fractal dimension can serve as an indicator of the number of factors affecting the system [8]. When the fractal dimension is less than 1.4, the system is affected by one or more forces moving the system in one direction. If the dimension is about 1.5, then the forces acting on the system are different, but more or less compensate for each other. If the fractal dimension is much larger than 1.6, the system becomes unstable and ready to move to a new state. Examining the dynamics, it is seen that during fairly stable periods and slow rises, the fractal dimension of the time series remains quite low, while in periods of abrupt changes, the total fractal dimension increased.

Conclusions

A set of methods of nonlinear dynamics has been applied, the joint application of which allows recognizing more than 75% of known situations in the juice purification department at the sugar factory. It is possible to increase the efficiency of decision-making on the management and formation of situationally important scenarios for the management of the technological complex of sugar production.

References

1. Ladanyuk, A., Shkolna, O., Kyshenko, V.: Automation of evaporation plants using energy-saving technologies. In: Szewczyk, R., Kaliczyńska, M. (eds.) SCIT 2016. AISC, vol. 543, pp. 220–226. Springer, Cham (2017). https://doi.org/10.1007/978-3-319-48923-0_27
2. Kyshenko, V.: The problems of technological monitoring in the control of industrial enterprise. In: Pressing Issues and Priorities in Development of the Scientific and Technological Complex, pp. 62–68. B&M Publishing, San Francisco (2013)
3. Korobiichuk, I., Ladanyuk, A., Zaiets, N., Vlasenko, L.: Modern development technologies and investigation of food production technological complex automated systems. In: ACM International Conference Proceeding Series, pp. 52–57 (2018). https://doi.org/10.1145/3185066.3185075
4. Novakovska, N., Kyshenko, V.: Fractal analysis of distillation unit time series in prediction and control problems. Ukr. J. Food Sci. **3**(2), 243–253 (2015)
5. Korobiichuk, I., Ladanyuk, A., Shumyhai, D., Boyko, R., Reshetiuk, V., Kamiński, M.: How to increase efficiency of automatic control of complex plants by development and implementation of coordination control system. In: Szewczyk, R., Kaliczyńska, M. (eds.) SCIT 2016. AISC, vol. 543, pp. 189–195. Springer, Cham (2017). https://doi.org/10.1007/978-3-319-48923-0_23
6. Korobiichuk, I., et al.: Synthesis of optimal robust regulator for food processing facilities. In: Szewczyk, R., Zieliński, C., Kaliczyńska, M. (eds.) ICA 2017. Advances in Intelligent Systems and Computing, vol. 550, pp. 58–66. Springer, Cham (2017). https://doi.org/10.1007/978-3-319-54042-9_5
7. Grassberger, P., Procaccia, I.: Characterization of strange attractors. Phys. Rev. Lett. **50**, 346–349 (1983)
8. Ladanyuk, A., Kyshenko, V., Smityuh, Y.: The biotech complexes control in conditions of situational. Ann. Warsaw Univ. Life Sci. **60**, 149–154 (2012)

Resource Efficiency Forecasting Neural Network Model for the Sugar Plant Diffusion Station

Nataliia Zaiets[1], Lidiia Vlasenko[2(✉)], Nataliia Lutska[3], and Vladimir Shtepa[4]

[1] National University of Life and Environmental Sciences of Ukraine, Kiev 03041, Ukraine
z-n@ukr.net
[2] Kyiv National University of Trade and Economics, Kyiv 02156, Ukraine
vlasenko.lidia1@gmail.com
[3] National University of Food Technologies, Kyiv 01033, Ukraine
[4] Polessky State University, 225710 Pinsk, Republic of Belarus

Abstract. The article proposes a model of the diffusion station of a sugar factory based on the neural network, which predicts the main technological indicators of resource efficiency. In contrast to existing solutions, forecasting is performed in real time, which allows one to increase the informative support of the operator-technologist on the enterprise quality measurements without additional workload on the industrial laboratory. The proposed model predicts the loss of sugar in the pulp and the amount of sugar in the diffusion juice with an error of less than 2%. The input variables of the forecast model are automatically measurable technological variables of the process, a total of eleven. On the basis of such information one can make a corresponding decision to change technological modes (temperature and consumption) at the station, thereby increasing the quality of beet processing. The article consists of two parts: analysis of process variables, including statistical and definition of input and output model variables; development and verification of the forecast model.

Keywords: Diffusion station · Neural network · Forecast · Sugar · Beet · Quality assessment

1 Introduction

Improvement in the efficiency of the sugar factory is directly related to operational assessments of the main departments functioning and forecasting their dynamics. The need for timely change of production parameters aimed at eliminating deviations and minimizing costs arising from the operation of the technological complex, with their detection in the early stages and anticipation of possible dynamics, which leads to increased efficiency of the system [1, 2]. Given that part of the technological information is measured in automatic mode, and the other – in the laboratory, there is a problem of combining data in a single model with different periods [3]. Therefore, to take into account the existing operational information that directly affects the quality of the original product, it

R. Szewczyk et al. (Eds.): AUTOMATION 2022, AISC 1427, pp. 151–161, 2022.
https://doi.org/10.1007/978-3-031-03502-9_16

is necessary to use new approaches in the management of the technological complexes subsystems based on the use of forecasting methods and algorithms [4]. It will be ensured by development of the diffusion station (DS) model on the basis of a neural network [5], the input variables of which are data collected from sensors, and the output – selected estimates of resource efficiency.

Removal of sugar from beet chips is one of the sugar production main processes, which largely determines the quality and quantity of sugar obtained and the efficiency of the plant as a whole. The efficiency of DS is determined by its productivity, sugar content in the diffusion juice and sugar losses in the pulp [6]. The same parameters determine the economic efficiency of the sugar extraction from beet chips process.

The prognostic task of determining the rational regime is to determine the optimum of the objective function, which can be both economic and technical: total losses of sugar during production, sugar yield, operating profit of the plant (technological). The economic indicators of the DS are directly influenced by qualitative technological variables, which in turn are the mode parameters of the station automated control system. The purpose of automated control is to maximize sugar yield, operating profit, or minimize total sugar losses in production, which is achieved by the introduction of global corrective actions that are typical for sugar mills: change in the productivity of the plant, change in the pumping of diffusion juice, change in the consumption of lime for juicing or change in water consumption for washing the filtered sludge. The non-stationarity of the process associated with the quality of processed beets is taken into account by recalculation of the coefficients and parameters based on the actual values of variables for the previous period, which diagnoses the need to calculate plant performance, which is reduced to the formation of diagnostic signs and messages, which, in turn, indicate the need for calculations and development of the recommendations for the organizational and technological process of sugar production. Such a correction lags behind at least a shift, and the placement of a system for forecasting key technological indicators will become a timely information support for the operator.

Object of the study is a process of forecasting the main indicators of the diffusion station of the sugar plant, which are crucial for the resource efficiency of production as a whole.

Subject of the study is the neural network model of the diffusion station to predict the resource efficiency of production as a whole.

The purpose of the work is the development of a neural network model of the diffusion station, which allows one to predict its resource efficiency and increases the informativeness of the main indicators in real time, thereby increasing the quality of beet processing on the DS.

2 Methods

Over the past five years, the world and Ukrainian sugar markets have experienced a decline in sugar production. It is due to many factors, staring from the impact of social advertising, which calls on humanity to reduce sugar consumption, ending up with reduction of the area for sugar beets cultivation in favor of alternative crops, drop in sugar prices in contrast to the high-cost production, low quality raw materials due to weather conditions and more [7]. In addition, it should be noted that every year there is an increase in the share of sugar cone compared to sugar beet. To make a profit and produce quality products, Ukrainian sugar factories have to increase resource efficiency in modern ways and methods.

The peculiarities of sugar production include a high level of energy consumption, seasonality of production and variable product quality both in different seasons and during the single season.

Therefore, to increase the sugar factory efficiency it is necessary to place modern intelligent methods. There are various approaches to improving the efficiency of sugar beet production, based on the use of traditional automation systems, adaptation methods, synergetic [8], scenario [9] approaches, the use of cognitive maps [10], robust regulators [11], statistical methods [12], but these methods either work with linearized models, or consider the process as quasi-stationary, or do not take into account all types of uncertainties acting on the object, or work with data distributed according to the normal law, and so on. Sugar production belongs to the continuous type, which is characterized by high complexity due to its non-stationary, distributed parameters, high degree of uncertainty, nonlinearity and inconsistency of changes in technological variables according to the normal distribution law.

Usually the efficiency of sugar production depends primarily on the efficient operation of the diffusion station. To ensure its efficient operation, it is necessary to make periodic adjustments of technological regulations on the basis of current values of technological indicators and expected values of individual indicators of the diffusion station, which determine the intermediate losses of sugar.

The main indicators of the diffusion station should include: sugar content in the pulp, sugar content in the diffusion juice and purity of the juice (1, 2). It should be noted that the sugar content in the dry matter of the diffusion juice and the non-sugar content in the diffusion juice are not measured substances, but calculated.

Sugar content in pulp, % by weight of beet, S_g:

$$S_g = \frac{1 - \frac{\alpha_{gv}}{\beta_0(\alpha_g + \alpha_{gv})} S_s \alpha \cdot 10^{-2}}{(\alpha - \alpha_{gv}\rho_0) \cdot 10^{-2}\, exp\left(Y \frac{\alpha \cdot 10^{-2} - 155,5q}{\alpha A + \alpha_{gv}\rho_0 \cdot 10^{-2}}\right)} \tag{1}$$

where α is the relative consumption of pumping for production, α_{gv} is the relative consumption of pulp water, α_g is the relative consumption of pulp, ρ_0 is the density of diffusion juice, A is the productivity of the beet plant, Y is the coefficient of diffusion installation, q is the coefficient of equation.

The sugar content in the diffusion juice, % by weight of the juice, S_{ds}:

$$S_{ds} = \frac{S_s - S_g - S_n}{10^{-2} \cdot \alpha}, \tag{2}$$

where S_s – sugar content in the chips, S_n – unaccounted for sugar costs in production.

To improve the operation of the sugar plant, it is advisable to forecast the main characteristics of the diffusion station. The quality of forecasting depends on competently selected data, the depth of the forecast, taking into account all the features of the technological process. It is often impossible to take into account all the actions that affect the object, and neural networks successfully cope with this problem because of the possibility of their learning. In addition, during training, it reveals complex relationships between input and output parameters, and therefore makes forecasting better.

Analysis of the sources showed that the use of neural networks to improve and predict the operation of the sugar factory is quite common. In particular, for the production of cane sugar, the following studies were performed using neural networks of different types: prediction of juice parameters [13], determination of optimal parameters of sugar cane grinding based on artificial neural networks [14], neural pH control model for lighting process [15], prediction of syrup crystallization process [16] and prediction of energy consumption during its passage [17].

For sugar beet processing, there are also a number of studies on the use of neural networks to solve various problems. In particular, [18] devoted to the comparison of methodology (RSM) and artificial neural networks (ANN) in the production of mixtures of oligosaccharides from sugar beet pulp, in [19] proposed an algorithm for monitoring the state of the evaporator station based on neural network, predicting the properties of purified diffusion juice [20].

As it can be seen, the vast majority of work is devoted to the production of sugar from cane rather than sugar beet, and the share of work devoted to predicting parameters based on neural networks is quite low. The authors also failed to find works devoted to forecasting the main performance indicators of the diffusion station of the beet sugar factory. Therefore, the work is relevant.

To develop a predictive model of the diffusion station, the following steps must be performed: process analysis, determination of process features – input/output variables; development of the forecast model.

2.1 Problem Analysis, Selection of Interdependent Variables

The main requirements for the automation system are the creation of conditions for the most complete extraction of sugar from chips, obtaining diffusion juice of a given concentration, ensuring a given productivity of the installation while ensuring maximum efficiency of the process. However, the operation of the automated control system directly depends on the mode parameters (tasks for regulators), which are determined on the basis of data from the production laboratory during each shift. Such determining variables are the loss of sugar in the pulp and the amount of sugar in the diffusion juice. At the same time, when changing the technological regime, the operator-technologist is guided by his own a priori knowledge about the process, a posteriori data, as well as his own vision of the forecast of the main indicators for the next shift. That is why the automated forecast of sugar content in diffusion juice and sugar losses in pulp will increase the resource efficiency of the diffusion station and the work of operational staff.

A typical scheme of the DS automation system provides: stabilization of the specific load of the apparatus, stabilization of the diffusion juice concentration, stabilization of temperature regimes in the zones of the diffusion apparatus; stabilization of the level in the main part of the device; measurement of chip consumption, diffusion juice, pulp and sulfated water; pH of the mixture of chips and water.

The extraction process is influenced by: temperature regimes in the zones of the apparatus (along its length), the quality of beet chips, pH level of the environment in the apparatus, the ratio of chip and water consumption, the quality of feed water, the persistence time of chips in the apparatus. Thus, the diffusion coefficient of sucrose in beet tissues is determined by the dependence:

$$D = \left(1{,}89 \cdot 10^{-8} R_e^2 t^{29}\right) / \tau^{0{,}54}, \tag{3}$$

where D is the diffusion coefficient of sucrose, m^2/s; R_e – equivalent chip radius, m; t is the temperature of active diffusion, °C; τ – duration of the process, s.

Figure 1a shows the effect of temperature on the diffusion coefficient in beet tissues, which shows that with increasing temperature, the diffusion coefficient increases. However, the increase in temperature to 80 °C leads to overheating of the chips, while the chips lose elasticity and results possible sticking. Thus, to obtain diffusion juice with the necessary technological qualities, it is necessary to quickly heat the chips to the temperature of protein denaturation and maintain the temperature of active extraction in the range of 72–75 °C. The approximate temperature regime of desiccating beet chips in an inclined screw diffusion apparatus is shown in Fig. 1b.

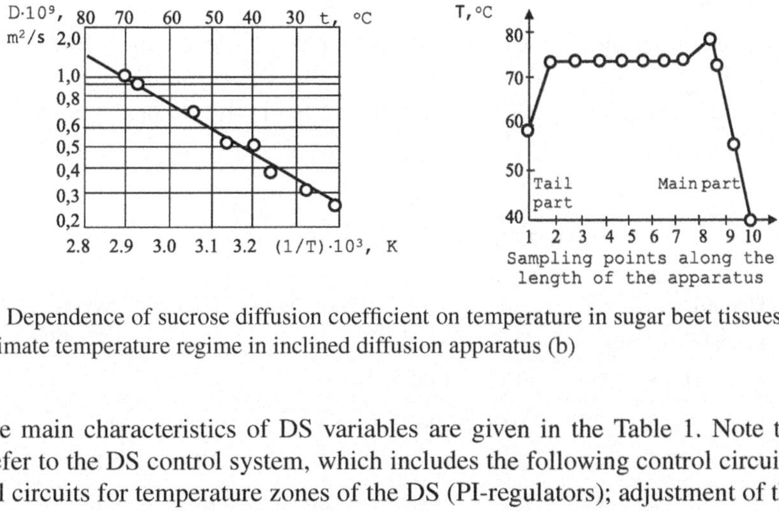

Fig. 1. Dependence of sucrose diffusion coefficient on temperature in sugar beet tissues (a) and approximate temperature regime in inclined diffusion apparatus (b)

The main characteristics of DS variables are given in the Table 1. Note that the data refer to the DS control system, which includes the following control circuits: four control circuits for temperature zones of the DS (PI-regulators); adjustment of the chip consumption and water consumption ratio (ratio regulator). Sampling length – about 3 days (sampling interval – 5 s).

As it can be seen from the result, the variables of costs can only be conditionally attributed to the normal law of distribution. The calculation of quantiles showed the following results – Table 2. A negative asymmetry value means that the data is skewed, and an excess value greater than 3 indicates rapid peaks compared to the normal distribution.

Table 1. Basic statistical characteristics of DS variables

Variable	Marking	mx	sx	[Min Max]	Trend	Median
Consumption of feed water by diffusion installation m^3/h	Var1	82.00	8.74	[65.24 90.56]	80.72	78.08
Consumption of juice by the diffusion installation m^3/h	Var2	81.38	8.80	[12.12 108.00]	83.40	82.12
Chip consumption, t / h	Var3	84.33	15.87	[0.00 99.63]	99.58	82.58
Water temperature after steam heater, °C	Var4	70.37	10.96	[54,39 79.52]	72.38	71.12
pH of feed water, units pH	Var5	4.13	1.00	[3.00 6.29]	4.10	3.95
Temperature in the I zone of the diffusion unit, °C	Var6	63.86	2.80	[54.48 69.53]	65.03	64.23
Temperature in the II zone of the diffusion unit, °C	Var7	63.58	2.23	[58.38 67.45]	64.73	64.03
Temperature in the III zone of the diffusion unit, °C	Var8	63.92	2.22	[59.02 67.45]	65.04	64.70
Temperature in the IV zone of the diffusion unit, °C	Var9	64.92	1.68	[60.70 68.48]	65.75	65.43
Diffusion juice temperature, °C	Var10	28.29	4.20	[20.70 52.24]	26.17	27.19
Voltages of the thyristor Coverter, V	Var11	170.10	12.76	[131.7 201.00]	175.10	171.50

Table 2. Statistical characteristics of costs

Costs	Quantiles					Asymmetry	Excess	Average absolute deviation
	Probability level							
	0	0.25	0.5	0.75	1.00			
Var1	0	73.2	78.1	80.4	90.6	−5.7	48.9	4.8
Var2	12.1	77.6	82.1	86.2	108.0	−2.5	17.2	5.8
Var3	0	0.25	82.6	86.8	99.6	−3.30	15.1	8.8

Thus, the standard mathematical expectation, standard deviation or variance cannot be characteristics of the data, as the data are not described by the normal distribution. In addition, these characteristics are sensitive to emissions. An adequate estimate of such data, respectively, can be the median, mean absolute deviation and interquantile range. However, the neural network structure of the model takes into account the variable nature of these variables.

As noted, the main assessments of the resource-efficient operation of DS are the loss of sugar in the pulp and the amount of sugar in the diffusion juice. The first estimate should be the lowest, the second - subject to maximization. In the course of work the variables influencing the resulted indicators by methods of the correlation analysis were defined. The values of the correlation coefficients between the input and predicted variables (Table 3) indicate a close relationship between the selected inputs and outputs, which confirms the use of the neural network model.

Table 3. Values of correlation coefficients between input and predicted variables

	Correlation coefficient					
	Train Var12	Validation Var12	Test Var12	Train Var13	Validation Var13	Test Var13
MLP 11-9-2	0.9756	0.9760	0.9685	0.98730	0.9756	0.9734
MLP 11-11-2	0.9774	0.9801	0.9735	0.9704	0.9741	0.9713
MLP 11-8-2	0.9699	0.9712	0.9675	0.9715	0.9738	0.9740
MLP 11-10-2	0.9819	0.9802	0.9757	0.9746	0.9759	0.9738
MLP 11-14-2	0.9843	0.9843	0.9820	0.9817	0.9822	0.9798

Thus, we obtain the model of DS, which is described by the dependence:

$$\overline{y} = f_{NN}(\overline{x});\ \overline{y} = \begin{bmatrix} \text{Var12} \\ \text{Var13} \end{bmatrix};\ \overline{x} = \begin{bmatrix} \text{Var1} \\ \text{Var2} \\ \cdots \\ \text{Var11} \end{bmatrix}, \tag{4}$$

where Var1 – Var11 – inputs of NN (Table 1); Var12 – sugar content in pulp, %; Var13 – sugar content in diffusion water, %.

For the proposed model f_{NN} for effective modeling in the package Statistica Neural Networks input data is automatically divided into three blocks: training, control, test. The presence of three blocks is not mandatory, however, the test block improves the quality of further work, as it allows you to make sure that there is no “retraining” (overfitting) of the network. At all stages during forecasting, much attention is paid to the depth of the forecast. The representativeness of the training sample for the neural network is provided by a set of input data in the form of time series (Var1–Var11) in the amount of 34 561 at and the original laboratory data (Var12, Var13) in the amount of 72 values (3 days in total).

2.2 A Model for Predicting Resource Efficiency

For efficient modeling in the Statistica Neural Networks package, the input data is automatically divided into three blocks: training, control, test. The presence of three blocks is not mandatory, however, the test block improves the quality of further work, as it allows you to make sure that there is no "retraining" (overfitting) of the network. The best result was demonstrated by the MLP 11–14-2 network (Table 4), which provided a performance of more than 98% in all samples, which proves the possibility of its further use.

Table 4. The results of solving the optimization problem of the synthesis of the DS model in the form of neural networks

	Training perfect	Validation perfect	Test perfect	Training error	Validation error	Test error	Training algorithm
MLP 11-9-2	0.9743	0.9758	0.9710	0.0006	0.0006	0.0006	BFGS 298
MLP 11-11-2	0.9740	0.9771	0.9723	0.0007	0.0006	0.0007	BFGS 360
MLP 11-8-2	0.9707	0.9725	0.9707	0.0007	0.0006	0.0006	BFGS 381
MLP 11-10-2	0.9783	0.9780	0.9747	0.0006	0.0005	0.0006	BFGS 513
MLP 11-14-2	0.9830	0.9832	0.9809	0.0004	0.0004	0.0005	BFGS 290

Table 5. The results of the selected NN

Samples (for target Var13)	MLP 11-14-2	Samples (for target Var13)	MLP 11-14-2
Minimum prediction (Train)	5.5013	Minimum residual (Validation)	−0.2888
Maximum prediction (Train)	6.0737	Maximum residual (Validation)	0.2789
Minimum prediction (Validation)	5.5013	Minimum residual (Test)	−0.3291
Maximum prediction (Validation)	6.1338	Maximum residual (Test)	0.1754
Minimum prediction (Test)	5.5013	Minimum standard residual (Train)	−15.1298
Maximum prediction (Test)	6.2954	Maximum standard residual (Train)	13.7241
Minimum prediction (Missing)		Minimum standard residual (Validation)	−13.7017
Maximum prediction (Missing)		Maximum standard residual (Validation)	13.2282
Minimum residual (Train)	−0.3196	Minimum standard residual (Test)	−15.0412
Maximum residual (Train)	0.2899	Maximum standard residual (Test)	8.0147

The results of the survey statistics, as well as fragments of the resulting model weight coefficients, are shown in Table 5, from which it can be seen that the resulting model has a high degree of adequacy. The latter shows that the model coefficients are significant.

3 Result and Discussion

According to the results of the use of the neural network (4), the dependencies presented in Fig. 2, 3 with an error of less than 2%. The analysis of materials allows us to draw a conclusion about the interdependence between input and output variables, which is significantly nonlinear.

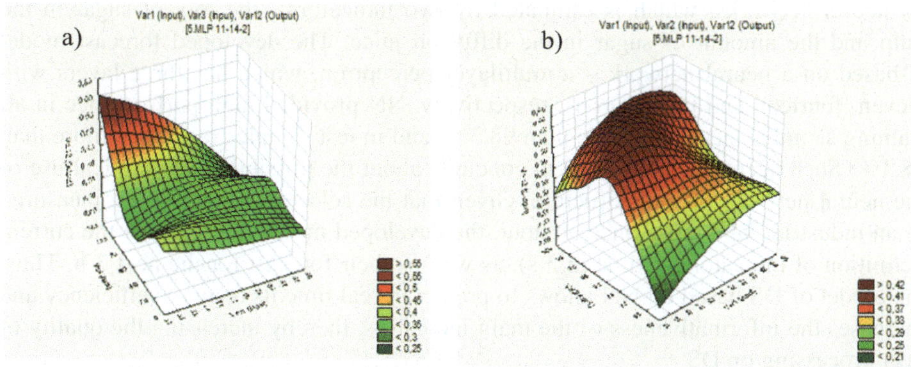

Fig. 2. Dependence of sugar content in pulp on: a) consumption of feed water and chips; b) consumption of feed water and diffusion juice.

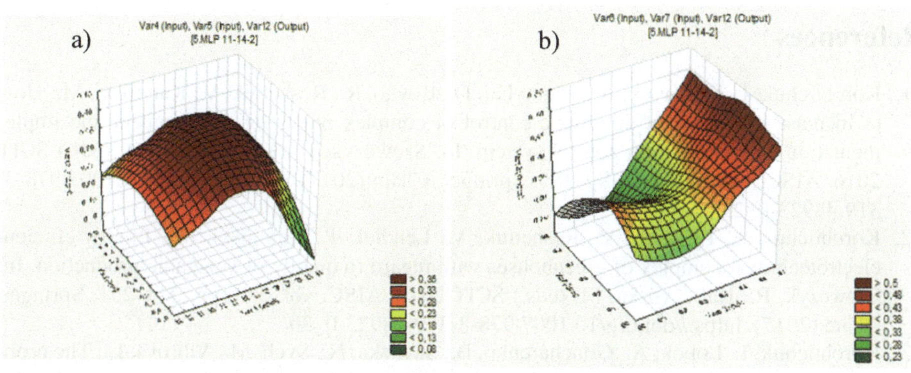

Fig. 3. Dependence of sugar content in pulp on: a) water temperature after steam contact heater and feed water pH; b) temperatures in I and II zones of the diffusion installation

Taking into account the forecast indicators obtained from the model of DS NN, a strategy is chosen for the DS regime parameters, in particular temperature and flow regimes. The latter directly become tasks for the corresponding automatic regulators of the station management system.

Given the established sufficient quality of operation of the NN model and the ability to generate the appropriate program code in C++ for use in real-time systems, DS neural model, namely a multilayer perceptron, it is advisable to predict the appropriate mode parameters of the station.

Improving the resource efficiency of sugar production becomes possible because of the high level of DS automated control, the mode parameters of which change in a timely manner based on the predicted values of the DS main indicators.

4 Conclusion

This paper proposes a model for predicting the energy efficiency of DS networks based on neural networks, which is estimated by two indicators - the loss of sugar in the pulp and the amount of sugar in the diffusion juice. The developed forecast model is based on a neural network – a multilayer perceptron, which has three layers with eleven, fourteen and two neurons, respectively. NN provided high performance in all training samples (accuracy more than 98.3%) and in test samples (accuracy more than 98.1%).Such accuracy allows one to conclude about the possibility of practical use of the neural network of such structure. Given that the selected indicators are measured in an industrial laboratory once an hour, the developed model will provide the current definition of indicators (1 time in 5 s), as well as their forecast for the next 3 h. Thus, the model of DS based on NN allows to predict in real time its resource efficiency and increases the informativeness of the main indicators, thereby increasing the quality of beet processing on DS.

Further research is aimed at developing an automated decision-making procedure to change the DS regime parameters, thus increasing the resource efficiency of sugar production.

References

1. Korobiichuk, I., Ladanyuk, A., Shumyhai, D., Boyko, R., Reshetiuk, V., Kamiński, M.: How to increase efficiency of automatic control of complex plants by development and implementation of coordination control system. In: Szewczyk, R., Kaliczyńska, M. (eds.) SCIT 2016. AISC, vol. 543, pp. 189–195. Springer, Cham (2017). https://doi.org/10.1007/978-3-319-48923-0_23
2. Korobiichuk, I., Lysenko, V., Reshetiuk, V., Lendiel, T., Kamiński, M.: Energy-efficient electrotechnical complex of greenhouses with regard to quality of vegetable production. In: Szewczyk, R., Kaliczyńska, M. (eds.) SCIT 2016. AISC, vol. 543, pp. 243–251. Springer, Cham (2017). https://doi.org/10.1007/978-3-319-48923-0_30
3. Korobiichuk, I., Lobok, A., Goncharenko, B., Savitska, N., Sych, M., Vihrova, L.: The problem of the optimal strategy of minimax control by objects with distributed parameters. In: Szewczyk, R., Zieliński, C., Kaliczyńska, M. (eds.) AUTOMATION 2019. Advances in Intelligent Systems and Computing, vol. 920, pp. 77–85. Springer, Cham (2020). https://doi.org/10.1007/978-3-030-13273-6_8
4. Kharlamenko, V., Ruban, S., Korobiichuk, I., Petruk, O.: Adaptive control of dynamic load in blooming mill with online estimation of process parameters based on the modified kaczmarz algorithm. In: Szewczyk, R., Kaliczyńska, M. (eds.) SCIT 2016. AISC, vol. 543, pp. 227–233. Springer, Cham (2017). https://doi.org/10.1007/978-3-319-48923-0_28
5. Tregub, V., Korobiichuk, I., Klymenko, O., Byrchenko, A., Rzeplińska-Rykała, K.: Neural network control systems for objects of periodic action with non-linear time programs. In: Szewczyk, R., Zieliński, C., Kaliczyńska, M. (eds.) AUTOMATION 2019. Advances in Intelligent Systems and Computing, vol. 920, pp. 155–164. Springer, Cham (2020). https://doi.org/10.1007/978-3-030-13273-6_16

6. Bakhmash, M.I., Ihnatev, M.I., Vityvskyi, I.A.: Buriakovyi tsukor tekhnolohii vyrobnytstva. Abetka-NOVA, Kamianets-Podilskyi (2004)
7. Krasnyak, O., Amons, S.: Sugar market: problems and prospects. Efektyvna Ekon. **1** (2020). https://doi.org/10.32702/2307-2105-2020.1.64
8. Kishenko, V., Zaika, V.: Synergetic synthesis of hierarchical control system of sugar factory technological complex. Eastern-Eur. J. Enterp. Technol. **4/2**(64), 46–51 (2013). https://doi.org/10.15587/1729-4061.2013.16658.
9. Ladanyuk, A.P., Shumygai, D.A., Boiko, R.O.: Situational coordination of continuous technological complexes subsystems. J. Autom. Inf. Sci. **45**(8), 68–74 (2013). https://doi.org/10.1615/JAutomatInfScien.v45.i8.70
10. Zaiets, N.A., Savchuk, O.V., Shtepa, V.M., Lutska, N.M., Vlasenko, L.O.: The synthesis of strategies for the efficient performance of sophisticated technological complexes based on the cognitive simulation modelling. Naukovyi Visnyk Natsionalnoho Hirnychoho Univ. **2**, 110–117 (2021). https://doi.org/10.33271/nvngu/2021-2/110
11. Lutskaya, N.N., Ladanyuk, A.P.: Problems features of the robust control of process plants. Part II. Examples of modeling of robust control systems. J. Autom. Inf. Sci. **48**(12), 62–69 (2016). https://doi.org/10.1615/JAutomatInfScien.v48.i12.60
12. Korobiichuk, I., Ladanyuk, A., Vlasenko, L., Zaiets, N.: Modern development technologies and investigation of food production technological complex automated systems. In: Proceedings of 2nd International Conference on Mechatronics Systems and Control Engineering, ICMSCE 2018, pp. 52–56 (2018). https://doi.org/10.1145/3185066.3185075
13. Coelho, A.P., Bettiol, J.V.T., Dalri, A.B., Fischer Filho, J.A., Faria, R.T., Palaretti, L.F.: Application of artificial neural networks in the prediction of sugarcane juice. Pol. Rev. Bras. Eng. Agríc. Ambient **23**(1), 9–15 (2019). https://doi.org/10.1590/1807-1929/agriambi.v23n1p9-15
14. Oktarini, D., Mohruni, A.S., Sharif, S., Yanis, M., Madagaskar: Optimum milling parameters of sugarcane juice production using artificial neural networks (ANN). J. Phys. Conf. Ser. **1167**(1), 012016 (2019). https://doi.org/10.1088/1742-6596/1167/1/012016
15. Lin, X., Yang, J.: HDP for the neutralized PH value control in the clarifying process of sugar cane juice. In: AIP Conference Proceedings, vol. 1127, pp. 162–173 (2009). https://doi.org/10.1063/1.3146188
16. Zhang, J., et al.: Monitoring sugar crystallization with deep neural networks. J. Food Eng. **280** (2020). https://doi.org/10.1016/j.jfoodeng.2020.109965.
17. Koc, B.A., Kocer, S., Tümer, A.E.: Artificial neural network models for predicting the energy consumption of the process of crystallization syrup in Konya sugar factory. Int. J. Intell. Syst. Appl. Eng. **1**(5), 18–21 (2017). https://doi.org/10.18201/ijisae.2017526691
18. Astray, G., Gullón, B., Labidi, J., Gullon, P.: Comparison between developed models using response surface methodology (RSM) and artificial neural networks (ANNs) with the purpose to optimize oligosaccharide mixtures production from sugar beet pulp. Ind. Crops Prod. **92**, 290–299 (2016). https://doi.org/10.1016/j.indcrop.2016.08.011
19. Ladanyuk, A., Kyshenko, V., Shkolna, O., Sych, M.: Development of the algorithm of determining the state of evaporation station using neural networks. Eastern-Eur. J. Enterp. Technol. **5**(2–83), 54–62 (2016). https://doi.org/10.15587/1729-4061.2016.79322
20. Zarifpoor, M., Hakimzadeh, V.: Investigating of artificial neural network potential to predict the properties of refined raw sugar beet juice by electrocoagulation process. Ukr. Food J. **7**(4), 682–691 (2018). https://doi.org/10.24263/2304-974X-2018-7-4-12

Wavelet Analysis of the Behavior of the Process of Rectification of Alcohol as an Object of Control

Yaroslav Smityuh(✉), Vasii Kyshenko, Anatoliy Ladanyuk, and Dmitry Krishchenko

National University of Food Technologies, Kiev, Ukraine
smityuh1@gmail.com, ladanyuk@ukr.net

Abstract. Mathematical and numerical modeling of time series of variables of technological process of alcohol rectification by means of wavelet analysis is considered in the work. The algorithm for processing the time series is to determine the stochasticity, chaos and analysis of special states of the system using wavelet transform. This complex application provides more complete information about the studied process, described by time series. The efficiency of the method of wavelet analysis of time is proved series of technological parameters of alcohol production to improve the accuracy of control and diagnosis of critical situations.

Keywords: Control object · Wavelet analysis · Time series · Alcohol production · Monitoring

1 Introduction

The technological process of distillation of a distillery is a complex technological object of management [1]. It is characterized by high levels of noise and interference, significant uncertainty, significant nonlinearity of relationships between technological parameters, significant non-stationary, stochastic and chaotic, problematic situational behavior due to bursts, gaps and other drastic changes in the state of the object [2]. These features complicate the management decision-making process: there is a need to find, justify and adopt real-time operational management strategies in accordance with the situation in the management facility, which is determined based on current and retrospective information about its behavior [3–5]. In this sense, in order to ensure the automatic implementation of applied control functions, methods of analysis of time series of technological variables of the control object are important [6, 7]. Traditional methods of statistical analysis of random variables and functions and Fourier spectral analysis are widely used to analyze random processes. Along with them in recent years, widespread methods of signal processing based on fractal and wavelet transforms. Wavelet analysis is one of the promising methods currently used for the analysis of time signals. Unlike Fourier spectrum analysis (spectral analysis), it has a number of advantages: it can be localized in both time and frequency; in comparison with the Fourier analysis has a lot of different basic functions, the properties of which allow to solve different problems [8, 9]. The purpose of these

R. Szewczyk et al. (Eds.): AUTOMATION 2022, AISC 1427, pp. 162–170, 2022.
https://doi.org/10.1007/978-3-031-03502-9_17

studies was to analyze the time series of technological variables of the distillation of alcohol, including wavelet transform, to establish the characteristic manifestations of the object of control, to provide automatic monitoring of technological situations of distillation, to organize effective strategies for controlling alcohol production in critical situations.

2 Materials and Methods of Research

Along with traditional methods, in recent years, methods of signal processing based on fractal and wavelet analysis have become widespread [4]. A distinctive feature of the latter is that they reveal the features of the local structure of a complex signal and identify its various properties, invisible in real time. The wavelet transform transforms additional information that is not available in its original form. An important characteristic of methods based on wavelet transforms is their universality. To date, there are two main objectives of the analysis of the system: the need to identify internal patterns in the behavior of time series and the forecast of periods of stability of the studied processes. Requirements for the quality of detection and study of these patterns are now being strengthened and these goals require that the series model be identified. Therefore, there is a need to develop new and modify existing algorithms for time series analysis in computer systems [10].

Considers mathematical and numerical modeling of time series using wavelet analysis. The main processing algorithm is based on the existence of a time series and is to determine the stochasticity, chaos and analysis of special states of the system using wavelet transform. This complex application provides more complete information about the studied process, described by time series. Wavelet transform is a mathematical tool for hierarchical decomposition of functions. With the help of wavelets, the functions are represented in the form of a composition of coarse low-frequency approximation and refining components (parts), which are missing in the approximation of the elements of the graph function. Regardless of the type of function (image, curve), the wavelet represents the function as a hierarchy of display levels with different accuracy of detail [11]. Wavelet transform is used in image pre-processing to reduce noise, analyze textures, highlight objects, and compress images.

The wavelet transform is well localized in time, which allows to detect the structure of the studied signal at certain time intervals. In addition, the use of multiscale decomposition makes it possible to analyze the process at different scales of time, as well as to identify a number of important features of signals associated with changes in its structure during the transition from one scale to another, with energy distribution. between scales, etc. [10]. A distinctive feature of wavelet transforms is that they, along with the global characteristics of processes, reveal the features of their local structure. Wavelet – analysis allows you to get local high-frequency and global large-scale information about the object and allows you to judge at what point in time there were certain components of the signal. An important characteristic of methods based on wavelet transforms is their versatility. Therefore, the relevance of the use of wavelet analysis of time series to automate the technological processes of bragorectification is beyond doubt [10]. The main processing algorithm is based on the existence of a time series and is to determine stochasticity, chaos and analysis of special states of the system using wavelet transform. This complex program provides more complete information about the process under study, described in time series. The analysis of wavelet coefficients allows to reveal nonstationarity of the studied process on any frequency-time scale, and thus to receive both qualitative, and quantitative characteristics of nonstationarity; to establish the level and nature of noise and disturbances, to diagnose the characteristic manifestations of the behavior of the object of control. The Fourier transform apparatus is quite simple to implement and provides a fairly clear interpretation of the results, but does not provide, for example, information about the frequency distribution over time, can not distinguish segments in signals with sharp changes in their level. Unlike the Fourier transform, which localizes frequencies but does not provide temporary identification of the process, the wavelet transform, which uses a self-organizing moving frequency-time window, detects as well as low-frequency ones thanks to a scaling function that performs rough approximation. signal by means of approximating coefficients and high-frequency characteristics of the signal, revealing the details of the signal based on detailed coefficients, at different time scales [10]. Wavelet analysis is to present the output signal, a function in the form of a linear combination of some basic functions – wavelets. Wavelet transform provides a two-dimensional scan of the studied one-dimensional signal, which makes it possible to analyze the properties of the signal both in time and frequency. Scale-temporal localization of wavelet transform is due to the principle of construction of the orthogonal basis of decomposition. In contrast to the Fourier transform, where harmonic functions are used as the basis, the basis of wavelet decomposition is localized in time, i.e. rapidly declining soliton-like wavelet – function:

$$\psi_{a,b} = \psi\left(\frac{t-b}{a}\right) \tag{1}$$

where t is the time, a is the scale factor and determines the amount of stretching of the wavelet, b has the dimension of time and determines the amount of shift of the wavelet function. There are currently a number of such features. The calculation of the wavelet transform is performed by convolution of the studied time sequence $f(t)$ with the wavelet function:

$$W(a,b) = \int_{-\infty}^{+\infty} f(t)\psi_{a,b}(t) \tag{2}$$

Wavelet transform forms an image as a vector of properties that reproduces the current state of the technological process. There are continuous and discrete transformation. Continuous wavelet transform is built using continuous scale transformations and wavelet transfers with arbitrary values of the scale factor a and the offset parameter b:

$$W(a,b) = |a|^{-1/2} \int_{-\infty}^{\infty} f(t)\psi^{*}(\frac{t-b}{a})dt \tag{3}$$

where the symbol * denotes the operation of complex coupling.

The two-parameter function $W(a, b)$ provides information about the change in the relative contribution of components of different scales over time and is called the spectrum of wavelet transform coefficients. Having the wavelet spectrum, you can calculate the total signal energy

$$E_f = \int f^2(t)dt = \int\int W^2(a, b)\frac{dadb}{a^2} \tag{4}$$

and the global spectrum of energy – the distribution of total energy on the scale (wavelet scan)

$$E_w = (a) = \int W^2(a, b)db \tag{5}$$

In practice, time series are set not by analytical functions, but by discrete acquisition of data determined on a finite time interval. In this case, it is assumed that for $t_k < t < t_{k+1}, f(t) = s_k, k = 1, 2, \ldots, n$ and formula (3) for the wavelet transform coefficients is modified as follows:

$$W(a, b) = \frac{1}{a^{1/2}}\sum_k S_k(\Phi_{k+1} - \Phi_k) \tag{6}$$

$$\Phi_k = |a|^{-1/2}\int_{-\infty}^{t_k} \psi^*(\frac{t-b}{a})dt \tag{7}$$

The choice of one or another class of wavelet functions is made taking into account the specifics of the research tasks and the characteristics of the signal, which helps to more fully identify and emphasize certain features of the analyzed time series. The wavelet transform spectrum of a one-dimensional signal represents a surface in three-dimensional space. Visualization of the spectrum is carried out by projection on a plane with isolines, which allows you to track changes in the coefficients at different scales over time, with a gradient of shades of blue and red between the lines, reflecting the frequency-time characteristics of the signal. Time is plotted on the abscissa, frequency – on the ordinate. In this paper, a variant of painting is chosen, in which the maximum area is white and the minimum area is black [11].

3 The Results of Research

The time series of the main technological variables of the distillation process of alcohol are analyzed: steam consumption, welding consumption, alcohol consumption, temperature on the control plates of the columns, temperature and pressure of the lower and upper part of the columns. Different types of wavelets were used in the analysis of time series characterizing the behavior of rectification processes: Mhat (Mexican hat), Morl (Morle), Gau (Gauss). Wavelet signal conversion was performed using the Matlab mathematical package. The results of the transformation are shown in the figures below. Two-dimensional arrays of coefficients are displayed in the form of point images,

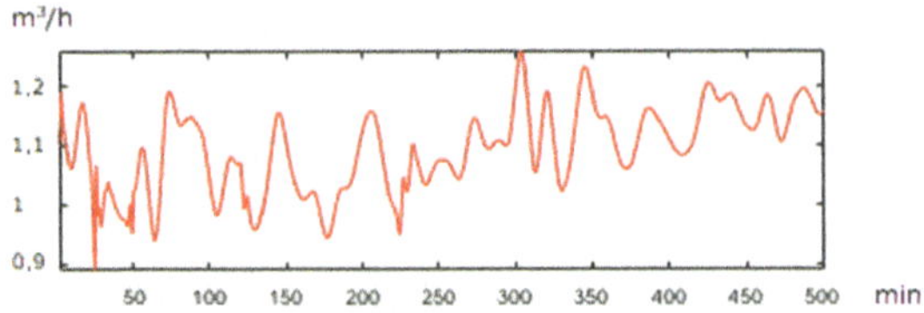

a)

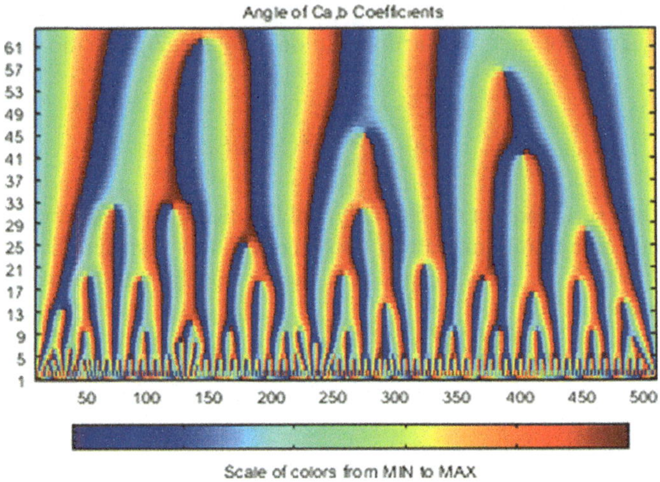

b)

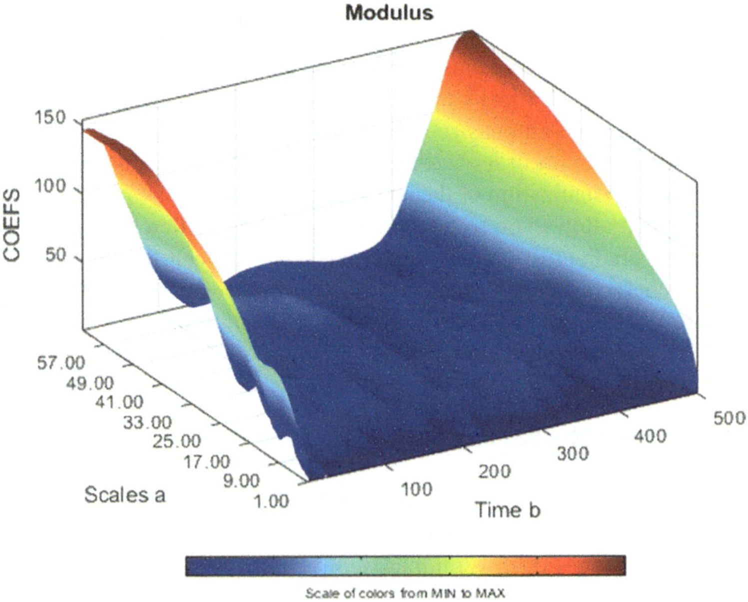

c)

Fig. 1. Wavelet-Gaussian transformation of the time series of alcohol consumption: **a** – time series; **b** – wavelet spectrogram; **c** – three-dimensional graph of wavelet coefficients

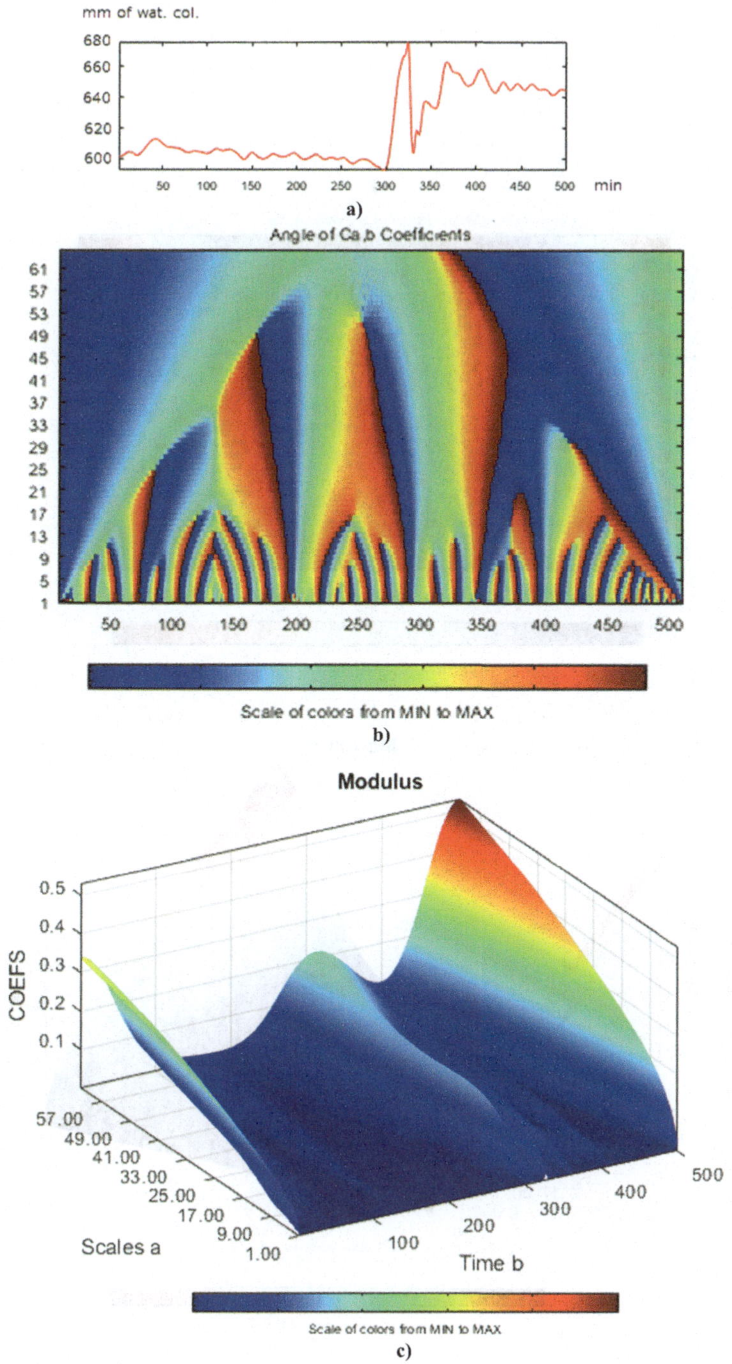

Fig. 2. Wavelet-Gaussian transformation of the time series of the pressure of the top of the meal column: **a** – time series; **b** – wavelet spectrogram; **c** – three-dimensional graph of wavelet coefficients

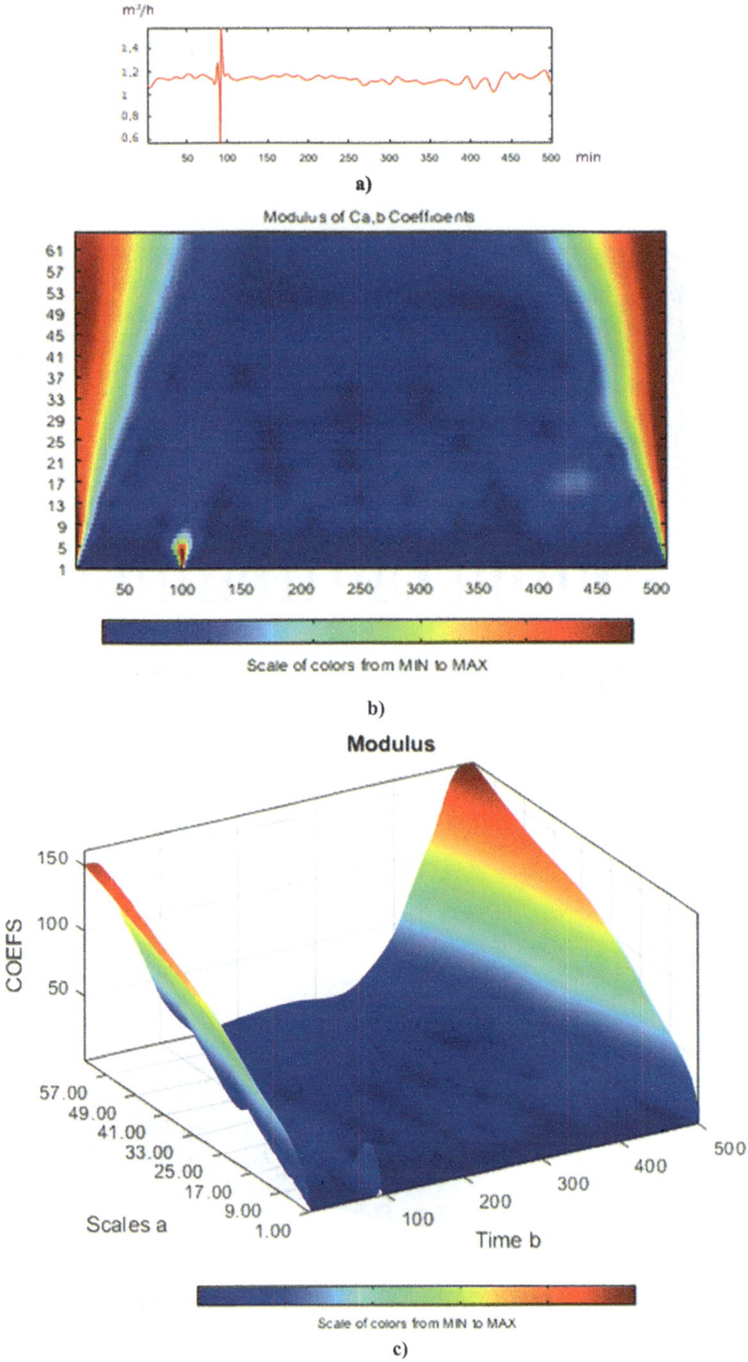

Fig. 3. Wavelet-Morle transformation of the time series of alcohol consumption: **a** – time series; **b** – wavelet spectrogram; **c** – three-dimensional graph of wavelet coefficients

and numerical values – shades of blue and red. In our case, 16 shades were used. The brightest parts of the image correspond to the maximum values of the coefficients [10].

In Fig. 1, 2 and 3 show the results of wavelet analysis of time series of a number of technological variables of the distillation process of alcohol, which demonstrates a clear fixation of the characteristic manifestations of the behavior of the object of control. Thus, in Fig. 1 in the time series of alcohol consumption there are periodic behaviors, and the wavelet spectrogram of the signal indicates the presence of high-frequency noise with their localization and characteristics of the frequency spectrum. In the time series of the pressure of the upper column (see Fig. 2) revealed the non-stationary nature of the studied process at certain frequency and time scales, which allows both qualitative and quantitative assessment of the non-stationary distribution of wavelet coefficients. Figure 3 highlights the area of abrupt emissions (bursts), which provides diagnosis of anomalies in the analysis of situations. One or another feature of the time series is emphasized by the level of wavelet coefficients: the higher the level of coefficients, the clearer this feature is established and the stronger its area in the frequency-time spectrum, which is almost impossible in Fourier analysis [9].

Conclusions

In this work, wavelet analysis allowed us to assess the structure of the studied signal, to observe separately the behavior of its various components. In addition, in contrast to the Fourier transform of wavelet decomposition allows you to analyze the signal in the form of components of its time-localized elementary processes, which is convenient for detecting local periodicities, nonstationarity, jumps, breaks, differences in steepness, etc. When processing data containing a multicomponent signal, ie a process containing several signals with separate characteristics, it is often necessary to separate and analyze these signals separately. As an effective approach can be used multi-scale approach technology, which allows based on the use of wavelet packets and clustering to make a fine and precise division of the frequency-time plane into cells so that you can simultaneously divide signals by frequency properties and location, which will automate the tasks of technological monitoring in the organization of effective strategies for situational management of the distillation of alcohol.

References

1. Wang, X., et al.: Review on etherification by reactive distillation. Huagong Jinzhan/Chem. Ind. Eng. Progr. **40**(4), 1797–1811 (2021). https://doi.org/10.16085/j.issn.1000-6613.2020-1982
2. Alves, J.L.F., da Silva Filho, V.F., Machado, R.A.F., Marangoni, C.: Ethanol enrichment from an aqueous stream using an innovative multi-tube falling film distillation column equipped with a biphasic thermosiphon. Process Saf. Environ. Protect. **139**, 69–75 (2020). https://doi.org/10.1016/j.psep.2020.03.039
3. Korobiichuk, I., Lobok, A., Goncharenko, B., Savitska, N., Sych, M., Vihrova, L.: The problem of the optimal strategy of minimax control by objects with distributed parameters. In: Szewczyk, R., Zieliński, C., Kaliczyńska, M. (eds.) Automation 2019. Advances in Intelligent Systems and Computing, vol. 920, pp. 77–85. Springer, Cham (2020). https://doi.org/10.1007/978-3-030-13273-6_8

4. Korobiichuk, I., et al.: Synthesis of optimal robust regulator for food processing facilities. In: Szewczyk, R., Zieliński, C., Kaliczyńska, M. (eds.) Automation 2017. ICA 2017. Advances in Intelligent Systems and Computing, vol. 550, pp. 58–66. Springer, Cham (2017). https://doi.org/10.1007/978-3-319-54042-9_5
5. Kharlamenko, V., Ruban, S., Korobiichuk, I., Petruk, O.: Adaptive control of dynamic load in blooming mill with online estimation of process parameters based on the modified Kaczmarz algorithm. In: Szewczyk, R., Kaliczyńska, M. (eds.) SCIT 2016. AISC, vol. 543, pp. 227–233. Springer, Cham (2017). https://doi.org/10.1007/978-3-319-48923-0_28
6. Kyshenko, V.: The problems of technological monitoring in the control of industrial enterprise. Pressing issues and priorities in development of the scientific and technological complex, pp. 62–68. B&M Publishing, San Francisco (2013)
7. Ladanyuk, A., Kyshenko, V., Smityuh, Y.: The biotech complexes control in conditions of situational. Ann. Warsaw Univ. Life Sci. **60**, 149–154 (2012)
8. Novakovska, N., Kyshenko, V.: Fractal analysis of distillation unit time series in prediction and control problems. Ukr. J. Food Sci. **3**(2), 243–253 (2015)
9. Korobiichuk, I., Ladanyuk, A., Shumyhai, D., Boyko, R., Reshetiuk, V., Kamiński, M.: How to increase efficiency of automatic control of complex plants by development and implementation of coordination control system. In: Szewczyk, R., Kaliczyńska, M. (eds.) SCIT 2016. AISC, vol. 543, pp. 189–195. Springer, Cham (2017). https://doi.org/10.1007/978-3-319-48923-0_23
10. Grinsted, A., Moore, J.C., Jevrejeva, S.: Application of the cross wavelet transform and wavelet coherence to geophysical time series. Nonlin. Processes Geophys. **11**, 561–566 (2004). https://doi.org/10.5194/npg-11-561-2004
11. Mallat S.: A Wavelet Tour of Signal Processing, p. 637. Academic Press (1999)

The Use of STM32f103xx Microcontroller to Create and Prototype Low Cost Laboratory Experiments

Svitlana Sharuda(✉), Oleh Sharuda, Sergey Kirichuk, Natalya Savitskaya, and Sergey Shved

National University of Food Technologies, Volodymyrska str., 68, Kyiv 01601, Ukraine
sveta.sharuda@gmail.com, {savitskanm,shved_sergey}@ukr.net

Abstract. This article describes the Home Lab Easy Kit project, which can be used to simplify the creation of an experimental setup and save time for other more important tasks, such as the logic of the experiment, calculations, etc. Efficiency is achieved by generated firmware for well known, cheap and accessible STM32F103C8T6 micro controller (MCU). There are number of boards accessible on the market which may be also referred as "Blue pill". This project allows linkage with generated C++ library that implements communication with micro controller and provide simple interfaces to MCU peripherals. This project allows to use some STM32F103C8T6 peripherals with Linux based microcomputers like Raspberry Pi using I2C bus.

Keywords: STM32F103C8T6 · Data science · Linux · I2C · CAN · 12-bit analogue to digital converter support · Multichannel software PWM support

1 Introduction

Modern industrial enterprise must have efficient product development. It's strategy must meet such goals as increasing competitiveness of products, reducing the time it takes for products to enter the market, and reducing overall production costs.

Engineering at enterprise often requires to test new solutions without risk for the business strategy. Unpredictable research and design lags is one of the most significant risks for production. In this case, engineers can use the simplified experimental setups. However, it is also not an easy task. There are many questions to be solved: design of experiment, the hardware selection, mechanics, etc.

One of these tasks is experiment automation. It is desirable to have one simple solution for as many cases as possible. There are a lot of system solutions for this problem are known. Commercial solutions provide good functionality, but any issues, incompatibilities may require long lasting customer support, which in turn increases the time of the experiment. Open-source solutions allow to make modifications to software, which is useful if some change is required. Enterprise may contribute open-source project providing advantages for production and community.

R. Szewczyk et al. (Eds.): AUTOMATION 2022, AISC 1427, pp. 171–182, 2022.
https://doi.org/10.1007/978-3-031-03502-9_18

Analysis of available open source solutions shows that microcomputers and microcontrollers are often used to create prototypes of experimental setups.

Costs, computational power, flexibility and other characteristics for the most popular platforms and solutions were analyzed in work [1]. Authors created an experimental setup and considered number of variations for data processing.

Next work [2] provide review and survey of platforms and different technologies for sophisticated and smart systems development for IoT applications.

Experimental setups built on a microcomputer and microcontroller can also be used remotely. The approach to remote experimentation, as well as hardware components and software technologies are described in the article [3].

There are also works that indicate the advantage of using experimental facilities for teaching students [4]. So, in the work [5], an experiment carried out by the authors is described, which proved the effectiveness of active learning using an experimental setup created under a number of conditions: low cost, easy to use, safe, rugged, clear instrumentation, flexible.

Many automation algorithms can also be implemented using a platform consisting of a microcomputer, microcontroller and control system, as described in the articles [6–8].

Authors didn't find other open-source solutions that provide firmware for MCU, communication protocol and generated library (which can be used with modern general purpose programming languages) as single solution. This is a reason why this project may be actual for enterprises, educational purposes and amateurs.

The object of research is the interface between a computer and a microprocessor device with peripherals.

The subject of the research is a software solution for the implementation of interconnections in an experimental setup.

The purpose of the work is to create an open-source project, with the use of which it is possible to implement an experimental installation with lower financial costs (using a Raspberry Pi [9], micro controller STM32F103C8T6 [10], sensors, required for a particular experimental setup, some commonly available electronic components), as well as with the ability debug and develop independently.

2 Key Components of Project

Project consist of three main parts: Customizer, Firmware and Software. JSON configuration file is written by user to describe peripherals to be used and configure them. This file describes user requirements to customizer and allows to generate software and firmware sources to meet user expectations.

Customizer generates C and C++ headers required for firmware and software parts. Also, customizer is capable to detect some issues in JSON configuration and to warn user.

Firmware part is CMSIS based project written on C. Once headers are generated by customizer it may be compiled and flashed into MCU. User is not required to change firmware sources, however it is not forbidden.

Software part consists of C++ library [11] based on JSON configuration file provided to Customizer. Also, there is monitor tool linked with this library that can be used to test software to firmware communication and firmware peripherals being used. Documentation is split onto three parts.

- This document serves as introduction. Also, it explains basic things like software and hardware requirements, JSON configuration file, etc.
- Software documentation is provided in doxygen form. It should be generated in order to get more information on classes, interfaces, etc. It is required if you plan to use this project.
- Firmware documentation is also provided in doxygen form. It is optional and incomplete. You can use it to understand how firmware works.

2.1 Hardware Requirements

For this project to work, you need a computer or microcomputer with Linux and possibility to connect and work with I2C bus [12]. Software implementations of I2C bus also should work. Note, I2C will not be used in clock stretching mode. It means all Raspberry Pi versions may be used (I2C stretching bug in BCM2835 processor shouldn't be a barrier). Also need STM32F103C8T6 micro controller (it may be one of "Blue pill" boards or self-made PCB with this micro controller) and STM32 ST-LINK v2 programmer or any other capable to work with STM32F103C8T6. Power supply should be selected carefully. It must provide stable voltage and enough current. Avoid using two or more power supplies in experimental setups, this may cause serious issues. Power supply topics is explained in many details in book [13].

2.2 Software Requirements

It is possible to use either microcomputer or PC as building machine. Below is a description of software dependencies required to use Home Lab Easy Kit project.

The following packages are used by Home Lab Easy Kit project:

- `Raspberry Pi OS Buster`: This is a new version of Raspberry Pi OS to be installed on Raspberry Pi microcomputer. Other Linux distributions may also be used and configured in this way, however some packages may be missing or have different names. Note, not all of the packages may be available for previous versions (before Buster) of Raspberry Pi OS. This project is not limited by Raspberry Pi, however it is difficult to write instructions for all possible microcomputers and operational systems. So, Raspberry Pi and `Raspberry Pi OS Buster` is basic configuration for instructions. Other configuration may require different configuration process.
- `CMake`: With version $\geq$3.10. Install `cmake` package.

- `ICU4C`: International Components for Unicode (for C/C++). It is official library from the Unicode Consortium. It is included in many Linux distributions, however in our case development files are required. Install `libicu-dev` package.
- `GNU Arm Embedded Toolchain`: It is a collection of packages required for bare-metal software development using 32-bit ARM processors. It also includes Cortex-M, used in STM32F103x. Install these packages: `binutils-arm-none-eabi`, `gcc-arm-none-eabi`, `gdb-multiarch`. Note, `tools` variable in firmware/toolchain.cmake should be modified to specify path to toolchain in order to configure project for CMake.
- `STM32F10x Standard Peripherals Library`: This is a library from STMicroelectronics which is required to build firmware. It can be downloaded from ST web site directly [14]. Unpack library and modify STDPERIF_PATH variable in firmware/toolchain.cmake.
- `OpenOCD`: Open On-Chip Debugger project required to debug firmware. Optionally it may be used to flash firmware, or to debug it from your favorite IDE. Install `openocd` package.
- `ST-LINK/V2 tools`: This software is required to flash and debug firmware. Install `stlink-tools` package.
- `i2c-tools`: are very useful during work with i2c bus. Install `i2c-tools` package.
- `doxygen`: Doxygen is required to generate documentation for firmware and software parts of the project. Install `doxygen` package.
- `ncurses`: This library is used for minimal terminal interface used by Home Lab Easy Kit monitor. Install `libncurses-dev` package.
- `python3`: Required to run Customizer, which is written using Python3. Python3 should be a part of the operational system.

3 Available Peripherals

3.1 Device Functions

Currently Home Lab Easy Kit support the following functionalities that represent peripherals available on STM32F103C8T6 (Table 1).

Table 1. Functions of peripheral devices

Name	Description	Exclusive	Remapping
INFODev	Firmware identification and information support	Yes	No
ADCDev	12-bit Analogue to Digital Converter support	No	No
DESKDev	Simple set of four buttons and one encoder	Yes	No
GPIODev	General purpose input/output support	Yes	No
IRRCDev	NEC standard IR remote control support	Yes	No
LCD1602ADev	Simple LCD1602a screen support	Yes	No
RTCDev	Real Time Clock support	Yes	No
SPWMDev	Multichannel software PWM support	Yes	No
StepMotorDev	Support for A4998 and DRV8825 stepper motor drivers	No	No
UARTDev	UART proxy (allow to communicate with UART devices connected to micro controller)	No	No
CanDev	CAN bus support (beta)	No	Yes

Exclusive feature means that this feature may be enabled in a form of single virtual device. Non-exclusive features may provide several virtual devices of similar functionality but using different peripherals and working independently.

3.2 Usage Instructions

Steps to start a project:

1. Install software dependencies.
2. Prepare JSON configuration file.
3. Run customizer.sh bash script with JSON configuration file passed as parameter.
4. Build firmware and flash MCU.
5. Build software and link generated library to your project.
6. Make all electrical connections for your project.

3.3 JSON Configuration File

JSON configuration file has the following structure (Table 2):

JSON configuration file has the following structure:

```
{
  "firmware" : {
    "mcu_model" : "stm32f103",
    "device_name" : "test_dev",
    "install_prefix" : "test_dev",
    "install_path" : "/opt",
    "i2c_bus" : {
      "clock_speed" : 100000,
      "buffer_size" : 512,
      "address" : 32,
      "requires" :  {"i2c" : "I2C2"}
    }
  },
  "devices" : {
    "DeskDevCustomizer" : { ... },
    "RTCCustomizer" : { ... },
    "IRRCCustomizer" : { ... },
    "UartProxyCustomizer" : { ... },
    "LCD1602aCustomizer" : { ... },
    "GPIODevCustomizer": { ... },
    "SPWMCustomizer": { ... },
    "ADCDevCustomizer": { ... },
    "StepMotorDevCustomizer": { ... }
  }
}
```

Each functionality is represented by corresponding key/object pair under devices key. The only exception is INFODev – it is not described in JSON file at all and is generated automatically based on information provided. Each key under devices must have one of these values: "DeskDevCustomizer", "RTCCustomizer", "IRRCCustomizer", "UartProxyCustomizer", 'LCD1602aCustomizer", "GPIODevCustomizer", "SPWMCustomizer", "ADCDevCustomizer", "StepMotorDevCustomizer". Internally they are called as customizers, to reflect their ability to customize firmware and software.

Each of these keys should have one or more key/object pairs, where key is a "name" of the device and object describes device itself. Exclusive features may be represented by single device only, non-exclusive features may have more devices described. Presence of all features is not required, so leave those customizers which are required for your project.

Note, "firmware" doesn't allow to select MCU frequency. For example, STM32F10x is configured to use 72 MHz clock and it's not possible to change (Table 3).

Most of the features require some peripherals from MCU. Such dependencies are described as JSON object with key equal to peripheral type and value with name of the

Table 2. The first `firmware` key describes basic firmware properties

Key	Value description	Possible values	Required
"`mcu_model`"	MCU model being used, currently the only MCU is supported	"stm32f103"	Yes
"`device_name`"	Name of the firmware represented by this JSON file	Reasonable name in quotes	Yes
"`install_prefix`"	Installation prefix for the library installation path	relative path (relative to "`install_path`")	Yes
"`install_path`"	Installation path for the library	Absolute path	Yes
"`i2c_bus`"	Describes i2c connection with microcomputer and MCU	Object, see below	Yes
"`clock_speed`"	I2C bus clock speed. This project was tested with standard frequency only (100 kHz)	100,000	Yes
"`buffer_size`"	Size of the buffer for commands being sent from microcomputer to MCU, in bytes	Number	Yes
"`address`"	I2C bus address. Must be valid 7-bit i2c bus address for your project	Number	Yes
"`requires`"	This object, here and after contain description of peripherals required	Object	Yes
"`i2c`"	Represents I2C peripherals being used for communication	"I2C1" or "I2C2"	Yes
"`devices`"	Describes required functionalities	Object	Yes

Table 3. Important settings that can't be changed by JSON configuration file

Parameter	Description	Value
F_{mcu}	MCU frequency	72 MHz
F_{gpio}	Maximum output frequency	50 MHz
D_{AHB}	AHB clock divider	1
D_{APB1}	APB1 clock divider	2
D_{APB2}	APB2 clock divider	1

peripheral. Most of the time name is the same as how peripheral is defined in CMSIS library.

Read information about the types of peripheral devices, possible values, etc. can be found on the project page [15].

4 Example Experimental Setup

As an example, experimental setup for light source radiation pattern scanning was built. This setup utilizes stepper motor used to rotate a source of light. Also a photo-transistor was connected to ADC as light sensor. Stepper motor is controlled by DRV8225 driver which is connected to blue pill module. Both ADC and DRV8225 support is provided by the project. Experimental setup is shown in Fig. 1.

Hardware part consist of:

1. Stepper motor Nema17 (SM42HT47) and DRV8825 stepper motor driver.
2. A source of light (bicycle flashlight) which is attached to the motor axis.
3. Photo-transistor (HPTC3C-44 J).
4. Terminal block used to connect photo-transistor, load resistor (10 kOhm). Terminal block allows to change load resistor quickly to adjust photo-transistor for the most kinds of light sources.
5. Microcomputer (Raspberry Pi Zero v1.3).
6. Blue Pill module with STM32F103C8T6 MCU.
7. Connection PCB board used for interconnection purposes and 3.3 V DC-DC converter. This PCB was made for project testing purposes, therefore there are three DRV8825 installed. If required this board allows to connect up to three stepper motor drivers, in our case single motor is enough.
8. Other devices like Power supply (not shown), USB hub, USB Ethernet adapter, MCU programmer and active cooling system.

Software part consist of:

1. JSON configuration file, written by used and used to describe which peripherals should be used by MCU.

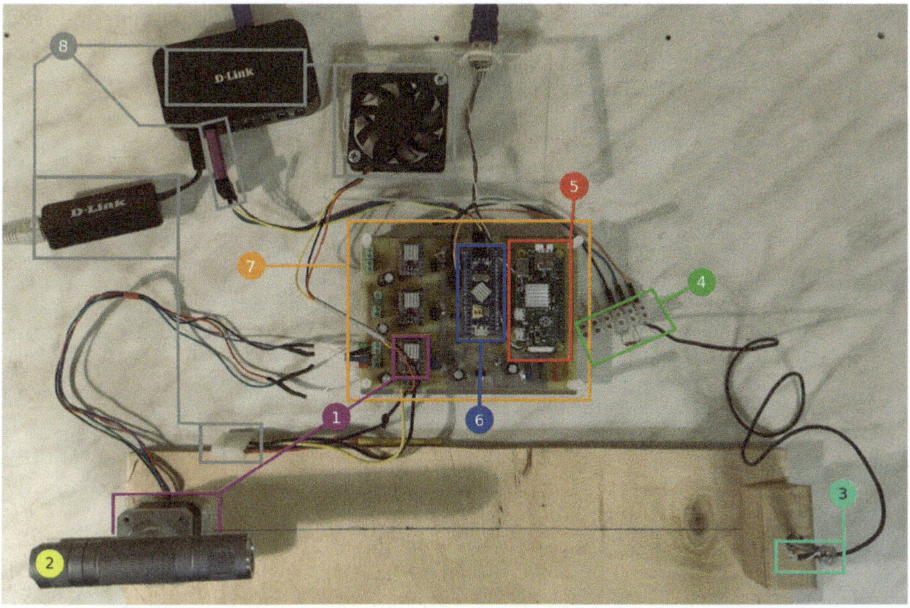

Fig. 1. Experimental setup for light source radiation pattern scanning

2. Firmware for STM32F103 MCU. Firmware is generated by the project using JSON configuration file.
3. C++ library generated by the project using JSON configuration file. This library should be installed in Linux operational system running on Raspberry Pi. It is possible to use cross-compiler as well.
4. C++ program linked with generated C++ library. This program initiates connection to I2C bus and firmware using several classes provide by the project. ADCDev and StepMotorDev classes are used to control ADC peripheral and DRV8225 stepper motor driver. Source code consist of 90 lines of C++ code which is relatively small program for this case.
5. Optional Python script to draw radiation pattern, 40 lines of code.

The experiment should be executed as following:

1. Operator adjust stepper motor axis to set light source direction to photo-transistor.
2. Operator starts a program with redirection of its output to a file (CSV file).
3. The program rotates source of light by 90° (counter clock wise)
4. Program sequentially rotates source of light by 1.8° (clock wise) and reads values from ADC. Each time MCU read 256 values. Angle and mean ADC value is put into output. Total scanning angle is 180°.
5. Program rotates source of light to initial position by rotation stepper motor by 90° (counter clock wise).

Video of the experiment is available on the website [16].

As result a CSV file is created and used to plot radiation pattern diagram on Fig. 2.

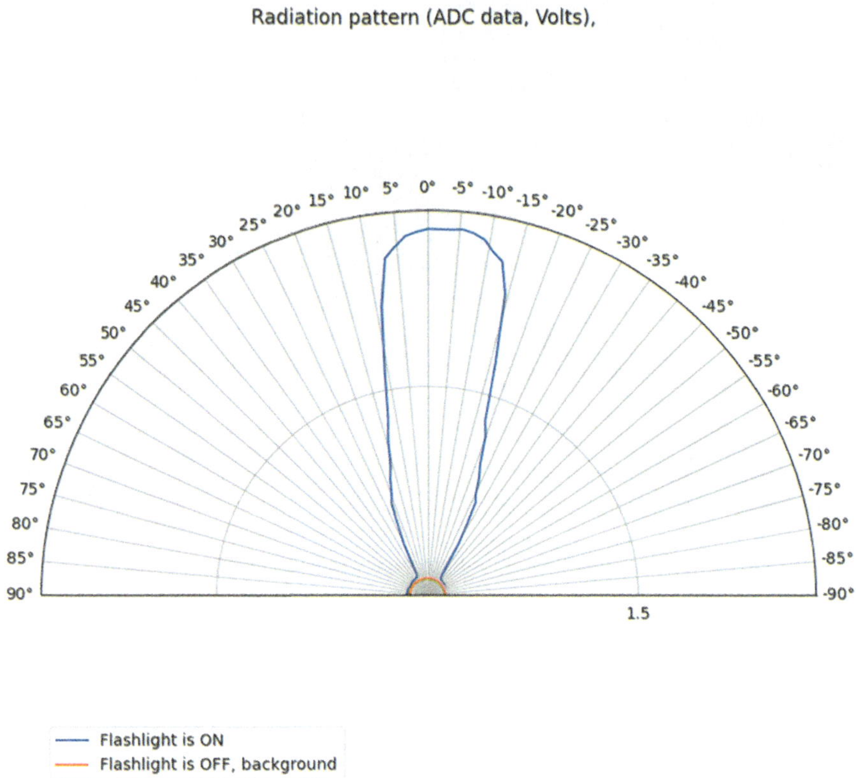

Fig. 2. Radiation pattern diagram

In this case data was analyzed using Pandas, Numpy and Matplotlib open-source libraries available for Python3. All these libraries may run on the very same Raspberry Pi alongside with diverse and rich software available on Linux. This is advantage over those projects that provide firmware only and do not provide possibility to build software used to process experimental data.

5 Result and Discussion

The result of the paper is overview of Home Lab Easy Kit project [12]. Low cost and powerful hardware may create efficient solutions only if combined with corresponding and diverse software.

The novelty of this work is an attempt to analyze experiment as cooperation of hardware and software. Home Lab Easy Kit project provide firmware and software (generated C++ library) which may significantly help.

Also, transfer of operations that require intensive CPU load and fast response time (interrupts) to microcontroller side (modern operating systems may require kernel mode driver development with serious programming skills demands). This project introduces I2C based interface capable to transfer data gathered by microcontroller to microcomputer running Linux with ability to use multiple libraries and software solutions for further data processing. The protocol is using control sum checks to detect failures during data transmission.

I2C bus allows to connect several BLUE PILLS to single microcomputer using the same lines. This may be used to efficiently extend experiment automation.

There is possibility to configure devices per experiment needs using simple JSON configuration file. Documentation is available on project's web page.

The advantages of the selected micro controller STM32F103C8T6: 32-bit ARM Cortex-M3 architecture, various peripherals, there is support for various interfaces that can be used by software.

The project supports ADC, PWM, stepper motors, GSM modem (AT commands, SMS, calls), which allows you to control the experiment remotely. Some peripherals may use configurable buffering to accumulate data making less data transfers, so the polling time of the microcontroller inputs can be increased (for example, for receiving messages from CAN bus).

6 Conclusion

This article describes a project that provides solution for experiments with a small project funding.

The developed solution has a number of advantages:

- Logic and heavy calculations may be moved from micro controller to computer, which performance is generally much better.
- It is possible to use other libraries for database access, data science, networking etc.
- Communication utilizes I2C in 7-bit addressing mode. In theory you can connect up to 112 micro controllers to single I2C port.
- Firmware architecture is intended to use MCU peripherals economically to allow as much peripherals to be used per micro controller as possible.

Further research is aimed to support SPI interface, use of USB as interface to software, generation of Python modules as alternative to C++.

References

1. Álvarez, J.L., Mozo, J.D., Durán, E.: Analysis of single board architectures integrating sensors technologies. Sensors **21**(18) (2021). https://doi.org/10.3390/s21186303
2. Ghate, R.A., Tilekar, S.K., Chavan, S.V.: Comparative study of intelligent and smart development platforms employed for internet of thing's applications. Turk. J. Comput. Math. Educ. **12**, 810–821 (2021)
3. Kalúz, M., Cirka, L., Valo, R., Fikar M.: ArPi Lab: a low-cost remote laboratory for control education. In: 19th IFAC World Congress, pp.9057–9062. IFAC Proceedings Volumes, Cape Town (2014). https://doi.org/10.3182/20140824-6-ZA-1003.00963

4. Korobiichuk, I., Dobrzhansky, O., Kachniarz, M.: Remote control of nonlinear motion for mechatronic machine by means of CoDeSys compatible industrial controller. Tehnički vjesnik/Tech. Gazette **24**(6), 1661–1667 (2017). https://doi.org/10.17559/TV-20151110164217
5. Meng, F., Van Wie, B.J., Thiessen, D.B., Richards, R.F.: Design and fabrication of very-low-cost engineering experiments via 3-D printing and vacuum forming. Int. J. Mech. Eng. Educ. **47**(3), 246–274 (2018). https://doi.org/10.1177/0306419018768091
6. Sobota, J., Pišl, R., Balda, P., Schlegel, M.: Raspberry Pi and Arduino boards in control education. In: 10th IFAC Symposium Advances in Control Education The International Federation of Automatic Control, vol. 46, no. 17, pp. 7–12. IFAC Proceedings Volumes, Sheffield, UK (2013). https://doi.org/10.3182/20130828-3-UK-2039.00003
7. Korobiichuk, I., Ladanyuk, A., Zaiets, N., Vlasenko, L.: Modern development technologies and investigation of food production technological complex automated systems. In: ACM International Conference Proceeding Series. 2nd International Conference on Mechatronics Systems and Control Engineering ICMSCE 2018, 21–23 February 2018, Amsterdam, Netherlands, pp. 52–57 (2018). https://doi.org/10.1145/3185066.3185075
8. Korobiichuk, I., et al.: Synthesis of optimal robust regulator for food processing facilities. In: Szewczyk, R., Zieliński, C., Kaliczyńska, M. (eds.) Automation 2017. ICA 2017. Advances in Intelligent Systems and Computing, vol. 550, pp. 58–66. Springer, Cham (2017). https://doi.org/10.1007/978-3-319-54042-9_5
9. Monk, S.: Programming the Raspberry Pi: Getting Started with Python.3rd edn. McGraw-Hill Education TAB, New York (2021)
10. Reference manual. https://www.st.com/resource/en/reference_manual/cd00171190-stm32f101xx-stm32f102xx-stm32f103xx-stm32f105xx-and-stm32f107xx-advanced-arm-based-32-bit-mcus-stmicroelectronics.pdf. Accessed 02 Jan 2022
11. Description of different C++ 11 standards. https://en.cppreference.com/w/cpp/11. Accessed 02 Jan 2022
12. I2C-bus Specification. https://www.i2c-bus.org/specification/. Accessed 02 Jan 2022
13. Horowitz, P., Hill, W.: The Art of Electronics, 2nd edn. Cambridge University Press, Cambridge (2015)
14. STM32F10x standard peripheral library. https://www.st.com/content/st_com/en/products/embedded-software/mcu-mpu-embedded-software/stm32-embedded-software/stm32-standard-peripheral-libraries/stsw-stm32054.html. Accessed 02 Jan 2022
15. HLEK. https://github.com/osharuda/home-lab-easy-kit. Accessed 02 Jan 2022
16. Radiation pattern of the flashlight using Home Lab Easy Kit project. https://youtu.be/8mFc6aBiLas. Accessed 31 Jan 2022

Theoretical Simulations of Scanning Electrochemical Microscope Positioning System

Mantas Makulavičius[1], Andrius Dzedzickis[1], Vytautas Bučinskas[1], Jurga Subaciute-Zemaitiene[1(✉)], and Inga Morkvenaite-Vilkonciene[1,2]

[1] Department of Mechatronics, Robotics, and Digital Manufacturing, Vilnius Gediminas Technical University, J. Basanaviciaus str. 28, Vilnius, Lithuania
jurga.subaciute-zemaitiene@vilniustech.lt

[2] Laboratory of Electrochemical Energy Conversion, State Research Institute Centre for Physical Sciences and Technology, Sauletekio 3, 10257 Vilnius, Lithuania

Abstract. This paper deals with the scanning electrochemical microscope microelectrode (SECM) positioning methods. Constant height, constant distance, and hopping intermittent contact methods were analyzed. Control algorithms are created, and the system is programmed in MATLAB Simulink environment using the existing mathematical SECM negative feedback model. Simulations were made using these models and the results were compared with a known object. Results show that both, constant distance and HIC methods provided better results than the constant height positioning method.

Keywords: Scanning electrochemical microscope · Ultramicroelectrode · Constant height · Constant distance · Hopping intermittent contact

1 Introduction

Scanning electrochemical microscope (SECM) is a electrochemical technique, which uses an ultra-microelectrode (UME) as a probe with a tip diameter ≤ 25 μm to scan over surface of interest by measuring the electrical current response [1, 2]. The probe can be positioned in the X, Y, and Z axes. The current response depends on the topography and the electrochemical activity of the surface [3, 4]. SECM modes of operation can be distinguished in constant height, constant distance, and hopping modes [5]. In the constant height mode (Fig. 1, a), where the UME is moved only in the X and Y axes, while the Z axis is at a constant height and remains stationary. This UME positioning method is suitable for estimating electrochemical activity at some distance from surface, since the structure topology at close distance has the influence in measurement result. However, positioning at this mode can damage the surface of the specimen or the UME itself [5]. This method is suitable when relatively large tip diameter probes are used [1]. Therefore, a smaller tip diameter UME can be selected to obtain a higher resolution. In this case, the constant height method becomes more complicated [6].

The constant distance mode (Fig. 2 b)) is when the UME maintains a constant distance from the measured object by moving in the X and Y axis and forcing the UME to follow

R. Szewczyk et al. (Eds.): AUTOMATION 2022, AISC 1427, pp. 183–191, 2022.
https://doi.org/10.1007/978-3-031-03502-9_19

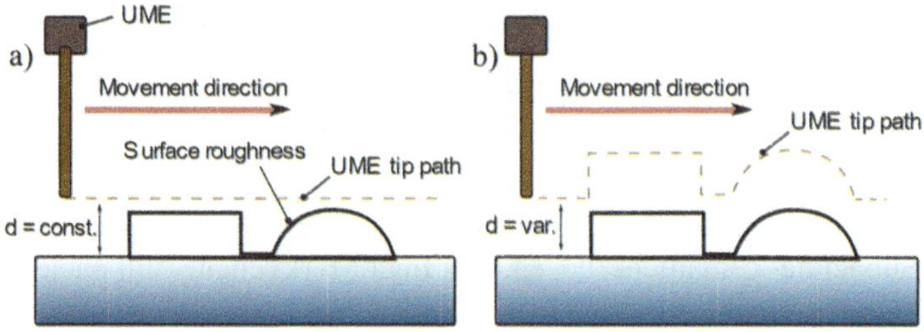

Fig. 1. Constant height a) and distance b) positioning methods

the contour of the surface. This is controlled using an electrochemically independent but distance-dependent analytical signal. These include tip position modulation (TPM), atomic force microscopy (AFM-SECM), scanning ion conduction microscopy (SECM-SICM), hopping positioning, or intermittent contact positioning (IC-SECM/HIC-SECM) [7]. One of the best ways to apply constant distance positioning is using a shear force-based feedback method [8, 9]. This requires additional equipment, but this method is widely accepted and applied commercially.

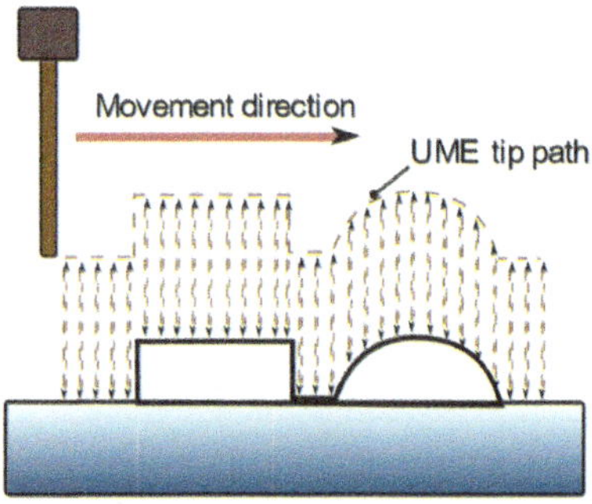

Fig. 2. HIC-SECM method

The IC-SECM (HIC-SECM) method (Fig. 2) allows the measured current to be used as feedback signal to control the UME distance to the surface [10]. Data is obtained as the UME approaches the surface and can be collected throughout the descent. This method is used in both SECM and SICM experiments. HIC-SECM is an efficient imaging method that allows not only the topography to be imaged, but also other data, and DC and AC signals are available. Using IC position feedback, the tip of the UME is stopped at the very surface, and the Z axis decay curves at each point in the X–Y plane are obtained [11].

Using constant distance positioning without intermission and with a sudden change in topography in the positive direction i.e. a sudden rise of the surface, the UME is not able to adjust the movement path and the tip or specimen itself can be damaged [11]. Such collisions are avoided by using HIC positioning, and both the specimen and the UME are protected regardless of the surface topography.

Based on the literature, it should be noted that the most efficient UME positioning method with the best results is the shear force feedback control, which requires additional equipment to generate and receive feedback [12]. Without that equipment, both UME positioning methods – constant distance and HIC are suitable based on the theoretical studies performed in this work.

2 Methodology

Mathematical model was described in the MATLAB software using the Simulink plug-in according to the [13]. An object (surface roughness) of known dimensions and topography is created for the mathematical model to perform the measurement. The 2D image of the object is compared with the original image of the object. Only negative feedback electrochemical mode was chosen for the system modelling. The dependence of the current on the distance to the measured surface is expressed as follows:

$$Ni_T^{ins}(L, R_g) = \frac{\frac{2.08}{R_g^{0.358}}\left(L - \frac{0.145}{R_g}\right) + 1.585}{\frac{2.08}{R_g^{0.358}}(L + 0.0023R_g) + 1.57 + \frac{lnR_g}{L} + \frac{2}{\pi R_g}\ln\left(1 + \frac{\pi R_g}{2L}\right)} \quad (1)$$

here Ni_T^{ins} is the measured current, R_g is the ratio of insulating and conductive radii of the UME tip, L is the distance to the surface.

The following steps are conducted as follows: firstly, the current vs. vertical distance dependence using this model was calculated; at the second step constant height, constant distance, and HIC control models were applied; at the third step measurement object used in the models was described; at the fourth step the measurement object was used for simulations in the Simulink environment using different control methods.

The measured object is generated from several curves (current vs. vertical distance dependences) and has more than one vertex at different heights (Fig. 3). Such an object is generated to create more variations, which would allow viewing clearer picture of the differences between positioning control methods, as well as creating more realistic measurement conditions. Due to the time-based principle of Simulink simulations, the graph is plotted as a function of height dependence on time t, but knowing the speed of UME movement, the t axis can be converted to μm [14]. In this case, the estimated scanning speed is 100 μm/s. Vertexes of the object are 15 μm and 40 μm, its total length is 85 μm. The most relevant information during the scan is the measured height of the object.

After object simulation, algorithms for the constant distance methods and hopping positioning method are made. The constant height method is excluded because the execution of this method is not comparable to other methods.

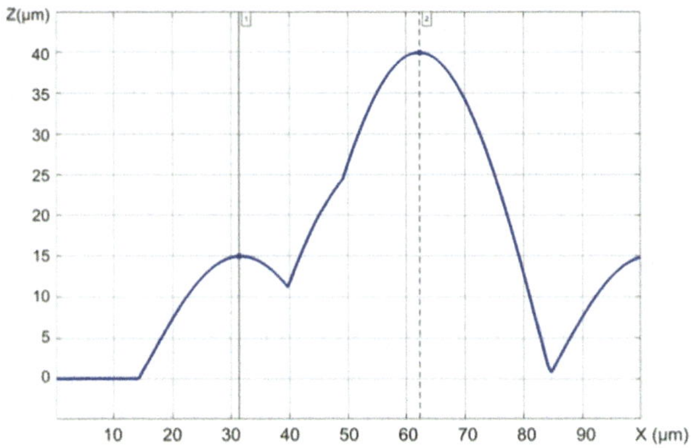

Fig. 3. Simulated measured object

3 Control Algorithms

3.1 Constant Distance Method

Firstly, the home position data of the device are determined (entered) as X axis travel speed (m/s), X axis total travel distance D (μm), constant distance d (μm) is maintained from the measured surface, distance change depends on the measured surface and the measured current variation. The program then converts the specified distance to the required current to be tracked using a reference current-distance dependence. The constant distance method allows the UME to track the topography of the surface using a feedback signal, which is obtained from the UME itself. With the response, the Z axis is adjusted to maintain the same distance. In this case, the measured current must ideally remain straight and any change in it must be corrected by using a displacement of the Z axis in one direction or the other.

The probe lowering step operates by first checking if the measured value is greater than the specified value. If it is true, then the Z axis is allowed to descend continuously until the values are equal. Once the equality is reached, the probe is at a set distance from the surface and the X-axis can begin to move according to the set criteria. If the value crosses the equality significantly, the probe returns to the home position to avoid damaging the probe and the process can be repeated automatically. The X axis moves at the set speed and the specified total distance. As the X axis moves in the specified direction, and arithmetic subtraction operation is performed with the measured value and the set value. According to [15], if there is a rise on the surface of the substrate at that location, the Z axis moves in the +Z direction and vice versa. The last part of the algorithm is the verification of the set total travel distance D. In comparison, the program stops only when the measured travelled distance is the same as the set one. The program stops even if the value becomes less than 0. This is possible if a stopping point in X axis is jumped due to errors or bad precision in the mechanics. In this case, it is assumed that stepper motors are rotating in a certain step and are used to control the X axis. Thus, it is possible to determine a distance whose stopping point would not coincide with the end of the motor step and the step would be completed beyond the stop point [16].

3.2 HIC Method

Firstly, the initial values are defined in the same way as it is in the constant distance method. The hopping step p was changed, while distance d between probe tip and surface was constant. The difference between step size p for motion trajectory and scanning at starting height H of the probe tip is presented in Fig. 4.

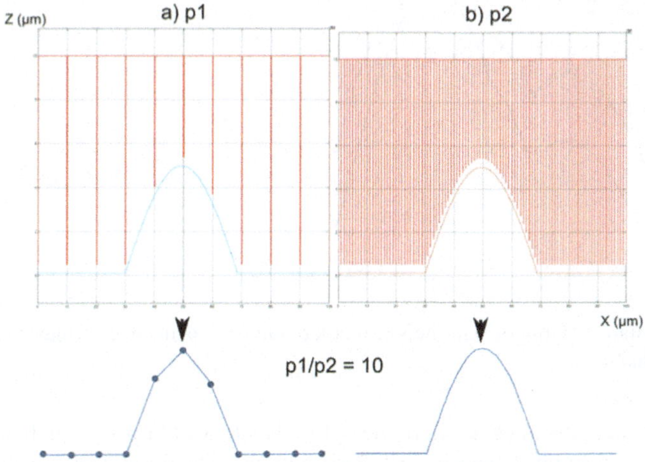

Fig. 4. Difference between a) too big and b) too small step sizes

The more points (smaller p), the more accurate the surface roughness can be determined, but the longer it takes to measure. It is noticeable how increasing the value of step p changes the quality of the result. The values of p shown in Fig. 4 differ by a ratio of 10. When the measured Z axis is lowered, the X axis remains stationary. When the measured value of the current becomes equal to the set value, the Z axis is stopped, and its position is read. That information is stored in the external memory and is also stored in the program as a reference point for the next cycle. After these steps, the Z axis rises in the +Z direction until the height H is reached and the probe is moved in X axis with a size of specified step p from the previous position and the process is executed in the same manner. Using both positioning methods, the measurement results are obtained by scanning the positions in the X and Z axes.

4 Simulation Results

For the simulation of constant height method, moving UME in the X direction is selected at the height of 40 µm. This height is chosen because the highest part of the simulated object is at a height of 40 µm and thus a single zero point can be formed. Moreover, with this can be noted, how well the exact points at different heights can be reconstructed. Figure 5 a) shows that the topography is reconstructed, but the lowest point is at a height of 12.07 µm. From this, it can be concluded that the system can work properly if a distance of 27.93 μm between the probe tip and the substrate surface is not exceeded. This result is in accordance with [15, 17]. In Fig. 4 b), the dependence of error on measurement distance between the probe tip and substrate surface and the bigger the measurement distance, the greater the error.

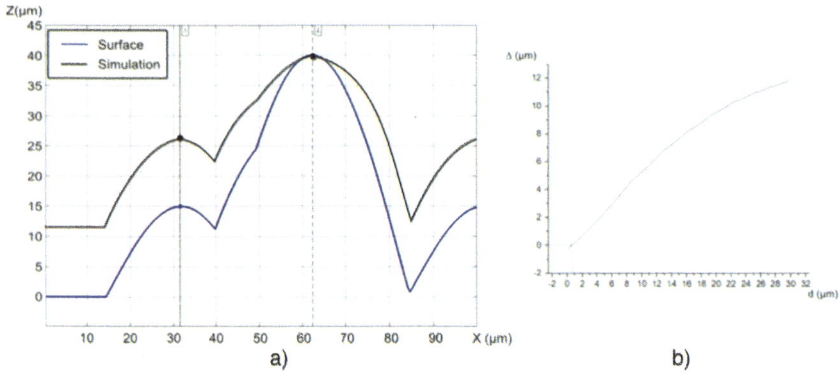

Fig. 5. a) Simulation with constant height method and b) dependence of measurement error on measurement distance.

For the constant distance method, the optimal distance of up to 2 μm has been chosen. With a smaller constant distance, collisions between the tip and the surface have been detected. Graphical results of using the constant distance method are shown in Fig. 6.

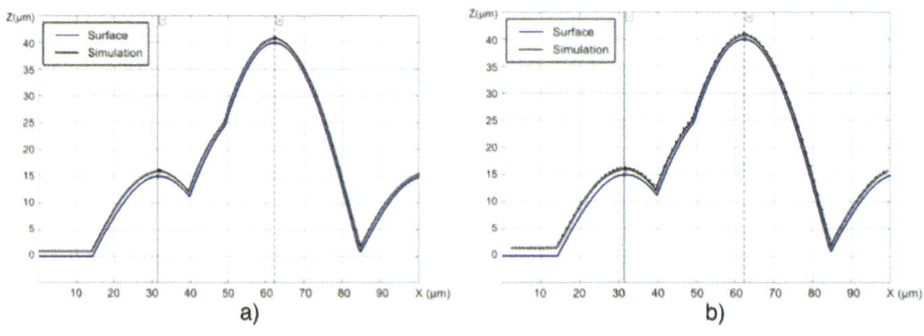

Fig. 6. a) Simulation with constant distance method; b) Simulation with HIC method.

For the HIC method, the hopping step has been set by 1 μm, the height H of the probe tip has been set at 50 μm. This distance is safe in this case, as it is known that the highest peak of the measured surface is 40 μm high. The supported distance from the surface is used in the same way as for the measurement of the constant distance mode (d = 2 μm). Obtained results with the HIC method are shown in Fig. 6 b).

After obtaining graphs from Fig. 5 to Fig. 6, numerical results are presented in Table 1.

The accuracy on the X-axis using the theoretical constant height model is high and reaches ±0.25 μm, but the accuracy on the Z-axis is variable. It is clear that the

Table 1. The tip coordinates of generated known surface and reconstructed surfaces using different methods

	Tip number (from the left)	Known surface		Obtained results		Differences	
		X (μm)	Z (μm)	X (μm)	Z (μm)	ΔX (μm)	ΔZ (μm)
Constant height	1st	31.46	15	31.71	25.93	0.25	10.93
	2nd	62.42	40	62.18	39.71	0.24	0.29
Constant distance	1st	31.46	15	32.08	15.71	0.62	0.71
	2nd	62.42	40	62.18	40.71	0.24	0.71
HIC	1st	31.46	15	32	15.70	0.54	0.70
	2nd	62.42	40	62.05	40.70	0.37	0.70

closer the UME is to the surface, the more accurate it can be measured. In this case, when the distance was 0, the error obtained does not exceed 0.3 μm in the Z axis. The accuracy of the X-axis remains constant no matter how close the UME is to the surface. Constant distance measurement results are better than constant height and the topography is displayed much better. However, the error in X with a smaller peak is bigger. For the HIC method, it is seen that the errors are very similar on the Z axis and larger on the X axis. The errors in the X axis depend on the selected step. Measurement speed of 10 μm/s has been chosen according to [16], the whole measurement for HIC took about 210 s, where at a constant speed the measurement of a constant distance took about 10 s.

In addition, with the use of real equipment, a mechanical error must be added for the existing theoretical error. The magnitude of the mechanical error depends on its accuracy. After the addition of mechanical error, it may be necessary to change the trajectory and increase the measurement height, thus losing the measurement quality. The system response time should also be evaluated additionally [18].

5 Conclusion

It has been observed that the topography method has a very limited topography and not all parts of the object are reproduced correctly, as well as a high risk of damaging the specimen. Constant distance and HIC methods solve the problem of specimen damage equally well, but theoretical studies have shown that measurement using the HIC positioning method takes several times longer to perform the same measurement but is more suitable for specimens with a very sharp topography.

Based on the results of theoretical research, constant distance and HIC positioning methods are suitable for measurements, as both greatly improve the measurement result compared to positioning at a constant height. Both positioning methods have errors not exceeding 1 μm.

Acknowledgements. This project has received financial support from the Research Council of Lithuania (LMTLT), Nr. P-LLT-21-6, State Education Development Agency of Latvia, Ministry of Science and Technology (MOST)) of Taiwan.

References

1. Whitworth, A.L., Mandler, D., Unwin, P.R.: Theory of scanning electrochemical microscopy (SECM) as a probe of surface conductivity. Phys. Chem. Chem. Phys. **7**, 356–365 (2005). https://doi.org/10.1039/B407397E
2. Morkvenaite-Vilkonciene, I., Ramanaviciene, A., Ramanavicius, A.: Redox competition and generation-collection modes based scanning electrochemical microscopy for the evaluation of immobilised glucose oxidase-catalysed reactions. Rsc Adv. **4**, 50064–50069 (2014). https://doi.org/10.1039/C4RA08697J
3. Sun, P., Laforge, F.O., Mirkin, M.V.: Scanning electrochemical microscopy in the 21st century. Phys. Chem. Chem. Phys. **9**, 802–823 (2007)
4. Polcari, D., Dauphin-Ducharme, P., Mauzeroll, J.: Scanning electrochemical microscopy: a comprehensive review of experimental parameters from 1989 to 2015. Chem Rev **116**, 13234–13278 (2016). https://doi.org/10.1021/ACS.CHEMREV.6B00067
5. Morkvėnaitė-Vilkončienė, I., Petkevičius, S., Keraitė, G., Šakalys, P., Lenkutis, T.: Positioning and control of scanning electrochemical microscopy. Moksl Lietateitis Sci. Futur. Lith **9**, 602–606 (2017). https://doi.org/10.3846/MLA.2017.1093
6. Sur, U.K.: Recent trend in electrochemical science and technology. Recent Trend Electrochem. Sci. Technol. (2012). https://doi.org/10.5772/1891
7. Nebel, M., Erichsen, T., Schuhmann, W.: Constant-distance mode SECM as a tool to visualize local electrocatalytic activity of oxygen reduction catalysts. Beilstein J. Nanotechnol. **5**, 141–151 (2014). https://doi.org/10.3762/bjnano.5.14
8. Etienne, M., Anderson, E.C., Evans, S.R., Schuhmann, W., Fritsch, I.: Feedback-independent Pt nanoelectrodes for shear force-based constant-distance mode scanning electrochemical microscopy. Anal. Chem. **78**, 7317–7324 (2006). https://doi.org/10.1021/ac061310o
9. Eckhard, K., Etienne, M., Schulte, A., Schuhmann, W.: Constant-distance mode AC-SECM for the visualisation of corrosion pits. Electrochem. Commun. **9**, 1793–1797 (2007). https://doi.org/10.1016/j.elecom.2007.03.035
10. McKelvey, K., Edwards, M.A., Unwin, P.R.: Intermittent contact-scanning electrochemical microscopy (IC-SECM): a new approach for tip positioning and simultaneous imaging of interfacial topography and activity. Anal. Chem. **82**, 6334–6337 (2010). https://doi.org/10.1021/AC101099E/SUPPL_FILE/AC101099E_SI_001.PDF
11. Lazenby, R.A., McKelvey, K., Unwin, P.R.: Hopping intermittent contact-scanning electrochemical microscopy (HIC-SECM): visualizing interfacial reactions and fluxes from surfaces to bulk solution. Anal. Chem. **85**, 2937–2944 (2013). https://doi.org/10.1021/ac303642p
12. Morkvenaite-Vilkonciene, I., Ramanaviciene, A., Kisieliute, A., Bucinskas, V., Ramanavicius, A.: Scanning electrochemical microscopy in the development of enzymatic sensors and immunosensors. Biosens. Bioelectron. **141**, 111411 (2019). https://doi.org/10.1016/j.bios.2019.111411
13. Cornut, R., Griveau, S., Lefrou, C.: Accuracy study on fitting procedure of kinetics SECM feedback experiments. J. Electroanal Chem. **650**, 55–61 (2010). https://doi.org/10.1016/j.jelechem.2010.09.007
14. Bard, A.J., Mirkin, M.V.: Scanning Electrochemical Microscopy. CRC Press, Boca Raton (2012)

15. Ballesteros Katemann, B., Schulte, A., Schuhmann, W., Katemann, B.B., Schulte, A., Schuhmann, W.: Constant-distance mode scanning electrochemical microscopy. Part II: high-resolution SECM imaging employing Pt nanoelectrodes as miniaturized scanning probes. Electroanalysis **16**, 60–65 (2004). https://doi.org/10.1002/elan.200302918
16. Kiss, A., Nagy, G.: New SECM scanning algorithms for improved potentiometric imaging of circularly symmetric targets. Electrochim Acta **119**, 169–174 (2014)
17. Schulte, A., Schuhmann, W.: Scanning electrochemical microscopy as a tool in neuroscience. Electrochem. Methods Neurosci. (2007)
18. Picatoste Ruilope, R.: Modelling and Control of Stepper Motors for High Accuracy Positioning Systems Used in Radioactive Environments. E.T.S.I. Industriales (UPM) (2014)

Sitting Posture Monitoring Using Velostat Based Pressure Sensors Matrix

Marius Sumanas(✉), Vaiva Treciokaite, Aurimas Čerškus, Andrius Dzedzickis, Vytautas Bučinskas, and Inga Morkvenaite-Vilkonciene

Department of Mechatronics, Robotics, and Digital Manufacturing, Vilnius TECH, J Basanaviciaus Str. 28, 03224 Vilnius, Lithuania
marius.sumanas@vilniustech.lt

Abstract. Working in an office environment leads to sedentary behavior. Prolonged sitting increases back pain and soreness, leading to spinal trauma – scoliosis. Therefore, this research aims to create an automatic tool to help identify the main human body characteristics responsible for good body mass distribution. For this purpose, an 8×8 pressure sensor matrix with signal acquisition system was built using Velostat. The commercially available composite piezoresistive material was selected as a pressure transducer due to its flexibility, versatility and low price. After system calibration, its functionality was validated by observing the sitting postures of 9 participants. Participant matrix map analysis showed differences in force distribution between sexes and subjects with and without spine trauma. This research will help monitor the office environment and increase workers' health and productivity.

Keywords: Sitting posture · Spinal trauma · Sensors matrix · Velostat · Multiplexer · BMI index

1 Introduction

Poor sitting posture can lead to spine, muscular disorders, hospitalization, or inhalable trauma [1]. Daily work in an office environment with prolonged static posture can lead to disorders of upper extremities – pain and functional impairment, deformation of soft tissues, disturbance of local blood supply and lymphatic circulation.

Over the previous decade, medical personnel evaluated human postures using visual detection and questionnaires to determine spinal cord disorders. This kind of subjective analysis is easy to carry out and gives fast results but does not provide hard data to prove the diagnosis [2]. Various pressure distribution sensing methods and systems have been developed as an alternative. Such systems can be integrated into clothing, cushions, and chairs to analyze: (i) human body pressure distribution [3]; (ii) absolute pressure amplitudes; (iii) sleeping posture; (iv) respiration rate [4]. Nevertheless, such systems have several huge disadvantages: they are expensive due to the standard regulations for medical equipment; the output data is complex and hard to interpret [5]. Due to that, cheaper alternative systems providing output data understandable without the interference of highly skilled doctors are in great demand [3, 6].

R. Szewczyk et al. (Eds.): AUTOMATION 2022, AISC 1427, pp. 192–201, 2022.
https://doi.org/10.1007/978-3-031-03502-9_20

A typical pressure distribution sensing system is made of sensors arrays placed between two sheets of a deformable material, hardware for data acquisition and software for data processing. Each sensor measures pressure applied on its active surface. Data processing software forms a pressure distribution map by combining data about sensor position in the array and actual pressure value. Commercially available systems usually use proprietary software [4]. Nevertheless, universal tools such as MATLAB also could be implemented [6].

A variety of measuring methods and pressure sensors could be used to measure the human body pressure distribution. However, small capacitive [7], resistive [8, 9], magnetic (Hall) [10] and piezoresistive sensors [11] are preferred due to their universality, straightforward integrity and simple data acquisition. This research aims to propose a methodology and equipment for human sitting posture monitoring to identify the physical properties of a human being and eliminate the risk of possible disorders caused by poor posture.

2 Experimental Research

2.1 Design and Analysis of Pressure Sensing Matrix

The most informative and straightforward information about sitting posture could be obtained using pressure sensing methods when sensors are integrated into the chair or clothes. Therefore, a piezoresistive flexible 350 × 350 mm pressure sensors matrix containing 64 (5 × 5 mm) active points (pixels) was built (Fig. 1b). The designed sensing matrix is based on commercially available piezoresistive material – Velostat (Adafruit Industries), and consists of three layers. The top and bottom layers are made of 0.5 mm thick sticky tape with attached 0.5 mm thick and 5 mm width copper tape strips acting as supply and readout electrodes. At the top layer, copper strips are placed horizontally 30 mm apart, at the bottom – vertically. The sheet of the Velostat placed between the top and bottom layers develops the ability to measure pressure acting in the junctions, where horizontal and vertical copper strips intersect.

The sensors readout circuit (Fig. 1a) is based on the Arduino Uno microcontroller and 8 channels CD74HC4051E multiplexer, allowing fast scanning. When the system is turned on, the microcontroller activates one of its outputs and provides a supply voltage to the first row of the matrix, while other rows remain connected to a common ground through internal pull-down resistors. When voltage starts to flow from the open row, it goes to the column with the least resistance. Meantime, a multiplexer connects each column one by one to the microcontroller's analogue input pin through a voltage divider circuit. After measuring the voltage in all eight columns, the same procedure is repeated with the next rows until values of all 64 sensors are obtained. The benefit of such a circuit is its ability to quickly collect data from many sensors using several microcontroller pins and send it to the computer as a simple 8 × 8 array containing integer values.

The main drawback of polymer-based piezoresistive sensors is their unstable characteristics and sensitivity to the environmental impact. Therefore, calibration of assembled sensor matrix is necessary. Calibration performed by placing the matrix on a flat surface, loading the sensors with known gradually increasing force (1–20 N), and measuring the

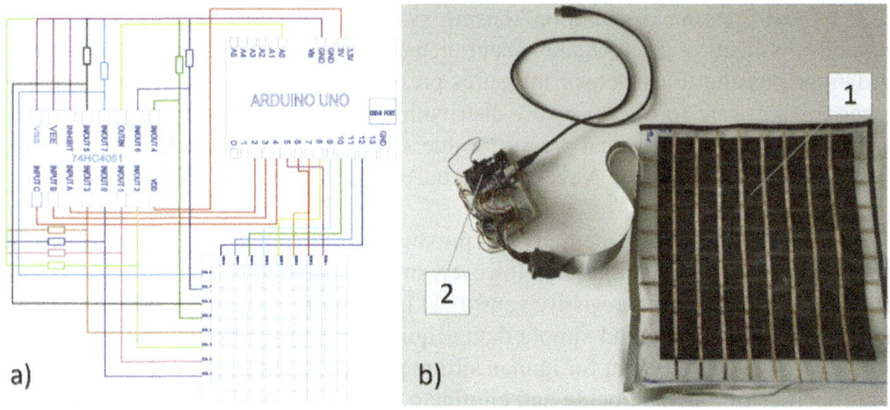

Fig. 1. a) Sensing matrix connection scheme, b) main parts of the system, 1- pressure sensors matrix, 2 - Arduino Uno microcontroller.

output voltage. The load was caused by placing 50 × 50 mm weights on the active sensor spot. The load variation step was 1 N. The output voltage values had been measured 2 min after load application to minimize the impact of sensors dynamics. The sensor resistance vs. force dependency (Fig. 2) was obtained by converting the digital value from 10-bit analogue–digital converter (ADC) to resistance by the following equation:

$$R_{(velostat)} = R_{(ref)} \frac{(1024 - V_{(ADC)})}{VV_{(DC)}}, \tag{1}$$

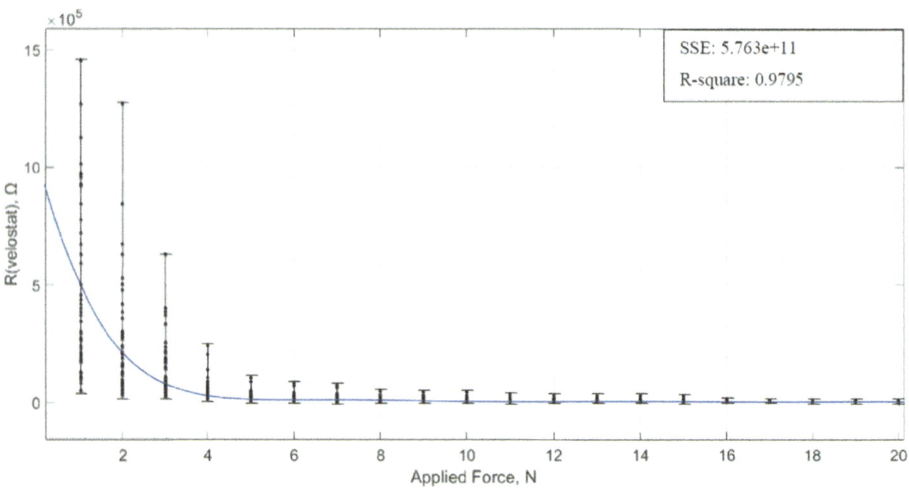

Fig. 2. Velostat resistance vs. applied force. SEE – data accuracy or the scatter of points about the line for a given value; R-square – statistical measure of fit that indicates how much variation of a dependent variable is explained by the independent variable(s) in a regression model

where: $V_{(ADC)}$ – value from 10-bit ADC, $R_{(ref)}$ – resistance of reference resistor (10 KΩ), $R_{(velostat)}$ – resistance of pressure sensor, $VV_{(DC)}$ – supply voltage.

Raw data presented in Fig. 2 was approximated (Eq. 2) to define the calibration curve for feature research.

$$R(x) = -0.04982x^7 + 4.335x^6 - 156.5x^5 + 3028x^4 - 3.379e^4 \cdot x^3 + 2.168e^5 \cdot x^2 - 7.41e^5 \cdot x + 1.054e^6, \quad (2)$$

Figure 2 shows the wide deviation of the results. This deviation might have occurred due to Velostast properties and manufacturing inaccuracies. Nevertheless, the deviation tends to reduce as the force increases. Therefore, the matrix is suitable for reliable monitoring of pressure distribution while a human sits on a chair.

2.2 Research Methodology

The purpose of the experimental research is to evaluate the functionality of the proposed method and developed hardware by analyzing human sitting posture (mass distribution) in an office environment. 9 participants: 6 males and 3 females between the ages of 24–42 years, were asked to sit in 7 most common postures (Fig. 3). They were:

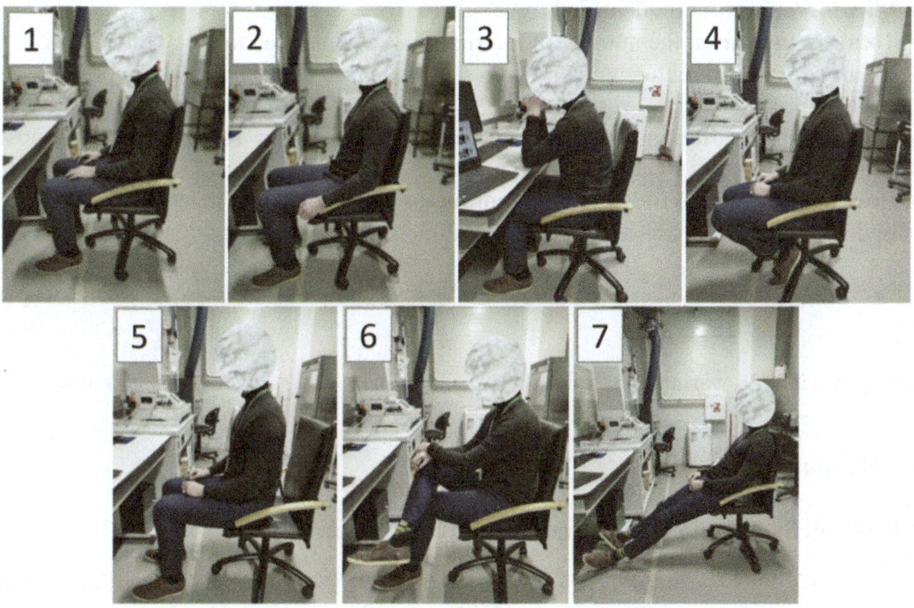

Fig. 3. Seven sitting postures: 1) Standard; 2) Lying-1; 3) Tilt forward; 4) Normal position feet under the chair; 5) Sitting on the edge; 6) Legs Crossed; 7) Lying-2

- **Standard**: sitting with supported back and feet on the floor. This posture is characterized by the body exerting pressure on the back and seat.

- **Lying-1**: (boredom). Subjects sit on the edge of the chair. The body exerts strong pressure on the upper part of the back and the front of the seat.
- **Tilt forward**: (attention). Subjects sit back, leaning forward, feet under the chair. The body exerts strong pressure on the front of the seat without pressure on the back.
- **Normal position**, feet under the chair (attention). Subjects sit with their back supported, sitting back with their feet completely flat. In this position, the body exerts very little pressure on support and strong pressure on the front of the seat.
- **Sitting on the edge** (attention): Subjects sits on the edge of the chair. No pressure is observed on the back, showing a homogeneous distribution on the seat.
- **Legs Crossed** (peace): subjects sitting normally with legs crossed. The weight is concentrated on the right or left side of the seat and slightly on the back edge.
- **Lying-2** (fatigue). In this position, a lot of pressure is exerted on both the seat and the backrest.

Experimental data presented as a pressure distribution map using MATLAB software. The participants were categorized by sex and health (subject with no spinal traumas versus subject with scoliosis); the following parameters for each case were calculated:

- **Average force**: mean average force acting on the whole matrix.
- **Max force**: maximum identified force value.
- **Total surface area**: the number of matrix points that reacted to the applied force.
- **Mean contact area**: the number of matrix points that reacted to average pressure in a range of deviation of 1 sigma.
- **Max force gradient**: max pressure vector in the matrix representing the most significant pressure change between points.
- **Pressure dispersion index (DI) (Fano factor)**: statistical coefficient explaining the scale of mass dispersion calculated according to Eq. 3.

$$DI = \frac{Variance}{F_{Avg}} \tag{3}$$

where: *Variance* – variance based on the entire matrix, F_{Avg} – average matrix force.

3 Research Results

3.1 Participant Data as Nominal for Matrix Comparison

9 people of 25 – 33 years old, 1.58 – 1.93 m height, and 56 – 93.5 kg weight participated in this research. One participant, a 25 years old man without any spinal trauma, was chosen as nominal for further analysis. Resulted pressure distribution map of the nominal subject is presented in Fig. 4. Statistical results for each posture are presented in Table 1.

The experimental data for each subject were grouped according to specified criteria and compared with nominal. Therefore, results of other participants and conclusions will be partially referenced to the results of the nominal subject.

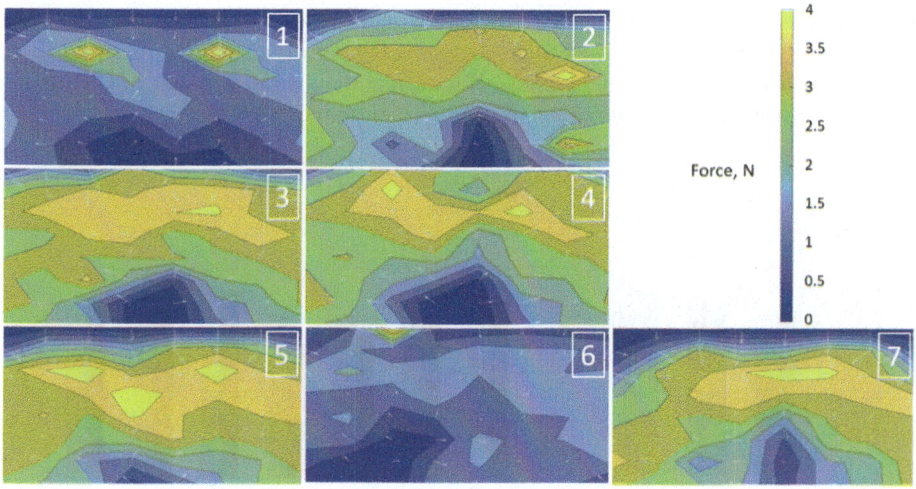

Fig. 4. Nominal results – 7 sitting postures: 1) standard; 2) lying; 3) tilt forward; 4) normal position – feet under the chair; 5) sitting on the edge; 6) legs crossed; 7) lying-2

Table 1. Nominal sitting posture results

Positions	Average Force	Max Force	Total surface area, Max (64)	Mean contact area	Max force gradient	Pressure dispersion index (DI)
1	2.292	8.947	53	48	21.97	3.744
2	2.417	4.874	55	47	20.47	1.019
3	2.620	4.100	62	46	12.95	0.491
4	2.554	4.326	62	49	13.03	0.400
5	2.591	4.452	56	45	16.03	1.039
6	2.483	8.879	61	50	32.73	1.340
7	2.385	4.309	56	44	16.72	1.035
Average	2.478	5.698	58	47	19.13	1.296
Min	2.292	4.100	53	44	12.95	0.400
Max	2.620	8.947	62	50	32.73	3.744
StDev	0.119	2.209	3.716	2.160	6.898	1.129

3.2 Result from Participants of Different Category

Comparison of Results by Gender

9 participants were grouped by their gender. Women weight was less than men, (Fig. 5 a), they produce bigger max and average force. The women force results are related to the Max force matrix results (Fig. 5b). When men sit on the chair, the distribution of force

was more even. When the sitting posture is based on the chair edge, where the mass must be concentrated in one area, men then surpass women with max force results – force distribution index values are the highest. But despite that, women manage to create the highest force on the matrix with the value of 7.14 N (men 6.95 N).

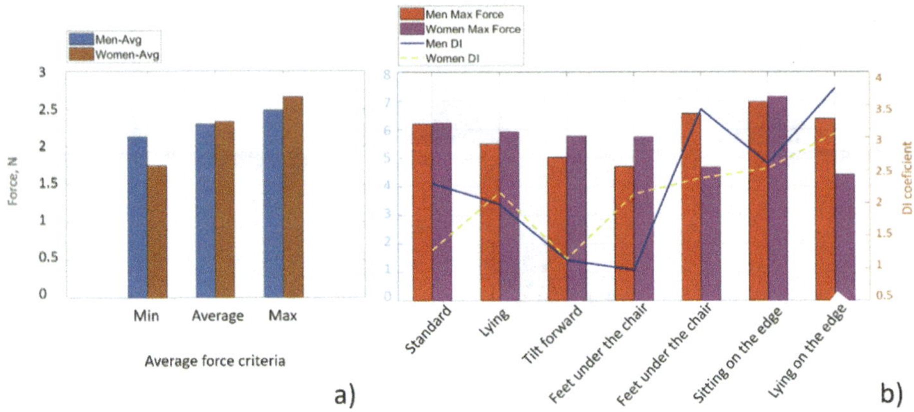

Fig. 5. a) Men & women averaged force b) Men & Women force & distribution index results.

From Fig. 5b, it is seen that with specific sitting postures (sitting on the edge, legs crossed, lying on the edge), the weight is focused more in areas where hip bones generate higher force values.

The results of women sitting with their feet under the chair are presented and compared with nominal in more detail (Fig. 6). Results show that while a nominal man can distribute his mass quite well, with a low DI index value of 0.400, the woman mass is focused much closer to the hip bone area. While the average matrix force is similar (2.55 N and 2.25 N), the woman generates the max force twice higher than the nominal man (8.859 N and 4.326 N, respectively). The assumption is that women lack muscle and fat layers for efficient body mass distribution in the wider area.

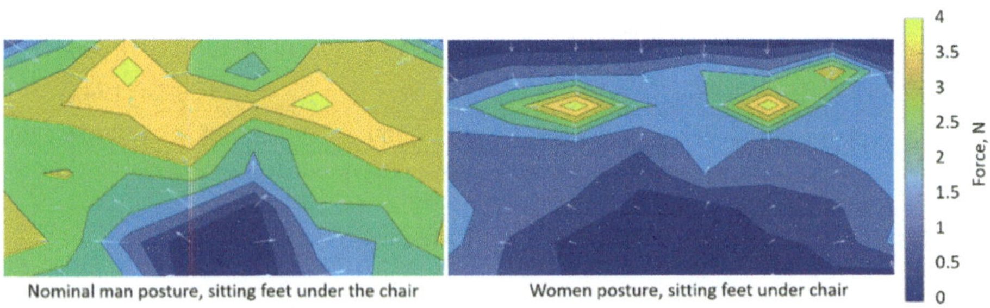

Fig. 6. Sitting feet under chair posture comparison between healthy & unhealthy subject

Comparison of Results by Healthiness

One participant had scoliosis – spinal trauma. His data was compared with nominal. From the matrix occupancy graph (Fig. 7a), it is seen that an unhealthy person's chair occupancy has a higher deviation. In standard sitting posture, an unhealthy subject occupies more chair area than a healthy, but the mean occupancy of the person with scoliosis, is smaller than the healthy one. Therefore, standard sitting posture was analyzed in more detail. From the Max force gradient graph (Fig. 7b), the peak of max force gradient was defined at the 6th sitting posture. Max force of a healthy person in all postures except the fourth one is higher. It might be due to a body mass index (BMI) difference.

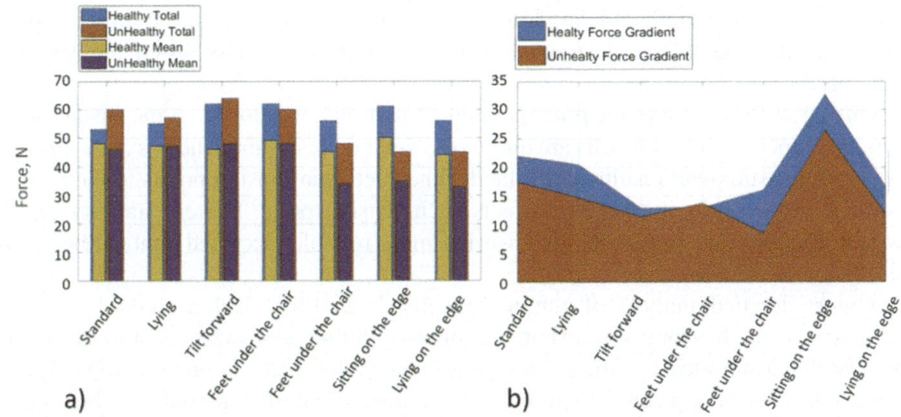

Fig. 7. Data comparison between healthy & unhealthy person: a) Matrix point occupancy; b) Max force gradient

Figure 8 shows a comparison of standard and legs crossed postures. Body weight force of a person with scoliosis is distributed more on the left side than nominal. Nominal

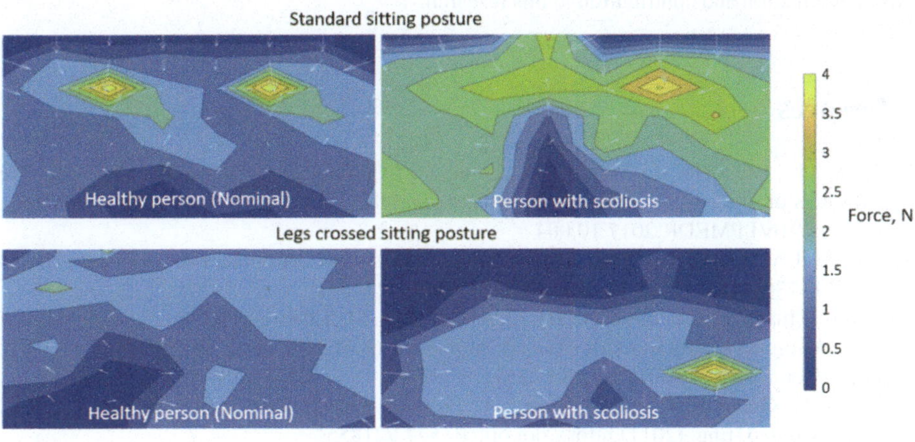

Fig. 8. Sitting posture comparison between healthy and scoliosis trauma having persons

persons mass in standard sitting posture is spread more evenly. When in legs crossed posture (right leg on the left), nominal man force density increased on the right hip area while the mass of person with scoliosis stayed focused on a left side.

4 Conclusions

Performed research provided valuable findings in the area of human sitting posture monitoring. Obtained results allow us to draw the following conclusions and guidelines for future research.

Obtained results prove the functionality of sitting posture monitoring methods based on pressure sensors installed in the chair seat. A homemade Velostat based sensor matrix with 64 sensitive points was developed, calibrated, and successfully tested, analyzing 7 sitting postures of 9 persons.

Sensing matrix calibration process showed that at lower loads, some instabilities occur, and repeatability of results are quite low. Nevertheless, this issue has a low impact on overall measurement results when the distance between sensing points in the matrix is sufficient and a significant load affects each sensing point. System stability under low load also could be improved by ensuring more reliable electrical contact, between Velostat and electrodes.

A more detailed analysis of obtained results showed that such a method not only creates a pressure distribution map but also allows defining poor sitting posture automatically. Each normal sitting posture corresponds to an individual combination of analyzed parameters. Therefore, each deviation from a defined combination could be threatened as a potential disorder. Nevertheless, it could also be an undefined posture; therefore, it is necessary to focus further research on developing smart data analysis algorithms capable of eliminating unreliable data.

Acknowledgements. This project has received financial support from the Research Council of Lithuania (LMTLT), Nr. P-LLT-21-6, State Education Development Agency of Latvia, Ministry of Science and Technology (MOST)) of Taiwan, also we are thankful to Rokas Dvarvytis for his active participation and contribution to this research.

References

1. Urda, J.L., Larouere, B., Verba, S.D., Lynn, J.S.: Comparison of subjective and objective measures of office workers' sedentary time. Prev. Med. Rep. **8**, 163–168 (2017). https://doi.org/10.1016/J.PMEDR.2017.10.004
2. Janicki, J.A., Alman, B.: Scoliosis: review of diagnosis and treatment. Paediatr. Child Health **12**, 771–776 (2007). https://doi.org/10.1093/PCH/12.9.771
3. Xu, W., Huang, M.C., Amini, N., He, L., Sarrafzadeh, M.: ECushion: a textile pressure sensor array design and calibration for sitting posture analysis. IEEE Sens. J. **13**, 3926–3934 (2013). https://doi.org/10.1109/JSEN.2013.2259589
4. Casey, V., Grace, P., Clarke-Moloney, M.: Pressure measurement at biomedical interfaces. Appl. Biomed. Eng. (2011).https://doi.org/10.5772/21855

5. Barba, R., De Madrid, Á.P., Boticario, J.G.: Development of an inexpensive sensor network for recognition of sitting posture. Int. J. Distrib. Sens. Netw. **2015**, 969237 (2015). https://doi.org/10.1155/2015/969237
6. Bao, J., Li, W.: Sitting posture recognition based on data fusion on pressure cushion. Artic. TELKOMNIKA Indones. J. Electr. Eng. (2013).https://doi.org/10.11591/telkomnika.v11i4.2329
7. Meyer, J., Arnrich, B., Schumm, J., Troster, G.: Design and modeling of a textile pressure sensor for sitting posture classification. IEEE Sens. J. **10**, 1391–1398 (2010). https://doi.org/10.1109/JSEN.2009.2037330
8. Ren, B., Liu, J.: Design of a plantar pressure insole measuring system based on modular photoelectric pressure sensor unit. Sensors **21**, 3780 (2021).https://doi.org/10.3390/S21113780
9. Ramirez, C., Concha, C., Valdes, B.: Non-invasive technology on a classroom chair for detection of emotions used for the personalization of learning resources. World Acad. Sci. Eng. Technol. **66**, 384–390 (2010). https://doi.org/10.5281/zenodo.1079802
10. .Barkallah, E., Freulard, J., Otis, M.J.D., Ngomo, S., Ayena, J.C., Desrosiers, C.: Wearable devices for classification of inadequate posture at work using neural networks. Sensors **17**, 2003 (2017). https://doi.org/10.3390/S17092003
11. Dzedzickis, A., et al.: Polyethylene-carbon composite (Velostat) based tactile sensor. Polymers (Basel) **12**, 1–17 (2020). https://doi.org/10.3390/polym12122905

Quinones-Mediated Microbial Biofuel Cell Based on Baker's Yeast

Juste Rozene[1(✉)], Katazyna Blazevic[1], Antanas Zinovicius[1], Vakare Guoba Sataite[1], and Inga Morkvenaite-Vilkonciene[1,2]

[1] Department of Mechatronics, Robotics, and Digital Manufacturing, Vilnius Gediminas Technical University, J. Basanaviciaus 28, 03224 Vilnius, Lithuania
juste.rozene@vgtu.lt

[2] Laboratory of Electrochemical Energy Conversion, State Research Institute Centre for Physical Sciences and Technology, Sauletekio 3, 10257 Vilnius, Lithuania

Abstract. The impact of the energy consumption increase on the environment is one big problem, and it makes alternative energy sources so significant. Biological fuel cells (BFCs) based on enzymes and microorganisms are devices capable to directly transform chemical to electrical energy via electrochemical reactions involving biochemical pathways microbial biofuel cells (MBFCs). The study shows the possibility to use yeast-based MBFCs as the microorganisms for the quinones caused toxicity. Quinones are a controversial substance that is very important to MBFCs electrons transfer process but can cause oxidative and replicative stress of the cell. This is why the main aim of our research was to improve charge transfer efficiency in Baker's yeast-based MBFCs by immobilizing quinones menadione (MD), 9, 10-phenantrenquinone (PQ), and MD/PQ on the anode surface. Metrohm μStat 400 Potentiostat/Galvanostat was used for the electrochemical measurements, which results are described in this paper.

Keywords: Microbial fuel cell · Sensor · Toxicity

1 Introduction

Although most electrical appliances are under development, energy consumption increases every year. Fossil fuels have supported the industrialization and economic growth of all the economies in the world over the past century. But because of using fossil fuels, the emission of greenhouse gases and the consequent increase in the average global temperature is the main concerns [1].

The impact on the environment is one big problem, and it makes alternative energy sources so significant. The last thirty years have been of particular interest in the biofuel cells, exploiting whole microorganisms as biocatalysts, attract an increasing research interest during the previous decade as a perspective technology for simultaneous electricity generation and wastewater purification [2]. Biological fuel cells (BFCs) based on enzymes [3] and microorganisms [4] are devices capable to directly transform chemical to electrical energy via electrochemical reactions involving biochemical pathways

R. Szewczyk et al. (Eds.): AUTOMATION 2022, AISC 1427, pp. 202–210, 2022.
https://doi.org/10.1007/978-3-031-03502-9_21

[5] microbial biofuel cells (MBFCs) [6]. For each microorganism applied as a biocatalyst, the clarification of the individual charge transfer mechanisms, is of significant importance [2].

During the last few years, the utilization of yeast, such as *Saccharomyces cerevisiae*, as biocatalysts in fuel cells has been reported. The yeasts are utilized for centuries in numerous biotechnological processes at the food industry, which generate huge amounts of wastewaters rich of organic matter. Biofuel cells are considered as an alternative for purification of such wastewaters [2]. The yeast as biocatalyst is ideal for this purpose because of its properties: (i) similarity to mammalian systems homology; (ii) predictable sensitivity/resistance to quinones; (iii) ability to survive in a variety of conditions; and (iv) easy availability and relatively low cost [7, 8].

The problem with yeast used in MBFCs is limited electron transfer efficiency from the cells towards the electrodes. To solve this problem, the redox mediator, a substance that can participate in intracellular reactions and transfer the charge towards the electrode are used [9]. The redox mediators should be chosen by certain characteristics, such as electrochemical activity, non-toxicity to microorganisms in MBFCs, ability to penetrate the cell's membrane, reduce and diffuse backwards while transferring electron, solubility and chemical stability [10, 11]. When microorganism is used as catalyst in biofuel cells, two redox mediators-based system is needed because lipophilic intracellular redox mediator penetrates microorganism's membrane but the charge transfer directly to the electrode is very weak. Second (hydrophilic) redox mediator cannot penetrate the cells' membrane, it acts in extracellular environment, and transfers the charge towards the electrode [12–14].

Toxicity of the redox mediator to the yeast cells should be minimized by evaluating its concentration in anode compartment of MFBC. Some studies demonstrate the utility of yeast as a probe to assess redox and electrophile-based toxicities [15, 16]. Microorganisms used in MBFCs can be useful in research related to the toxicity of the drinking water. Yeast, as the unicellular organisms with respiratory function, can be used as early toxicity warming device [17]. Oxygen concentration in the solution is the primary indicator of oxidative stress and the death of the cells caused by quinones [18, 19].

Lipophilic 2-methyl-1, 4-naphthoquinone (menadione, MD) and 9,10-phenanthrenequinone (PQ) can work intracellularly and transfer the charge from the cell [20, 21] to the second redox mediator, such as ferricyanide [12]. Both quinones cause oxidative stress – the situation when the disproportion of oxygen reactive species (ROS), between its production and collection in cells appears. A biological system is able to detoxify these reactive products: superoxide radicals (O^{2-}), hydrogen peroxide (H_2O_2), hydroxyl radicals (OH), and singlet oxygen (O_2) [22]. Also, replicative stress can be caused by MD, and the replication process will be affected [23].

The main aim of our research was to improve charge transfer efficiency in Baker's yeast-based MBFCs by immobilizing quinones MD, PQ, and MD/PQ on the anode surface.

2 Materials and Methods

2.1 Materials

Chemicals used in experiments were purchased from Merck, Riedel and Sigma–Aldrich companies. Dried yeast culture was purchased from food supplier "Dr. Oetker Lietuva" (Vilnius, Lithuania).

0.05 M phosphate buffer solution was prepared in distilled water with CH_3COONa, NaH_2PO_4, Na_2HPO_4, and 0.1 M KCl to increase solutions conductivity. Potassium ferricyanide were prepared in distilled water. 9, 10-phenantrenquinone (PQ) was dissolved in 97% ethanol and menadione (MD) was prepared in distilled water.

2.2 Preparation of Yeast

YPD medium was prepared by mixing 1 g of YPD-broth with 20 ml distilled water to get medium with YPD-broth 50 g/l concentration. 500 mg of dried yeast was introduced to prepared medium. Yeast cell suspension in YPD medium was incubated till yeast reaches Logarithmic Growth Phase (20–24 h) in shaking incubator at 200 rpm.

2.3 Anode Preparation

Graphite electrode (ø3 mm) was sanded with paper with three different grinding bead sizes and washed with distilled water and 97% ethanol. The 2.5 μL drop of redox mediator (MD, PQ, or MD/PQ) with concentration of 3.75 mM were immobilized directly on the dry electrode.

2.4 Electrochemical Measurements

Electrochemical measurements were performed using a Metrohm μStat 400 Potentiostat/Galvanostat (Utrecht, Netherlands), and DropView software. All experiments were carried out at ambient temperature (at 20 °C), while stirring in-between measurements, under aerobic conditions. Measurements were performed in three electrodes electrochemical cell, where the graphite electrode was connected as a working electrode, the platinum electrode as a counter electrode and Ag/AgCl (3 M KCl) electrode as a reference electrode (Fig. 1). Cyclic voltammograms (CV) were recorded at the scan rate of 0.1 V/s in the range from –0.6 V to 0.8 V.

For this research, a system of three electrodes was used. It consists of interconnected reference (Ag/AgCl), counter (platinum) and working (graphite) electrodes immersed in the phosphate buffer solution (Fig. 1). Electron transfer mediator – PQ, MD, or mix of both (MD/PQ) by a ratio (1:1) was immobilized on working electrode (anode). In all cases, 2.5 μL volume of 3.75 mM concentration droplets were immobilized directly on the dry anode. Afterwards, the yeast suspension was added to the solution, and CV were recorded.

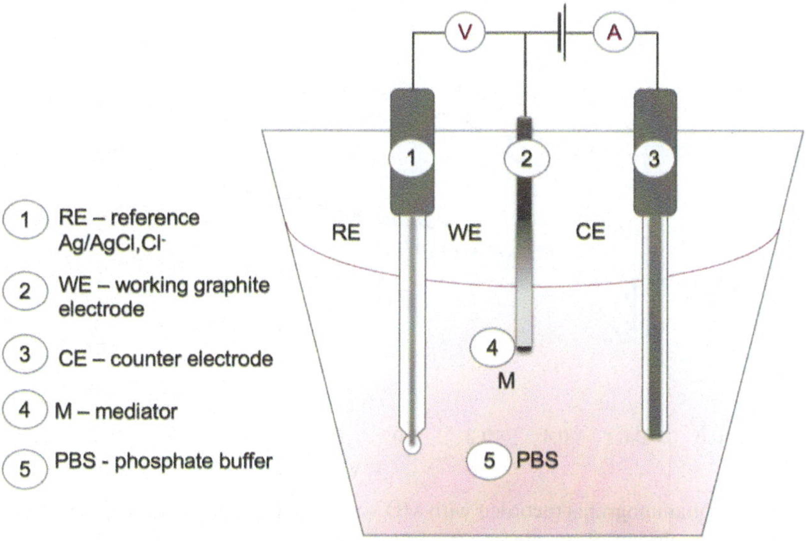

Fig. 1. Three electrodes electrochemical cell. RE – reference (Ag/AgCl) electrode, WE – working graphite electrode, CE – counter platinum electrode, M – immobilized mediator, PBS – phosphate buffer

3 Investigation of Toxicity Sensor

3.1 Experiments with Mediator-Modified Anode at Different Yeast Concentrations

To determine electron transfer efficiency towards the electrode, cyclic voltammetry was applied. This electrochemical method illustrates the redox processes in yeast cells and allows to evaluate the generation of electric current at certain electric potential. CVs were recorded while yeast concentration was being incrementally increased. In this experiment, MD was immobilized on the anode, and the concentration of yeast added in phosphate buffer solution was being increased till 2 g/ml (Fig. 2). The current density decreased from 0.55 to 0.4 mA/cm^2 at –0.25 and –0.15 V potentials while increasing yeast concentration from 0.04 to 2 g/ml.

Performing the same measurements with PQ immobilized on the anode and the increasing yeast concentration, the results were relevant. The highest current density appeared at the lowest concentration of yeast (Fig. 3). The current density decreased from 6 to 4.5 mA/cm^2 at –0.25 and –0.15 V potentials while increasing yeast concentration from 0.04 to 2 g/ml.

Comparing the results of CV obtained using only one of two lipophilic redox mediators (MD or PQ), 12 times higher values of current density were received while using PQ. In experiments with MD maximal current density reached 0.55 mA/cm^2, and 6 mA/cm^2 with the PQ. The results are consistent with previous our studies: the use of MD [24] and PQ [12] improves the electron transfer processes.

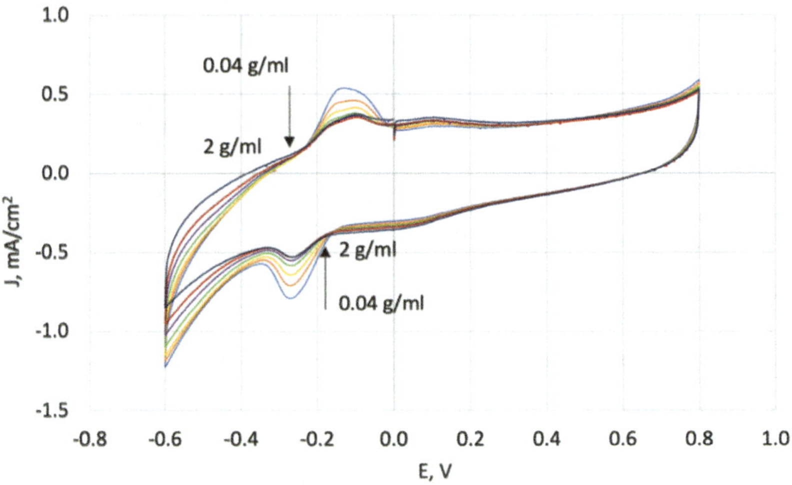

Fig. 2. Cyclic voltammograms recorded with MD – modified anode at different yeast concentrations in the buffer solution. Scan rate was 0.1 V/s

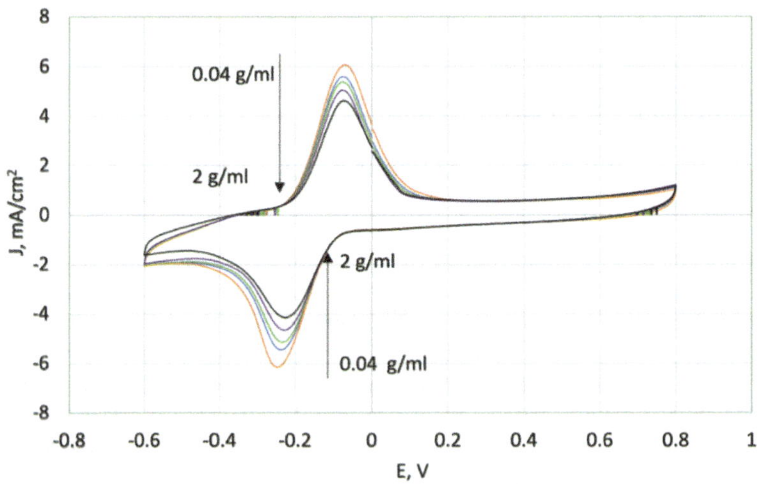

Fig. 3. Cyclic voltammograms recorded with PQ – modified anode at different yeast concentrations in the buffer solution. Scan rate was 0.1 V/s

3.2 Experiments with Two-Redox Mediators-Based System

Once the optimal yeast concentration was determined, the experiments to determine optimal hydrophilic (potassium ferricyanide) mediator concentration were performed with a two-mediator system.

At first, cyclic voltammetry measurements were performed with PQ-modified anode while increasing potassium ferricyanide concentration (Fig. 4).

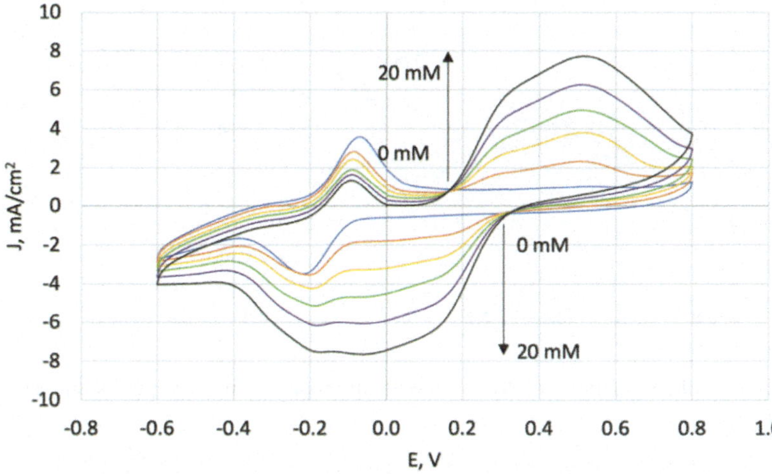

Fig. 4. Cyclic voltammograms of PQ – modified anode at different potassium ferricyanide concentrations

In this case, the result with PQ mediator is more complex, because at the –0.1 V potential, the current density value decreased after the potassium ferricyanide was added. That means that the concentration of the reduced form of PQ became lower. The oxidation–reduction peaks of potassium ferricyanide were detected at –0.1 and 0.53 V potentials. The gain of current density value is equal to $\Delta 6.6$ mA/cm^2 after the concentration of potassium ferricyanide was changed from 0 to 20 mM.

A similar situation was detected after the measurements with MD were carried out (Fig. 5). MD oxidation–reduction peaks were obtained at –0.14 and –0.25 V potentials. However, since the concentration of lipophilic MD mediator was stable, values of both (reduced and oxidized) forms are similar and decreasing. That means that the reaction is reversible. However, potassium ferricyanide peaks are sharper than in the previous case with PQ-modified anode (Fig. 4), this phenomenon can be explained as shorter reaction time. The gain of current density value is equal to $\Delta 4.4$ mA/cm^2 after the concentration of potassium ferricyanide was changed from 0 to 20 mM.

In the last case (Fig. 6), two lipophilic mediators were mixed with 1 by 1 ratio and immobilized on the graphite electrode. The result was expected to be better than using each one individually. However, the result was unexpected, because it has qualities of both lipophilic mediators. Even though current density is bigger than with MD (Fig. 5), it's smaller than with PQ (Fig. 3). The gain of current density value is equal to $\Delta 4.7$ mA/cm^2 after the concentration of potassium ferricyanide was changed from 0 to 20 mM. As in all cases, the increase of the hydrophilic potassium ferricyanide concentration makes an impact to the current density value.

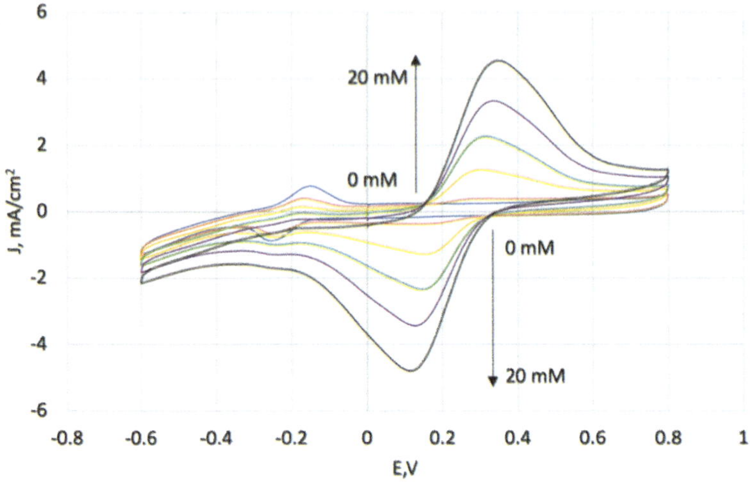

Fig. 5. Cyclic voltammograms of MD – modified anode at different potassium ferricyanide concentrations

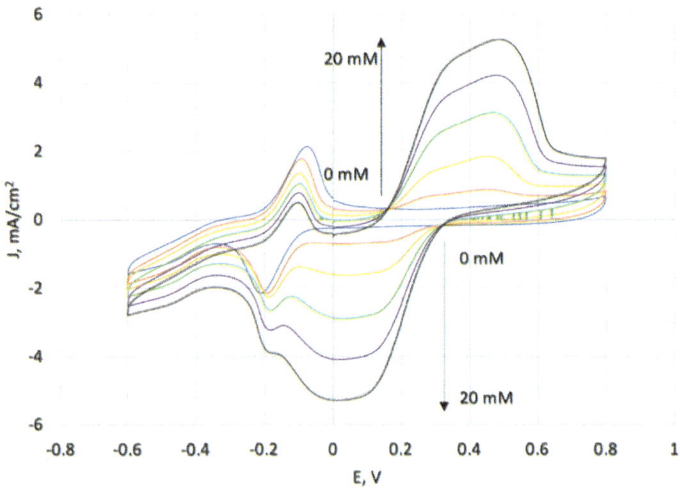

Fig. 6. Cyclic voltammograms of mix (PQ/MD) – modified ratio (1:1) anode at different potassium ferricyanide concentrations

4 Conclusions

The MD, PQ, and MD/PQ modified anode-based microbial biofuel cells was developed, evaluated and compared. Yeast and potassium ferricyanide concentrations were changed. The highest current density received at the lowest yeast concentration (0.04 g/ml) using both mediator-modified anodes (0.55 mA/cm^2 with MD; 6 mA/cm^2 with PQ). The values obtained using these two lipophilic mediators showed significant difference: current density value received at measurements with PQ was 12 times higher. Current density increased, while one of the lipophilic redox mediators (PQ or MD) were immobilized, and the concentration of hydrophilic potassium ferricyanide was increased too. An exceptional drop in current density is visible at –0.1 V potential where the redox of the mediator form is being consumed. The current density value was two times higher while using PQ (8 mA/cm^2 with PQ and 4.5 mA/cm^2 with MD). The current density increased while the mix of lipophilic mediators (PQ/MD) were immobilized on the graphite electrode, and the concentration of hydrophilic potassium ferricyanide was increased, but the results were lower than using only PQ (5 mA/cm^2). In further our research, improved construction of MBFC and different modifications of anode will be used.

Acknowledgements. This project has received funding from European Social Fund (project No 09.3.3.-LMT-K-712-22-0325) under grant agreement with the Research Council of Lithuania (LMTLT).

References

1. Trapero, J.R., Horcajada, L., Linares, J.J., Lobato, J.: Is microbial fuel cell technology ready? An economic answer towards industrial commercialization. Appl. Energy (2017). https://doi.org/10.1016/j.apenergy.2016.10.109
2. Hubenova, Y., Mitov, M.: Extracellular electron transfer in yeast-based biofuel cells: a review. Bioelectrochemistry **106**, 177–185 (2015). https://doi.org/10.1016/j.bioelechem.2015.04.001
3. Ramanavicius, A., et al.: Biofuel cell based on glucose oxidase from Penicillium funiculosum 46.1 and horseradish peroxidase. Chem. Eng. J. **264**, 165–173 (2015). https://doi.org/10.1016/j.cej.2014.11.011
4. Bruzaite, I., Rozene, J., Morkvenaite-Vilkonciene, I., Ramanavicius, A.: Towards microorganism-based biofuel cells: the viability of saccharomyces cerevisiae modified by multiwalled carbon nanotubes. Nanomaterials **10**, 954 (2020)
5. Bullen, R.A., Arnot, T.C., Lakeman, J.B., Walsh, F.C.: Biofuel cells and their development. Biosens Bioelectron **21**, 2015–2045 (2006)
6. Ramanavicius, A., Kausaite, A., Ramanaviciene, A.: Biofuel cell based on direct bioelectrocatalysis. Biosens. Bioelectron. **20**, 1962–1967 (2005). https://doi.org/10.1016/j.bios.2004.08.032
7. Rodriguez, C.E., Shinyashiki, M., Froines, J., Yu, R.C., Fukuto, J.M., Cho, A.K.: An examination of quinone toxicity using the yeast Saccharomyces cerevisiae model system. Toxicology (2004). https://doi.org/10.1016/j.tox.2004.04.016
8. Baronian, K.H.R.: The use of yeast and moulds as sensing elements in biosensors. Biosens. Bioelectron. **19**, 953–962 (2004)
9. Kisieliute, A., et al.: Towards microbial biofuel cells: improvement of charge transfer by self-modification of microoganisms with conducting polymer – Polypyrrole. Chem. Eng. J. **356**, 1014–1021 (2019). https://doi.org/10.1016/J.CEJ.2018.09.026
10. Babanova, S., Hubenova, Y., Mitov, M.: Influence of artificial mediators on yeast-based fuel cell performance. J. Biosci. Bioeng. (2011). https://doi.org/10.1016/j.jbiosc.2011.06.008
11. Qiao, Y., Li, C.M., Bao, S.J., Bao, Q.L.: Carbon nanotube/polyaniline composite as anode material for microbial fuel cells. J. Power Sources (2007). https://doi.org/10.1016/j.jpowsour.2007.03.048

12. Morkvenaite-Vilkonciene, I., Ramanaviciene, A., Ramanavicius, A.: 9,10-Phenanthrenequinone as a redox mediator for the imaging of yeast cells by scanning electrochemical microscopy. Sens. Actuators B Chem. **228**, 200–206 (2016). https://doi.org/10.1016/j.snb.2015.12.102
13. Allen, R.M., Bennetto, D.H.P.: Microbial fuel-cells electrons oxidation mediator [(Reduced) (Oxidant Fuel, Reduced Oxidant Ion-exchange. Appl. Biochesmestry Biotechnol. **39** (1993)
14. Hubenova, Y.V., Rashkov, R.S., Buchvarov, V.D., Arnaudova, M.H., Babanova, S.M., Mitov, M.Y.: Improvement of yeast−biofuel cell output by electrode modifications. Ind. Eng. Chem. Res. **50**, 557–564 (2010)
15. Morkvenaite-Vilkonciene, I., et al.: The improvement of the accuracy of electromagnetic actuator based atomic force microscope operating in contact mode and the development of a new methodology for the estimation of control parameters and the achievement of superior image quality. Sens. Actuators A Phys. **287** (2019). https://doi.org/10.1016/j.sna.2019.01.015
16. Rozene, J., Zinovicius, A., Kacinskaite, B., Bučinskas, V., Ramanavicius, A., Morkvenaite-Vilkonciene, I.: Microbial fuel cell-based toxicity sensor. In: Szewczyk, R., Zieliński, C., Kaliczyńska, M. (eds.) Automation 2020: Towards Industry of the Future. AUTOMATION 2020. AISC, vol. 1140, pp. 379–388. Springer, Cham (2020). https://doi.org/10.1007/978-3-030-40971-5_35
17. Stein, N.E., Hamelers, H.M.V., Van Straten, G., Keesman, K.J.: On-line detection of toxic components using a microbial fuel cell-based biosensor. J. Process Control **22**, 1755–1761 (2012). https://doi.org/10.1016/j.jprocont.2012.07.009
18. Jamieson, D.J.: Oxidative stress responses of the yeast Saccharomyces cerevisiae. Yeast **14**, 1511–1527 (1998)
19. Breitenbach, M., Weber, M., Rinnerthaler, M., Karl, T., Breitenbach-Koller, L.: Oxidative stress in fungi: its function in signal transduction, interaction with plant hosts, and lignocellulose degradation. Biomolecules **5**, 318–342 (2015)
20. Rodriguez, C.E., Sobol, Z., Schiestl, R.H.: 9, 10-Phenanthrenequinone induces DNA deletions and forward mutations via oxidative mechanisms in the yeast Saccharomyces cerevisiae. Toxicol Vitr **22**, 296–300 (2008)
21. Yamashoji, S.: Different characteristics between menadione and menadione sodium bisulfite as redox mediator in yeast cell suspension. Biochem. Biophys. Rep. (2016). https://doi.org/10.1016/j.bbrep.2016.03.007
22. Pizzino, G., et al.: Oxidative stress: harms and benefits for human health. Oxid. Med. Cell. Longev. **2017** (2017)
23. Ren, X., Santhosh, S.M., Coppo, L., Ogata, F.T., Lu, J., Holmgren, A.: The combination of ascorbate and menadione causes cancer cell death by oxidative stress and replicative stress. Free Radic. Biol. Med. **134**, 350–358 (2019)
24. Petroniene, J., et al.: Evaluation of redox activity of human myocardium-derived mesenchymal stem cells by scanning electrochemical microscopy. Electr. Anal. (2020)

Robotics

Neural Dynamic Programming with Application to Wheeled Mobile Robot

Zenon Hendzel[1] and Maciej Kołodziej[2(✉)]

[1] Faculty of Mechanical Engineering and Aeronautics, Department of Applied Mechanics and Robotics, Rzeszow University of Technology, Rzeszow, Poland
zenhen@prz.edu.pl

[2] Doctoral School of Engineering and Technical Sciences at the Rzeszow University of Technology, Rzeszow, Poland
d515@stud.prz.edu.pl

Abstract. In this paper we propose a new application sensor-based navigation method for navigation of wheeled mobile robot, based on neural dynamic programming (NDP). We discuss a sensor-based approach to path design and control of wheeled mobile robot (WMR) in an unknown 2-D environment with static obstacles. A strategy of navigation is developed including two main behaviors: a reaching the middle of a collision-free space behavior, and a goal-seeking. The NDP navigator which is the main theme of this paper, can fuse behaviors so that the mobile robot can go for the goal position without colliding with obstacles for the concave and convex obstacles. This solution is based on the reinforcement learning (RL) method of actor-critic architecture for continuous-time, and does not require pre-learning and working on-line. Computer simulations have been conducted to illustrate the performance of the proposed solution by a series of experiments.

Keywords: Neural networks · Neural dynamic programming · Behavior control · Wheeled mobile robot

1 Introduction

One of the most challenging problems in behavioral control involves the learning the appropriate mixture of elementary behaviors in order to execute complex tasks. There are a lot of studies on trajectory generation for robots using various approaches e.g. [1, 2, 5, 7, 9, 10]. Numerous behavior co-ordination mechanisms have been proposed. For a detailed overview, discussion, and comparison of behavior co-ordination mechanisms see [2]. We discuss a sensor-based approach to path design and control of wheeled mobile robot WMR in an unknown 2-D environment with static obstacles. A strategy of navigation is developed including two main behaviors: a reaching the middle of a collision-free space, and a goal-seeking. The NDP navigator, which is the main theme of this paper, can fuse low-level behaviors so that the mobile robot can go for the goal position without colliding with obstacles. Method of NDP, which is placed within the framework of optimal control [4, 7, 12], utilize two parametric structures (in this

R. Szewczyk et al. (Eds.): AUTOMATION 2022, AISC 1427, pp. 213–222, 2022.
https://doi.org/10.1007/978-3-031-03502-9_22

work-neural network (NN)), called the actor and the critic. The actor consists of a parameterized control law. The critic approximates a value function and captures the effect that the control law will have on the future cost.

This solution is based on the RL method of actor-critic architecture [3, 4, 7, 9, 11] for continuous-time, continuous state and does not require pre-learning and working on-line. The output of the navigation level is fed into a neural tracking controller that takes into account the dynamics of the WMR. The NDP architecture uses a high-level NN adaptive critic element (ACE), to generate the reinforcement signal to optimize the associative search element (ASE), which is applied to generate a given trajectory of motion realized by the tracking control system.

This subject of the article is an extension of the existing solutions in the field of mechatronics, especially mobile robotics. The remainder of the paper is organized as follows. In Sect. 2, the kinematics equations and the dynamic equations of motion of a WMR are given. Section 3 is the main topic of the article. A strategy of navigation is developed including two main behaviors: a reaching the middle of a collision-free space behavior, and a goal-seeking. The NDP navigator can fuse behaviors so that the mobile robot can go for the goal position without colliding with obstacles. Assumed structure of the neural network is discussed, and its description is given. The RL algorithm, which was used to solve optimal behavioral control, was also given. In addition, the structure of the higher-level control system is discussed. Section 4 includes results of NDP tests, obtained after numerical simulations. Section 5 resumes the results of the research.

2 Kinematics and Dynamics of a WMR

The mechanical structure of the mobile robot, in the measurement environment is shown in Fig. 1. The model consists of frame 4, two road driving wheels 1 and 2 and a free rolling castor wheel 3. Inthe word co-ordinates a posture is defined as $[x_A, y_A, \beta]^T$, where (x_A, y_A) is the position of the point A, and β is the heading angle of the robot with respect to absolute co-ordinates (x, y). In addition the following notations were used: l, l_1, l_2 – dimensions resulting from WMR's geometry, G – selected goal, x_1, y_1 – axes of the movable system of coordinates connected with point *A* in such a way so that axis x_1 is identical to the symmetry axis of the WMR frame, AG – straight line passing through the points A and G, ψG – angle between the line AG and axis x_1. Described mobile robot is equipped with eight an ultrasonic sensors divided into three group. A group is composed of: three, two and three neighboring sensors: d_{Li}, d_{Ri} – measurements of distance to obstacles, for the left and right side of the WMR frame, $i = 1, 2, 3$, d_{Fi} – measurement of distance to an obstacle in front of the WMR frame, $i = 1, 2$. The distance of WMR to the goal is determined as the length of the line segment |AG|. Each sensor covers an angular view.

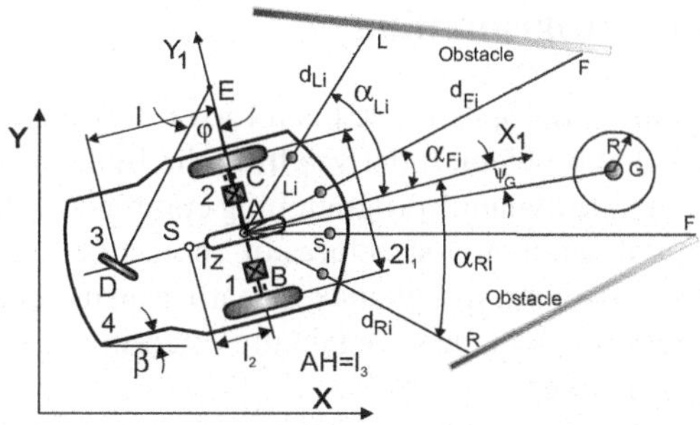

Fig. 1. The schematic of mobile robot

which are oriented by angles α_{Li}, α_{Fi}, α_{Ri} respectively. A dead zone with a radius R around point G is also marked. The algorithms generating behavioral control signals are based on feedback from the environment. The mobile robot's kinematics is defined by [6]

$$\begin{bmatrix} \dot{x}_A \\ \dot{y}_A \\ \dot{\beta} \end{bmatrix} = \begin{bmatrix} V_{Am}\cos(\beta) & 0 \\ V_{Am}\sin(\beta) & 0 \\ 0 & \omega_m \end{bmatrix} \begin{bmatrix} u_v \\ u_\beta \end{bmatrix} \tag{1}$$

with the maximum linear V_{Am} and angular ω_m speeds and u_v is the multiplying coefficient applied to the maximum linear velocity of point A of the robot and u_β is the multiplying coefficient applied to the maximum angular velocity of the frame. That coefficients form a vector $u_B = \left[u_\upsilon, u_\beta\right]^T$ generated by ACE_ASE navigator. Tacking α_1, α_2 – angles of the self-turn of the propel wheels as independent co-ordinates, then from velocity vectors of points A, B, C and (1) we receive angular velocities of the robot's driving wheels expressed in mobile base co-ordinates

$$\dot{\alpha}_1 = \frac{V_A}{r_1} + \dot{\beta}\frac{l_1}{r_1} \tag{2}$$

$$\dot{\alpha}_2 = \frac{V_A}{r_2} - \dot{\beta}\frac{l_1}{r_2} \tag{3}$$

where $V_A = V_{Am}u_v$, $\dot{\beta} = \omega_m u_\beta$, $r_1 = r_2 = r$ is radii of the wheels. Equations (2) and (3) define a prescribed motion trajectory $x_d = [\alpha_1, \dot{\alpha}_1, \alpha_2, \dot{\alpha}_2, \ddot{\alpha}_1, \ddot{\alpha}_2]^T$ which is determined in a higher-level path planner in real time, which is carried out by the neural control system standing at the control object. To describe the movement of a mobile robot model, Maggie's equations were used [6]. In the analyzed case the Maggie's equations have the following form:

$$\begin{bmatrix} a_1 + a_2 + a_3 & a_1 - a_2 \\ a_1 - a_2 & a_1 + a_2 + a_3 \end{bmatrix} \begin{bmatrix} \ddot{\alpha}_1 \\ \ddot{\alpha}_2 \end{bmatrix} + \begin{bmatrix} 0 & 2a_4(\dot{\alpha}_2 - \dot{\alpha}_1) \\ -2a_4(\dot{\alpha}_2 - \dot{\alpha}_1) & 0 \end{bmatrix} \begin{bmatrix} \dot{\alpha}_1 \\ \dot{\alpha}_2 \end{bmatrix} + \begin{bmatrix} a_5 \operatorname{sgn}\dot{\alpha}_1 \\ a_6 \operatorname{sgn}\dot{\alpha}_2 \end{bmatrix} = \begin{bmatrix} M_1 \\ M_2 \end{bmatrix} \tag{4}$$

where a_j is a vector of the mobile robot parameters, which results from the system geometry, weights distribution and motions resistance, $u = [M_1, M_2]^T$ is a vector of the moments propelling driving wheels. The objective of mobile robot control is to select the control vector $u = [M_1, M_2]^T$ so that the mobile robot follows a prescribed motion trajectory, $x_d(t)$ as depicted in the Fig. 3.

3 Neural Dynamic Programming

Neural Dynamic Programming [3, 4, 11, 12] is a forward dynamic programming approach, which solve signaled problem. It derives from the Bellman's backward dynamic programming, but uses iterate dynamic programming combined with a parametric structures like NN. NN are the parametric structures of choice, because they easily handle large-dimensional input and output spaces and can learn in an incremental fashion. Based on the architecture of reinforcement learning control algorithm [3, 4, 8, 11] we employ NN as two parametric structures.

3.1 Linear-In-The-Parameter Neural Nets

A two layer NN is depicted in Fig. 2a, where there are two layers of neurons, with one layer having m neurons feeding into a second layer having r neurons, with linear activation functions on the output layer. The output of the two-layer NN is given by the equation [8]

$$y = W^T S\left(V^T x\right) \tag{5}$$

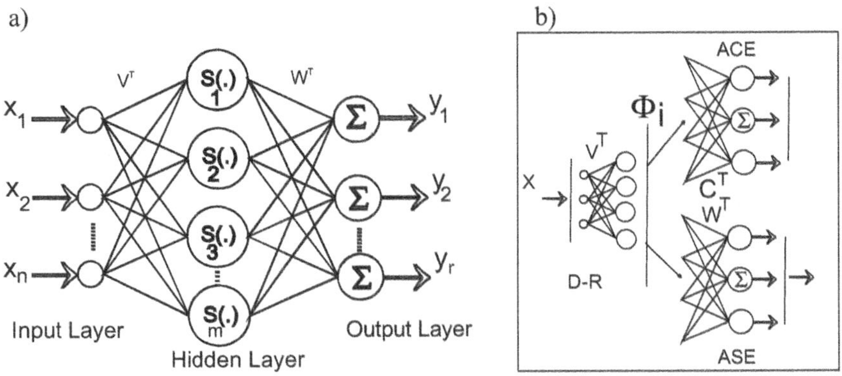

Fig. 2. a). Neural network scheme, b) ACE-ASE model

If the first-layer weights V are predetermined by some a priori method, then only the second-layer weights W are considered to define the NN. Define the fixed function $\varphi(x) = S(V^T x)$ so that such a one-layer NN has the equation

$$y = W^T \varphi(x) \tag{6}$$

This NN is linear in the NN parameters W so that it is easier to train by tuning the weights. It is shown in [8] that, selecting the matrix V in (5) randomly the resulting function $\varphi(x) = S(V^T x)$ is a basis, so that the linear in the NN parameters has the universal approximation property. In this approach, S(.) can be the standard sigmoid functions. In our solution presented NN is separated into three parts; one is corresponding to the decoder (D-R), the task of which is to scale the input signals to the network, shown in Fig. 2b and other is corresponding to the ACE and ASE respectively. Thus, the outputs of ASE and ACE can be represented as $W^T\varphi$ and $C^T\varphi$, respectively.

3.2 RL Algorithm

Our task is to determine the law $u_B \in U \subset R^2$ of controlling the system kinematics (1), that maximizes the expected reinforcement $r(t) = [r_1, r_2]^T$ for a certain period in the future. We define the value function as [3, 4, 11]

$$V_j(t) = \int_t^{\infty} \frac{1}{\tau_c} e^{\frac{\tau - t}{\tau_c}} r_j(\tau) d\tau,\ j = 1, 2 \tag{7}$$

where τ_c is constant. A local consistency condition for the value function is as follows

$$\tau_c \dot{V}_j(t) = V_j(t) - r_j(t) \tag{8}$$

Let P(t) be the prediction of the value function. If the prediction is correct, then the following equation should be satisfied

$$\tau_c \dot{P}_j(t) = P_j(t) - r_j(t) \tag{9}$$

If the prediction is not accurate, prediction error is determined as

$$\hat{r}_j(t) = r_j(t) - P_j(t) + \tau_c \dot{P}_j(t) \tag{10}$$

The critic is a parametric structure $P_j(t)$ with parameters c_i that is used to approximate prediction of the value function and generates a signal in the form

$$P_j(t) = C^T \varphi(t) \tag{11}$$

Critic weight adaptation is performed according to the relation

$$\dot{c}_i(t) = \gamma \hat{r}(t) \bar{e}_i(t) \tag{12}$$

and activity path $\bar{e}_i(t)$ is described as follows

$$\tau_c \dot{\bar{e}}_i(t) = \varphi_i(t) - \bar{e}_i(t) \tag{13}$$

where γ denotes the learning rate. The purpose of this paper is to develop the ASE output which is of the form

$$u_B = \eta[S(W^T \varphi(t) + u_E)] \tag{14}$$

where η – scaling factor of actor output and signal u_E is defined differently for elementary behaviors, and S(.) is a sigmoid types. The law of actor weight adaptation takes the form

$$\dot{w}_{ki}(t) = \alpha \hat{r}_k(t) u_{Ek} \varphi_i(t) \tag{15}$$

where α denotes the learning rate. In the following, the ACE_ASE navigator analysis will be discussed.

3.3 ACE-ASE Navigator

To solving the trajectory tracking problem for non-holonomic mobile robot with considering the vehicle dynamics, it is assumed that the current given trajectory is generated at each time step, by NDP navigator which generates vector of coefficients $u_B = [u_\upsilon, u_\beta]^T$, based on the environment information. A strategy of navigation is developed including two main behaviors: a reaching the middle of a collision-free space, and a goal-seeking. The NDP navigator, can fuse this behaviors so that the mobile robot can go for the goal position without colliding with obstacles. We assume that low-level of the control system (tracking control) has been well designed [12] and in the further part of the article we will omit the implementation of the desired motion trajectory by the lower tracking control system. Diagram of the navigator and controller architecture is depicted in Fig. 3. The individual decision of different behaviors are fused into combined decision. In the proposed NDP navigator low-level modules are denoted as obstacle avoider (OA) and the upper one refers to reaching the goal (GS). Each module receives different signals and produce output signals. Let us assume a given configuration of the A point of a mobile robot $q = [x_A, y_A, \beta]^T$ and given coordinates of the G point $q_G = [x_G, y_G, 0]^T$. Let us define the error of the given and final configuration as $\tilde{q} = q_G - q = [\tilde{x}, \tilde{y}, \tilde{\beta}]^T$. Let's define the error in polar coordinates $q_b = [\delta_{AG}, \psi_G]^T$, where $\delta_{AG} = \sqrt{\tilde{x}^2 + \tilde{y}^2}$, $\psi_G = \arctan2(\tilde{y}, \tilde{x}) + \tilde{\beta}$, $\psi_G \in [-\pi, \pi]$. The input signals to the GS module are δ_{AG} and ψ_G signals. The module GS determines the action, $u_{B1} = [u_{vGS}, u_{\beta GS}]^T$ for the behavior of goal seeking, while the module OA determines the action $u_{B2} = [u_{vOA}, u_{\beta OA}]^T$ for the behavior of obstacle avoidance, on the basis of the d_{Li}, d_{Fi}, d_{Ri} signals, received from the ultrasonic sensors.

These two behavioral modules work independently and their actions are fused by ASE to produce action u_B for the navigation. The reinforcement signal plays an important role in NDP and we express it as in the following vector form

$$r(t) = [-(u_{\beta OA} - u_{\beta GS})^2, -(u_{vOA} - u_{vGS})^2]^T \tag{16}$$

so we want to maximizing it. The auxiliary signal u_E in (14), that is referred to as the exploration process, is determined as follows. Let us suppose that we establish the following linear combination for both navigation actions

$$u_{E1} = \omega[a_1 u_{vGS} + a_2 u_{vOA}] \tag{17}$$

$$u_{E2} = \omega[a_1 u_{\beta GS} + a_2 u_{\beta OA}] \tag{18}$$

where $a_1 + a_2 = 1, 0 \le a_1, a_2 \le 1, 0 < \omega << 1$. Thus, values of a_1 close to unity means that the mobile robot is almost unconcerned about avoiding obstacles, whereas the opposite case of a_2 approaching unity implies that the mobile robot's main goal is to avoid obstacles. After testing different switching functions, we have found that the rule

$$a_2 = 1 - \exp[-\kappa(d_{(.)\,min} - \rho)] \tag{19}$$

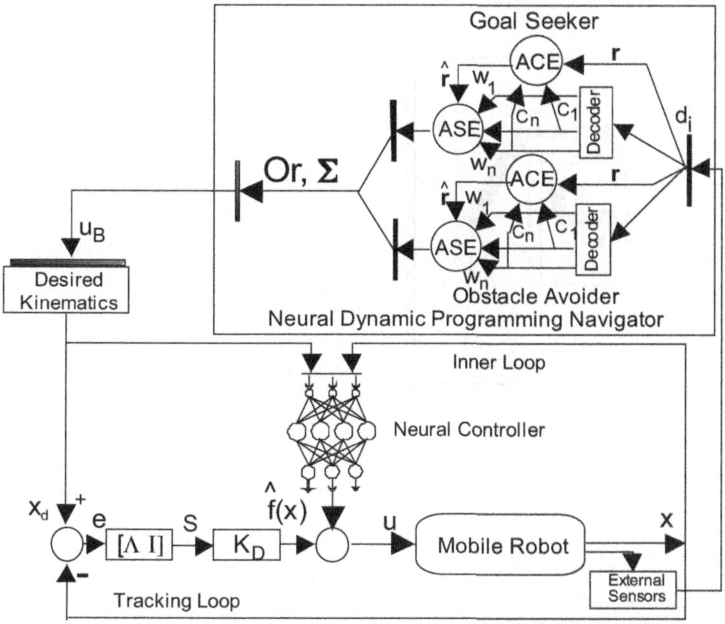

Fig. 3. ACE_ASE navigator and tracking controller

where κ, ρ are parameters adjusted by experimentation to get the best solution, is very efficient when suitably tuned. The final multiplying coefficient applied to the Eq. (1) is generated by Eq. (14), where $u_E = [u_{E1}, u_{E2}]^T$.

4 Simulation Results

In this section, we will present simulations of the proposed behavioral control for WMR using NDP. The parameters of the neural ACE_ASE navigator as shown in the Fig. 2 is as follows: $\omega_1 = 3$, $\omega_2 = 8$, $\eta_1 = \eta_2 = 0.1$, $\kappa = 7$, $\rho = 0.08$, $\gamma = 10$, $\tau_c = 0.1$. The maximum distance d_{max} measured by the sensors was 6[m]. All initial weights v_i of the decoder (D-R) shown in Fig. 2b, were set randomly. The parameters set $\{V_m, \omega_m\}$ are equal to $\{0.4[m/s], 0.3[rad/s]\}$ for test. The mobile robot had received mission to reach a given goal position G, depicted in Fig. 4a, from given start position with reaching the middle of a collision-free space. The environment was considered as fixed and completely unknown. The mobile robot starts with $(x_A, y_A, \beta) = (2.5, 5, 0)$ in the test. R is the distance of influence of the goal shown in Fig. 1.

It is assumed that no obstacle exists in the circle of R = 1[m] diameter. The successive activation of the different behaviors can be observed in Fig. 4a. Figure 4b shows the prediction $P_2(t)$ of the value function $V_2(t)$ and determined control signals as depicted in the Fig. 4c, confirm that maximizes the expected reinforcement $r(t) = [r_1, r_2]^T$ as shown in Fig. 5a and 5b. Furthermore, Fig. 4d shows auxiliary signal u_{E2} which make up 10% of control signal u_β. This confirms that proposed ASE element can learns. The weights, for neural actor element which produces control signal u_β, are depicted

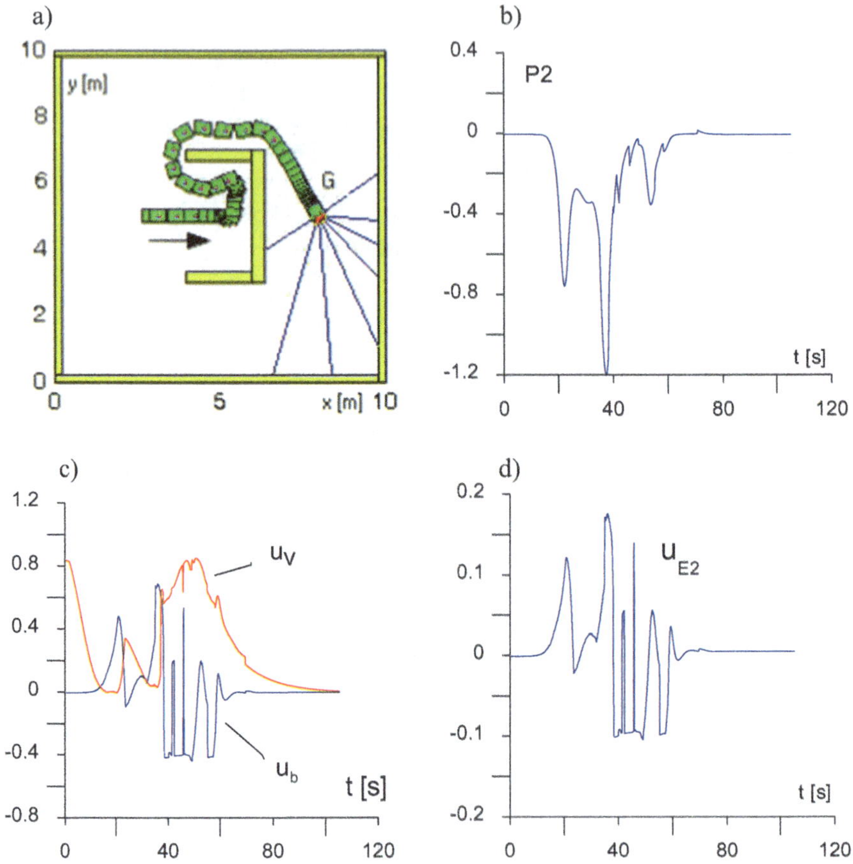

Fig. 4. The numerical results of the proposed solution

in Fig. 5d. Figure 5c shows that more than one behavior is active at the same time, that permits a soft transition. Moreover, it was found that all signals are bounded. The following example, shown in Fig. 5, demonstrates the abilities of the NDP controller in another workspace. The basic driving tasks are to drive the WMR toward the target and to avoid collision with obstacles. The successive activation of the different tasks can be observed in Fig. 6a.

On the other hand, in Fig. 6b there are control signals determined from the higher level of the hierarchical system, which determine the given trajectory of movement for the lower level of the system.

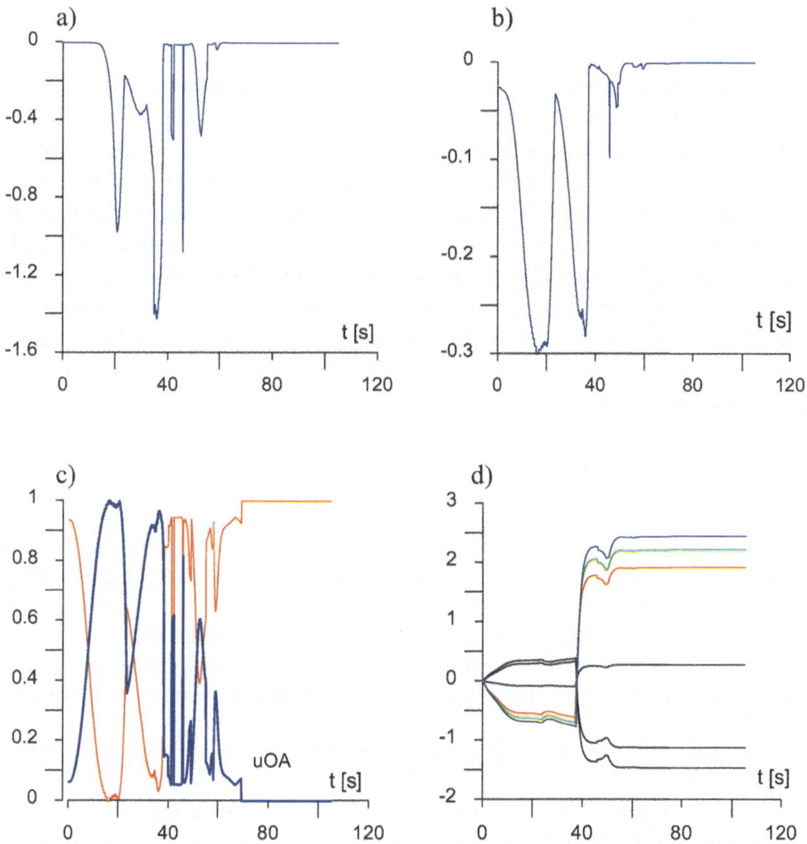

Fig. 5. Simulation results of reinforcement learning, behaviors and the weights

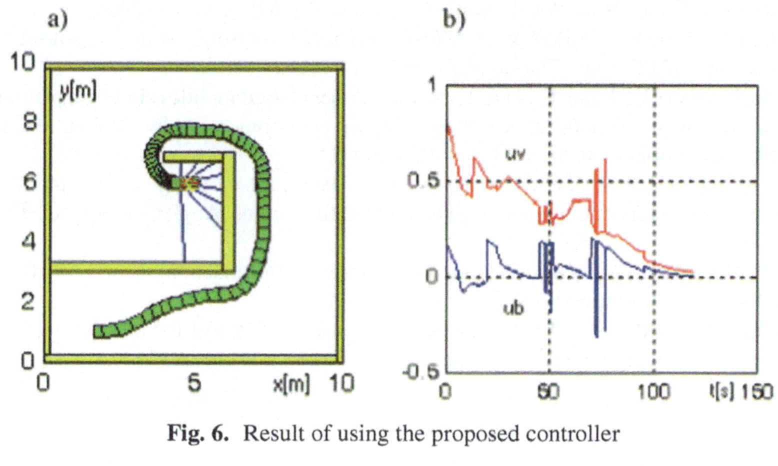

Fig. 6. Result of using the proposed controller

5 Conclusion

In this work we propose a new sensor-based navigation method for navigation of wheeled mobile robot, based on neural dynamic programming. In our consideration the learning machine is RL. We have observed that NDP based on neural network model with actor-critic architecture can solve paths planning in complex unknown environment. A strategy of reactive navigation was developed including two main behaviors: a reaching the middle of collision-free space behavior, and a goal-seeking. Compared with traditional NDP designs and RL approaches, the proposed solution does not require pre-learning, working on-line, and can prevent from time-consuming trial-and error learning, which is not suitable for navigation of an autonomous mobile robot. Simulation results for the proposed NDP navigator demonstrate that the control objectives can be achieved effectively and successfully.

References

1. Ali, M.A., Mailah, M.: A simulation and experimental study on wheeled mobile robot path control in road roundabout environment. Int. J. Adv. Rob. Syst. **16**(2), 1729881419834778 (2019)
2. Arkin, R.C.: Behaviour-Based Robotics. The MIT Press, Cambridge (1998)
3. Barto, A.G., Sutton, R.S., Anderson, C.W.: Neuronlike adaptive elements that can solve difficult learning control problems. IEEE Trans. Syst. Man Cybern. **SMC-13**(5), 834–846 (1983)
4. Doya, K.: Reinforcement learning in continuous time and space. Neural Comput. **12**(1), 219–245 (2000)
5. Fiorato, N., Yasutaka, F., Roberto, O.: A LSTM neural network applied to mobile robots path planning. In: IEEE 16th International Conference on Industrial Informatics (INDIN), pp. 349–354 IEEE (2018)
6. Giergiel, J., Żylski, W.: Description of motion of a mobile robot by Maggie's equations. J. Theor. App. Mech. **43**, 511–521 (2005)
7. Hendzel, Z.: Collision free path planning and control of wheeled mobile robot using Kohonen selforganising map. Bull. Polish Acad. Sci. Techn. Sci. **53**(1), 39–47 (2005)
8. Levis, F.L., Liu, K., Yesildirek, A.: Neural net robot controller with guaranteed tracking performance. IEEE Trans. Neural Netw. **6**(3), 703–715 (1995)
9. Chang, L., Shan, L., Jiang, C., Dai, Y.: Reinforcement based mobile robot path planning with improved dynamic window approach in unknown environment. Auton. Robot. **45**(1), 51–76 (2020). https://doi.org/10.1007/s10514-020-09947-4
10. Ramanathan, K.C., Mohan, M., Dhanraj, J.A.: Backward motion planning and control of multiple mobile robots moving in tightly coupled formations. Appl. Comput. Sci. **17**(3), 60–72 (2021)
11. Sutton, R.S., Barto, A.G.: Reinforcement Learning: An Introduction, second edn. Cambridge MIT Press, Cambridge (2018)
12. Szuster, M., Hendzel, Z.: Intelligent Optimal Adaptive Control for Mechatronic Systems. Springer, Heidelberg (2018). https://doi.org/10.1007/978-3-319-68826-8

Predicting Dynamics of a Rehabilitation Exoskeleton with Free Degrees of Freedom

Piotr Falkowski[1,2](✉)

[1] ŁUKASIEWICZ Research Network – Industrial Research Institute of Automation and Measurements PIAP, Warsaw, Poland
piotr.falkowski@piap.lukasiewicz.gov.pl
[2] Warsaw University of Technology, Warsaw, Poland

Abstract. The challenges of the modern world accelerated the development of many fields of technology, including rehabilitation robotics. The focus of the robot-aided motor therapy is put on providing remote, home treatment. According to the conducted research, also, functional kinesiotherapy is a leading trend for the future. Therefore, designing a mechatronic device enabling such a workout is needed. The presented study is based on a concept of an upper extremity exoskeleton with free degrees of freedom. However, it is applicable for any exoskeleton with non-controlled joints. The paper corresponds to the problem of a control system for a not fully controlled robot; especially, as the mass parameters of the user's limb remain unknown. Investigated prediction control is based on a recurrent neural network model of the system. The dynamics simulations are all performed without the usage of a physical device. Nevertheless, the outcomes of the trials prove, that such a control approach is suitable for the exoskeletons with free degrees of freedom. The results may be the base to adjust basic parameters of the neural networks used for similar applications.

Keywords: Artificial neural networks · Exoskeleton · Multibody dynamics · NARX · Rehabilitation robotics

1 Introduction

The COVID-19 pandemic was a key driver to changes in various areas; in an approach towards contemporary medicine, among others [1–3]. It particularly exposed weaknesses of current rehabilitation. The technological advancement in robotics is developed enough to organise physiotherapy, which should not require a therapist in the same room [4]. So far, it is not practised yet, due to the restrictive safety standards. Moreover, kinesiotherapy, dedicated mostly to people suffering from neurological and musculoskeletal disorders, is not available for everyone. Mostly, people from rural areas have limited access to professionals and specialist clinics located in big cities. Also, physiotherapists meet problems while performing their jobs. They are mostly connected with an objective assessment of progress, keeping patients motivated and the physical conditions required to work with heavier and paralysed people. A solution for these problems is

R. Szewczyk et al. (Eds.): AUTOMATION 2022, AISC 1427, pp. 223–232, 2022.
https://doi.org/10.1007/978-3-031-03502-9_23

implementing rehabilitation robots into physiotherapeutic practice. Due to their mechanical constructions, they may perform the same movements many times in a repeatable way and support a patient's extremity with force. Moreover, they are resistant to the impact of biological viruses and may be used remotely by a therapist [5]. However, this requires advanced safety systems and control algorithms; so to be compatible with users of different anatomic parameters.

78% of 130 questioned Polish physiotherapists pointed out function-oriented therapy as the most promising trend for the future. Such treatment is based on multi-joint mobilisation to perform typical physical activities. During a function-oriented workout, a patient follows a given trajectory with their extremity, so as to mimic everyday movements. Thanks to such an approach, not only effects are visible earlier, but also genuine neurological patterns are recalled. Moreover, the motivation of a patient determined to perform their favourite activities is easier to keep [6]. Nevertheless, combining function-oriented rehabilitation with robots is a challenge. Not only need they to activate more degrees of freedom (DOFs) at the same time, but also to measure more motion parameters, control a complex dynamic system in real-time, and assess the performance of a user, considering correlation with the anatomical patterns. This task is normally performed by a physiotherapist based on their visual observations and experience. Therefore, it is important to adequately model an assessment function in a mathematical way [7].

As to keep the exoskeleton lightweight and do not restrain the natural motion of the extremity, free degrees of freedom may be used. Thanks to this, fewer motors are involved. However, then the control system needs to compensate for the motion of a patient in real-time. This may be realised with the use of predictive control [8]. The paper addresses the problem of building the model based on recurrent artificial neural networks (ANNs) dedicated to this approach. To have an applicable value, the test sets are built to represent a short-time calibration, while the test set reflects a long usage within continuous therapy series. The following paper is a simulation proof of concept for using a predictive control for an exoskeleton with free DOFs and training the ANNs model based on a set obtained during a short calibration.

2 Methodology

A considered rehabilitation system is based on an exoskeleton. Thanks to adjustments it is suitable for users of anatomical dimensions between the 5th percentile of the Polish women population (5W) and the 95th percentile of the Polish men population (95M). These include not only lengths of arm, forearm and hand, but also their perimeters and mass parameters.

All the works are organised within MatLab R2020b and no physical model is used for validation. An exoskeleton is modelled with the *rigidBodyTree* structure as a multibody dynamics (MBD) system. It contains six rigid bodies (only two with mass parameters and transition transform not equal to zero), two point masses, and two joints to enable shoulder and elbow mobility (Fig. 1). The bodies' 4 and 6 mass parameters are calculated according to the Steiner's theorem for the considered sub-bodies remaining fully constrained one to another (extremity's and exoskeleton's elements). Corresponding rotation axes of the exoskeleton fully overlap the ones from the extremity model. It is described with physical parameters presented in Table 1 [9, 10].

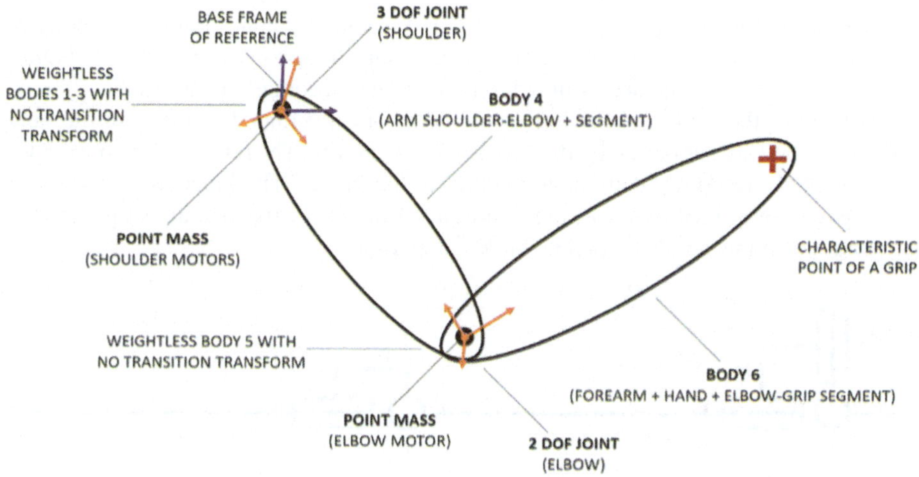

Fig. 1. Schematic rigidBodyTree structure used for the MBD simulation

Table 1. Biomechanical parameters of extremities

Body	Parameter	Value		
		W5	M50	M95
Arm	Mass [kg]	1.500	2.349	2.970
	Moment of inertia xx [kg m^2]	0.109	0.269	0.413
	Moment of inertia yy [kg m^2]	0.395	1.254	1.861
	Moment of inertia zz [kg m^2]	0.420	1.240	2.011
	Length [mm]	286	349	383
	Radius of centre of mass [mm]	125.84	153.56	168.52
Forearm	Mass [kg]	0.775	1.209	1.535
	Moment of inertia xx [kg m^2]	0.061	0.128	0.186
	Moment of inertia yy [kg m^2]	0.317	0.822	1.144
	Moment of inertia zz [kg m^2]	0.326	0.842	1.175
	Length [mm]	189	235	273
	Radius of centre of mass [mm]	81.27	101.05	117.39
Hand	Mass [kg]	0.500	0.780	0.990
	Moment of inertia xx [kg m^2]	0.024	0.125	0.159
	Moment of inertia yy [kg m^2]	0.067	0.125	0.159
	Moment of inertia zz [kg m^2]	0.054	0.103	0.130
	Length (to the grip) [mm]	115	117	125
	Radius of centre of mass [mm]	67.16	87.22	98.04

Electric motors are assumed to be fully and accurately controllable by setting an input torque for the system. The output of the controlled system, the DOFs' rotation angles of the exoskeleton, are assumed to be fully and accurately measurable.

As three of the DOFs require higher moments of force than the others for the same joints [11], they are supported by the exoskeleton's motors. The remaining two were left unconstrained by adding plain open bearings to the model. The kinematic structure of the exoskeleton in its based position is presented in Fig. 2. Its mechanical parameters and ranges of joints are described in Table 2 and Table 3.

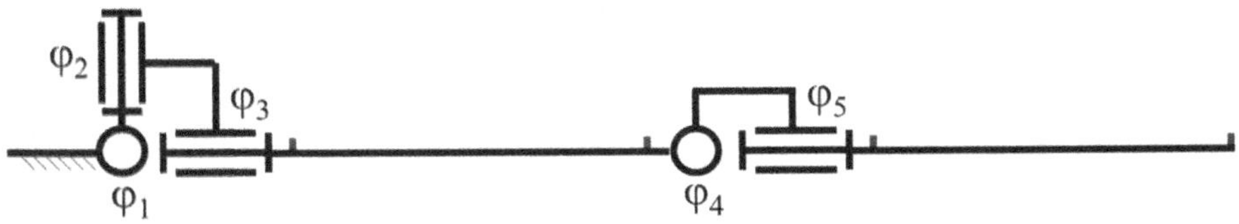

Fig. 2. Kinematic structure of an exoskeleton with corresponding rotations (red - attachments to the extremity)

Table 2. Parameters of the exoskeleton's elements

Body	Parameter	Value (M50)
Shoulder Motors	Mass [kg]	1.58
Shoulder – Elbow Segment	Mass [kg]	0.608
	Moment of inertia xx [kg m^2]	$7.4 \cdot 10^{-3}$
	Moment of inertia yy [kg m^2]	$7.4 \cdot 10^{-3}$
	Moment of inertia zz [kg m^2]	$0.16 \cdot 10^{-3}$
Elbow Motor	Mass [kg]	0.315
Elbow – Grip Segment	Mass [kg]	0.612
	Moment of inertia xx [kg m^2]	$7.58 \cdot 10^{-3}$
	Moment of inertia yy [kg m^2]	$7.58 \cdot 10^{-3}$
	Moment of inertia zz [kg m^2]	$0.16 \cdot 10^{-3}$

The exoskeleton may be described with the typical MBD equations with the non-identified mass matrices [12]. As the ANN should predict the behaviour of the system based on prior records and current input signals (torques), it has to involve a recurrent approach [13]. For this reason, a one-layer nonlinear autoregressive exogenous (NARX) model has been chosen [14]. Different structures of NARX ANNs, various in terms of the number of neurons, activation functions and internal parameters N_u [steps] and N_y [steps] were trained and tested (their considered values are presented in Table 4). The performance quality metric used for ANNs were a Mean Square Error *MSE* and Pearson correlation coefficient *R*.

Training set input $\boldsymbol{X}_{tr}$ contains a time series with 12000 samples. Every sample is a vector $3N_u$ x1 constructed as presented in formula (1), where $\boldsymbol{\tau}_i$ is a 3×1 vector of

Table 3. DOFs of a modelled exoskeleton

Rotation	Joint	Driven/Free	min./max
φ_1	Shoulder	Driven	−140°/90°
φ_2		Driven	−50°/130°
φ_3		Free	−50°/130°
φ_4	Elbow	Driven	0°/140°
φ_5		Free	−80°/80°

Table 4. Considered parameters of the ANNs

N_u	N_y	Neurons	Activation
5, 10, 20	3, 5, 10, 20, 30	10, 50, 100	Sigm, Lin

input motor torques in the ith (from 1 to 12000) time sample and N_u is a delay of control. The validation set consists of 10–15% of the training set.

$$X_{tr,i} = \begin{bmatrix} \tau_i \\ \vdots \\ \tau_{i-N_u} \end{bmatrix} \quad (1)$$

As to prepare a well-performing training set, first random signals $\boldsymbol{\varphi}$ (time-series of rotations of five DOFs) were generated for 1200 samples, as a simulation of 2 min calibration with a 0.1 s sample time. Then, an inverse dynamics simulation was run with a random set of mass parameters to compute torque $\boldsymbol{\tau}$ time series. The derivatives of signals $\boldsymbol{\varphi}$ were gathered from kinematics calculations.

Training set output $\boldsymbol{Y}_{tr}$ contain of 12000 samples. Every sample is a vector 15 × 1 constructed as presented in the formula (2). The $\boldsymbol{Y}_{tr,i}$ output corresponds to the $\boldsymbol{X}_{tr,i}$ input, as they are measured for the same time sample.

$$Y_{tr,i} = \begin{bmatrix} \varphi_i \\ \varphi_i \\ \varphi_i \end{bmatrix} \quad (2)$$

As all the values stay within similar ranges, they are not being normalised. The aim of training with such a set is to prove the possibility of modelling the forward dynamics with ANNs for a predictive control system.

3 Results

The results of the most important considered ANNs are presented in Table 4; where N_y is a delay of output. All of these were trained with the Levenberg-Marquardt algorithm within a maximum time of 5 h each. As the ANN 2 has not completed even one epoch by this time, it was stopped and not evaluated in the further stages. The assessment of performance included analyses of quality metrics, response and error plots (see example in Fig. 3). None of the responses significantly outstand from the desired time series. The quality of an ANN was assessed based on the minimum MSE values, as the R values were all approximately equal to 1 (see Table 5). However regression graphs were also analysed to validate whether the observations lay on the desired regression line (see example in Fig. 4); thus, to confirm whether an ANN is not overtrained and if all the width of the targets is estimated accurately.

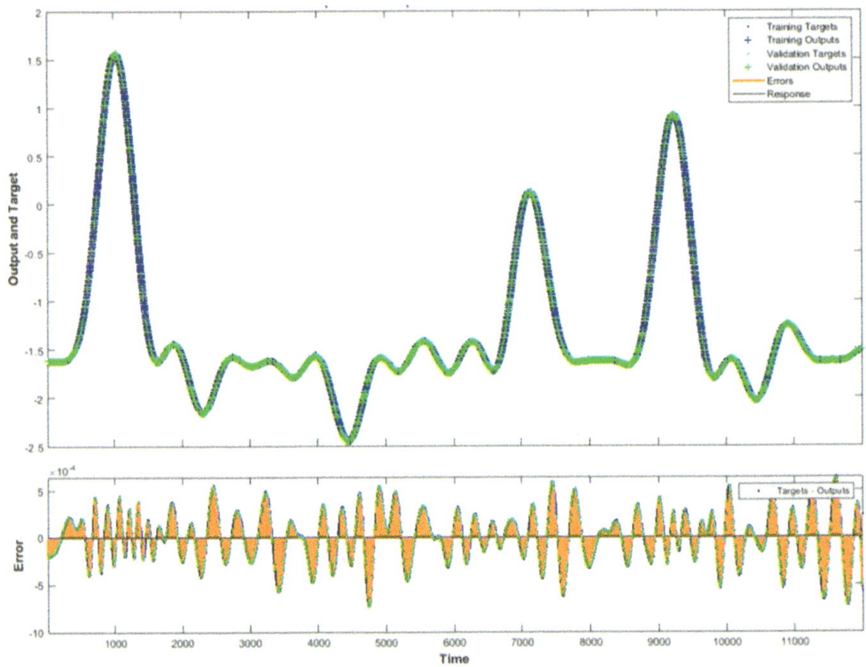

Fig. 3. Response of the ANN 3

The results obtained with the most important considered structures, mentioned before are presented in Table 6. Due to the poor performance, the ANN 1 was not considered an applicable structure. The ANN 2 was not analysed due to the too long computation time. Apart from these two, the rest of the presented ANNs' performance is relatively similar.

Table 5. Results of training of the chosen ANNs

No.	N_u	N_y	Neurons	Activation	MSE	R	Epoch
1	20	20	10	Sigm, Lin	1.28e-10	1	469
2	5	30	100	Sigm, Lin	0.384	–	1
3	5	3	50	Sigm, Lin	2.34e-7	1	20
4	5	3	100	Sigm, Lin	7.59e-12	1	64
5	5	10	50	Sigm, Lin	2.31e-7	1	14

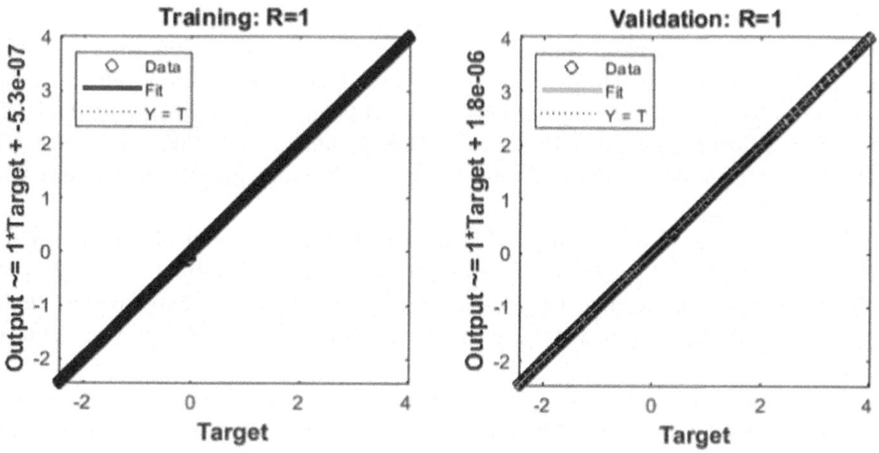

Fig. 4. Regression of the ANN 3

Table 6. Results of testing some of the ANNs

No.	MSE	R	Result
1	1.18e-1	0.23	Overfitted
2	–	–	Undertrained (too long time of training)
3	4.22e-4	0.99984	Well performing
4	1.70e-3	0.99936	Well performing
5	7.81e-4	0.99971	Well performing

4 Discussion

As proved with the experiments, NARX ANNs give a possibility to predict the behaviour of such a dynamic system, based on past information about its kinematics and applied inputs (here, the torques). Moreover, a 2-min calibration covering most of the device's workspace with the 0.1 s sampling is enough to precisely predict dynamics during the 10-h workout.

As predicted, the bigger neural networks give generally better results. However, they severely slow down the training process. Therefore, much smaller N_u and N_y may be used without the significant loss of precision. The quality of predicting first, second and fourth DOFs are relatively accurate for most of the N_u and N_y settings if the number of hidden neurons is high enough. However, change in the prediction of parameters related to the third and fifth DOFs, which are uncontrollable by the system, is clearly visible for different settings of N_y.

The outcomes are accurate even for relatively small delays N_u and N_y. As could be predicted, previously applied control signals does not affect behaviour on a system as much as the current ones. However, as some of the DOFs are non-controlled, the data on the previous states (here, the configurations and kinematics parameters) are relevant to predict the further movement of the exoskeleton. Hence, the higher values of N_y are desirable. However, they need to be raised significantly much over 10, which may slow down the learning process.

The largest relative errors may be observed for the prediction of rotation in free DOFs (see example in Fig. 5). Nevertheless, they are still acceptable for the control of a rehabilitation exoskeleton, as bigger inaccuracies may appear due to the elasticity of connections with the extremity. If necessary, reduction of these errors may be achieved by increasing N_y.

Application of such ANNs is especially beneficial, due to the fact that the mass parameters related to the robot's dynamics are not measurable. Moreover, the moments of inertia related to particular DOFs, so the motors, are dependent on the extremity's configuration. Hence, if not ANN, a decoupled model would have to be used. Such an approach is much more labour-consuming and does not guarantee accurate results, as has to be based on the estimated mass parameters of the user. Moreover, a control system not involving an ANN model cannot be effectively used for the exoskeleton with free DOFs, due to the lack of information on the non-controlled joints.

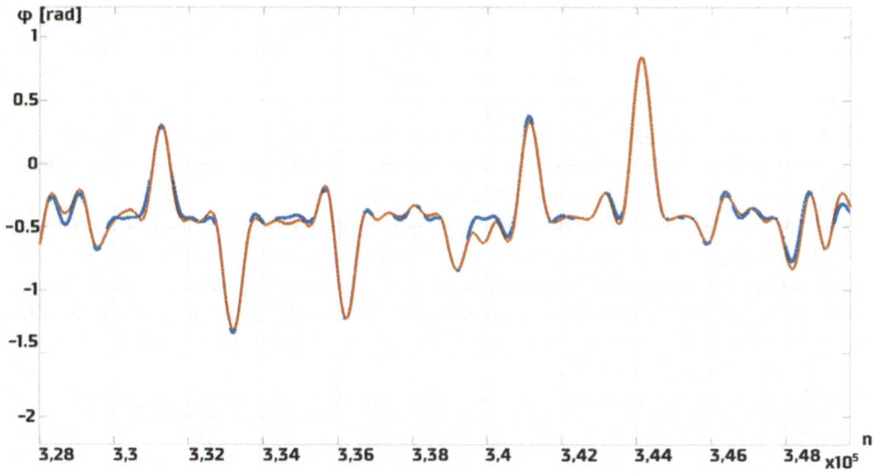

Fig. 5. Real (blue) and predicted (orange) φ_3 for ANN 3

5 Summary

In conclusion, the investigation proves that NARX ANN may be used for the long time prediction of the dynamics of an exoskeleton with free DOFs. Such an approach requires only a short calibration, which covers different combinations of input signals within their full scope.

Presented ANNs models 3–5 perform well. Based on assumed metrics, model 3 performs the best; however, model 5 if trained longer, could achieve the best performance. Nevertheless, as the error from all three models is acceptable, it is best to get one that requires the least time to train (model 3).

To achieve more statistically reliable results, further research should be conducted with the multiplication of training more configurations of ANNs. Also, different methods of training shall be considered. Finally, the simulated system should be replaced with the real one, to feed the ANNs with the data collected from experimental trials. Such a change will possibly decrease the performance of the proposed structures, as the inputs will be affected with noises of relatively high values (e.g. external forces, backlash within exoskeleton's joints or inaccuracies of measurements).

References

1. Dwivedi, Y.K., et al.: Impact of COVID-19 pandemic on information management research and practice: transforming education, work and life. Int. J. Inf. Manag. **55** (2020). https://doi.org/10.1016/j.ijinfomgt.2020.102211
2. Jnr, B.A.: Use of telemedicine and virtual care for remote treatment in response to COVID-19 pandemic. J. Med. Syst. **44**(7) (2020). https://doi.org/10.1007/s10916-020-01596-5
3. Bettger, J.P., Resnik, L.J.: Telerehabilitation in the age of COVID-19: an opportunity for learning health system research. Phys. Ther. **100**(11), 1913–1916. Oxford University Press (2020). https://doi.org/10.1093/ptj/pzaa151

4. Akbari, A., Haghverd, F., Behbahani, S.: Robotic home-based rehabilitation systems design: from a literature review to a conceptual framework for community-based remote therapy during COVID-19 pandemic. Front. Robot. AI (2021). https://doi.org/10.3389/FROBT.2021.612331
5. Manjunatha, H., Pareek, S., Jujjavarapu, S.S., Ghobadi ,M., Kesavadas, T., Esfahani, E.T.: Upper limb home-based robotic rehabilitation during COVID-19 outbreak. Front. Robot. AI (2021). https://doi.org/10.3389/FROBT.2021.612834
6. Thant, A.A., et al.: Effects of task-oriented training on upper extremity functional performance in patients with sub-acute stroke: a randomized controlled trial. J. Phys. Ther. Sci. **31**(1), 82–87 (2019). https://doi.org/10.1589/jpts.31.82
7. Nordin, N., Xie, S.Q., Wünsche, B.: Assessment of movement quality in robot- assisted upper limb rehabilitation after stroke: a review. J. NeuroEng. Rehabil. **11**(1), 1–23 (2014). https://doi.org/10.1186/1743-0003-11-137
8. Tahamipour, Z.S.M., Hosseini Sani, S.K., Akbarzadeh, A., Kardan, I.: An assistive strategy for compliantly actuated exoskeletons using non-linear model predictive control method. In: Proceedings of the 26th Iranian Conference on Electrical Engineering (ICEE), pp. 982–987 (2018). https://doi.org/10.1109/ICEE.2018.8472699
9. Batogowska, A., Słowikowski, J.: Atlas antropometryczny dorosłej ludności Polski dla potrzeb projektowania III. Warszawa: Instytut Wzornictwa Przemysłowego, Zakład Wydawnictw (1994)
10. McConville, J.T., Clauser, C.E., Churchill, T.D., Cuzzi, J., Kaleps, I.: Anthropometric relationships of body and body segment moments of inertia, December 1980. https://doi.org/10.21236/ADA097238
11. Guenzkofer, F., Bubb, H., Bengler, K.: Maximum elbow joint torques for digital human models. Int. J. Hum. Factors Model. Simul. **3**(2) (2012). https://doi.org/10.1504/IJHFMS.2012.051092
12. Sariyildiz, E., Oboe, R., Ohnishi, K.: Disturbance observer-based robust control and its applications: 35th anniversary overview. IEEE Trans. Ind. Electron. **67**(3), 2042–2053. Institute of Electrical and Electronics Engineers Inc. (2020). https://doi.org/10.1109/TIE.2019.2903752
13. Boyko, N., Pobereyko, P.: Basic concepts of dynamic recurrent neural networks development. ECONTECHMOD Int. Q. J. Econ. Technol. Model. Process. **5**(2), 63–68 (2016)
14. Xiu, Y., Zhang, W.: Multivariate chaotic time series prediction based on NARX neural networks, pp. 164–167 (2017). https://doi.org/10.2991/eame-17.2017.40

Intuitive Robot Programming and Interaction Using RGB-D Perception and CNN-Based Objects Detection

Bartłomiej Kulecki(✉)

Institute of Robotics and Machine Intelligence, Poznan University of Technology, Ul. Piotrowo 3A, 60-965 Poznan, Poland
bartlomiej.kulecki@put.poznan.pl

Abstract. Robots programming requires knowledge about a specific robot programming language and control interfaces. Also, re-programming the robot to perform a slightly different task requires domain-specific knowledge. These requirements limit the application of robots in small-scale flexible production plants where the robot should be re-programmed frequently. This paper presents a human-robot interface for programming a robotic arm with an RGB-D camera for pick-and-place tasks using natural language. We propose a novel architecture of the system that utilizes voice recognition, pointer, and CNN-based object detection modules to intuitively program robot motions. The created system allows the operator to program motion trajectories or choose an object for grasping using a visual pointer and voice commands.

Keywords: Human-robot interface · Robot programming · 3D perception

1 Introduction

Robotic arms are widely used for pick-and-place tasks. In repetitive tasks, robots are faster and more efficient than humans. Industrial robots are programmed once, and then they perform the same tasks for months or years. In contrast, robots used in flexible manufacturing should adapt to the changing production process. The robots equipped with RGB-D cameras can adapt to changing positions, sizes, and types of manipulated objects. Unfortunately, programming this kind of robotic system may be difficult for many human operators. It requires specific domain knowledge about the dedicated software and the programming language. The goal of this research is to develop a system that allows programming robots by operators using a set of voice commands. Additionally, we use a

This work was supported by the National Centre for Research and Development (NCBR) through project LIDER/33/0176/L-8/16/NCBR/2017 and Poznan University of Technology project number 04/45/SBAD.

R. Szewczyk et al. (Eds.): AUTOMATION 2022, AISC 1427, pp. 233–243, 2022.
https://doi.org/10.1007/978-3-031-03502-9_24

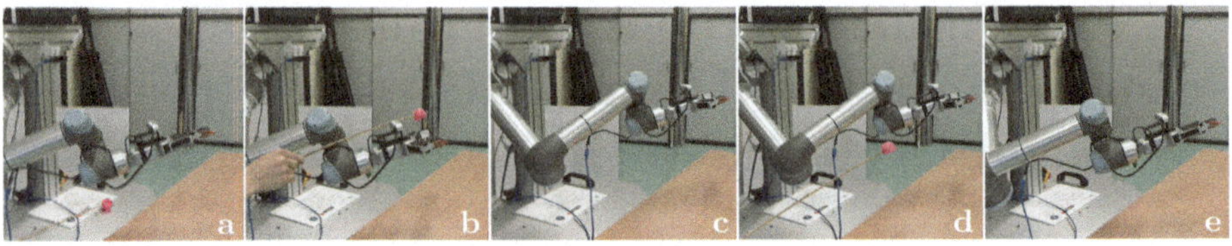

Fig. 1. Example of programming a simple trajectory: the robot moves to the initial position and saves the configuration (command *gtic*, *sic*) (a), the gripper opens (*og*, *sgc*), the operator uses the pointer and the command *gt* to set the goal position (b), the robot moves to the desired position, the configuration is saved (*src*) (c), the next goal is set by the pointer (*gt*) (d), the robot moves and the gripper closes (*cg*), the configuration and the trajectory are saved (*src*, *sgc*, *st*) (e)

visual marker and neural network-based object detector to define a sequence of the robot's motions.

In this research, we explore the problem of programming the cooperative robot (Fig. 1) using natural language. The main goal of the system is the interpretation of the operator's voice commands. Another challenge is to interpret the intentions of the operator from the camera images. The system produces a list of commands that are stored in the memory and later executed in the loop by the robot.

To deal with the problem of robot programming using natural language we utilize a visual pointer, voice commands, and neural-based objects detection. The operator uses the pointer (Fig. 1) to specify trajectory points or select objects to be gripped. Voice commands allow the system to trigger the desired robot actions such as opening or closing the gripper, moving to pose, or performing a pick-and-place task with the selected object. We also describe a grasp planning method based on the point cloud analysis. Finally, we show how the neural-based object detection module enhances the robot programming interface and improves human-robot interaction.

2 Related Work

Many human-robot interfaces use vision as a channel of communication. In application, where a robot has to grasp an object chosen by the operator, the pointing method is needed. A popular approach is the usage of a laser pointer and detecting the laser spot on the RGB image [6]. This solution is often applied in Wheelchair Mounted Robotic Arm (WMRA) systems [4]. The simple detection process and quick selection of objects from a long distance are the main advantages of using a laser pointer. However, using this method can be problematic to detect the laser spot on transparent or reflective objects. Other ways of indicating objects are pointerless methods like hand or gesture detection. One of the simplest solutions is hand detection based on the color of the skin [16]. A more advanced solution was proposed in [3] where the authors utilized the Leap Motion sensor and estimate the operator's hands' position and orientation with the Kalman and particle filters. In another work [11] we can find a method that uses RGB-D images to detect hand gestures and track whole body parts, which enables the operator to select objects on the floor with his whole arm.

One of the most natural ways to interact with the robot is verbal communication. In [10] a system is presented that controls a mobile robot with voice commands. The MFCC (Mel Frequency Cepstral Coefficients) are extracted as audio signal features. The DTW (Dynamic Time Warping) method, which is based on pattern matching, is applied for speech recognition. A similar and very popular approach is the utilization of the HMM (Hidden Markov Model) [1]. The voice commands recognition system, which is sensitive to the context of the user's utterance, was made with Fuzzy Neural Network in [2]. A commonly used solution to the issue of speech recognition is using ready-made software for this purpose, e.g. Microsoft or Google API [12], CMU Sphinx [14], or the wit.ai environment [12].

Another extensive issue that makes robots more intelligent is object recognition. The current state-of-the-art methods are based on the Convolutional Neural Networks (CNN). The examples are YOLO [13] and SSD [9]. The first one predicts the position of the bounding boxes on the image and classifies the object inside this region. A similar strategy is applied in the Single Shot Detector. It uses more convolutional layers than YOLO and operates on feature maps with various resolutions using additional computational resources. Another way of detection is provided by the Mask R-CNN [5] which performs segmentation inside the bounding box to determine which pixels belong to the detected objects.

In this research, we utilize recent advances in machine learning for objects detection and grasping to make the human-robot interface more efficient and intuitive for a human operator.

3 Robot Control System

All methods presented in this paper are implemented on our mobile-manipulating platform Robot 4.0 [7]. For our experiments, we use only the upper part of the robot - the Universal Robots UR5 robotic arm presented in Fig. 1. During experiments, we put various objects on the table in front of the robot. The robot is equipped with the Intel Realsense D435 camera (mounted on the wrist), Microsoft Kinect Xbox One (mounted on the head) for 3D environment perception, and the two-fingered OnRobot RG6 gripper. The Robot Operating System (ROS) integrates all modules into a common framework.

3.1 System Architecture

The architecture of the proposed system is presented in Fig. 2. The input of the system is the user's action - voice commands and pointer position. The pointer position is obtained from the RGB-D images. These images are also utilized to perform objects detection and scene segmentation. Recognized voice commands, pointer pose, objects' labels, and objects' models are used to plan the trajectory executed by the robot controller.

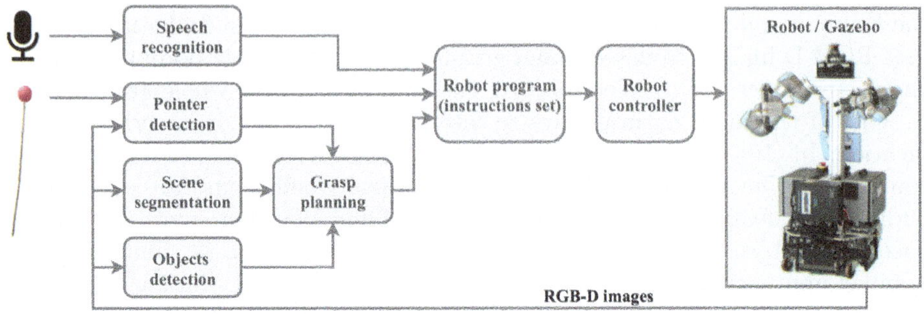

Fig. 2. Architecture of the proposed robot programming system

3.2 Scene Segmentation

In order to manipulate the objects in the robot's workspace, we need to obtain information about the environment, especially about the poses of the manipulated objects [7]. For this purpose, we process the point cloud obtained from the RGB-D camera. Processing a single frame includes ROI (region of interest) cropping and detecting the table plane using the RANSAC method. The detected plane is removed from the point cloud to extract points representing the objects. Because data obtained from a single perspective does not contain information about the full 3D shape of the object, the scanning procedure is executed. The robot collects scans from various poses around objects and accumulates point clouds. In the next stage, we perform Euclidean segmentation (based on Euclidean distance between points) to find instances of objects in the point cloud. The position is estimated for each cluster representing the object by calculating the centroid. Then, the Oriented Bounding Box (OBB) fitting is performed. We use Principal Component Analysis (PCA) to find three directions of the object's point cloud with the greatest variance. These directions correspond to the edges of OBB. Example OBB are presented in Fig. 3b.

3.3 Objects Grasping

For grasp planning, we use the method that fits the OBB to the point cloud representing the object [7]. The centroid of the object is used as a gripper target position. Orientation of the gripper depends on the OBB dimensions and alignment. If the height of the object's OBB is smaller than the width and length, the grip is performed vertically from the top. If the height of the box is larger than the remaining dimension, the robot tries to grasp the object horizontally from the side. The distance between the end-effector and the TCP depends on the gripper opening. We estimate gripper opening using the OBB dimensions. While closing the gripper on the object the force feedback is used. If the reaction force is larger than the threshold, the gripper stops the motion. This strategy compensates for the inaccuracies of the perception system, stabilizes the grasp, and increases the success rate of the grasping method.

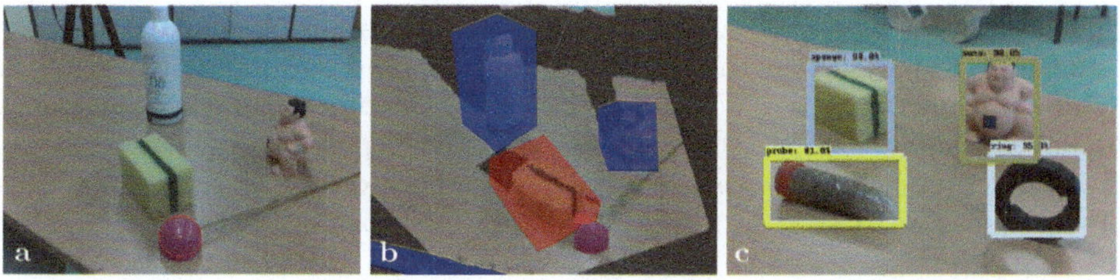

Fig. 3. Example result of pointer detection (a), object detection (c), and Oriented Bounding Boxes with chosen object marked in red (b)

3.4 Pointer Detection

The main task for the pointer detection module is to find the pointer on the image and estimate the 3D pointer position. The input of this module is the RGB image and the point cloud. We find the pointer on the image using the color and the shape. In the first step, we convert the RGB image to the HSV space. Then, we create a mask that extracts the pixels from the image that are inside the specific Hue-Saturation-Value ranges corresponding to the pointer color. Next, the resulting image is converted to grayscale, thresholded and blurred. After that, we utilize Hough transform to find the circle whose center is the position (x_{px}, y_{px}) of the pointer on the image. An example image with a marked pointer is presented in Fig. 3a. To obtain the 3D position of the pointer, the point cloud data is used. First, we calculate index of the point in the cloud:

$$index = x_{px} + y_{px} \cdot cloud.width \tag{1}$$

The index value allows us to find the XYZ values that represent the estimated coordinates of the pointer in the 3D space.

The obtained pose of the pointer is utilized to check which object is pointed by the operator, as shown in Fig. 3b. For this purpose, the Euclidean distance between the pointer and each object is calculated. If the minimum distance is smaller than the required threshold value, the selected object is found.

3.5 Object Detection

An alternative way to select the target object that we implemented is to include its name in a voice command. Therefore, an object detection procedure is required. A Single Shot Detector (SSD) based on the neural network Inception V2 architecture [15] has been used to find bounding boxes on the images. The detector was pre-trained on the MS COCO dataset [8]. We prepared a set of 10 objects, including a sumo figurine, a probe, a dispenser, and a sponge. The images (1500 samples) were collected and manually labeled with bounding boxes and object classes to create a training dataset. Example result of inference of the neural network used in this work is presented in Fig. 3c.

To find the object selected by the operator the recognized voice command is compared to detected objects' names. If the user's utterance contains any of the detected objects' names, the 3D pose of the object is estimated. The pose estimation process utilizes the position of the bounding box on the image, depth data, and camera matrix. When the position of the object is known, it is used to find associated OBB. We compare detected object position with the positions of the determined OBB in the same way as the position of the visual pointer as described in Sect. 3.4.

3.6 Voice Commands Recognition

The part of the system responsible for converting speech to text is very important from the point of view of the user's communication with the robot. We decided to use an of-the-shelves module designed for this purpose. The SpeechRecognition library allows us to easily use various engines for speech-to-text processing, such as Google Speech Recognition, Microsoft Azure Speech, Wit.ai, or CMU Sphinx. The sound from the microphone is recorded in the loop with a defined interval (5 s by default). The recording is saved to a wave file and then passed to the voice recognizer from the SpeechRecognition library. In the next step, the audio is sent to the server with the selected engine (Google Speech Recognition). If the processing on the server is successful, the resulting text is received as a response, and then sent to the main controller of our system.

In the proposed system, voice commands are used to trigger the appropriate action of the robot. The set of available commands and associated actions are presented below. Each command was implemented with a few equivalent versions which contain synonyms:

1) ***go to initial configuration (gtic)*** - the robot moves to predefined initial config.
2) ***open / close hand (og/cg)*** - the robot's gripper opens/closes
3) ***save robot configuration (src)*** - saving the current configuration of the robot arm (all saved robot configurations are listed)
4) ***save init configuration (sic)*** - saving the initial config. of the robot arm
5) ***save gripper configuration (sgc)*** - saving the actual config. of the gripper
6) ***save trajectory (st)*** - saving the robot's trajectory (list of saved robot configurations) and clearing the configurations list
7) ***go there (gt)*** - the robot arm moves to the position defined by the visual pointer; the pointer determines the given position of the wrist; only the translation of the end effector is taken into account
8) ***look down/front (ld/lf)*** - setting the end effector pitch angle
9) ***clear/execute program (cp/ep)*** - the motion program is cleared or executed
10) ***inspection / scanning*** - the robot starts scanning the environment (the table and objects)
11) ***grasp*** - the robot plans and executes the trajectory to selected object and closes the gripper
12) ***grasp + object name*** - the robot plans end executes the trajectory to selected object (the name of the object is included in the command) and closes the gripper

13) ***put down/release*** - the robot plans end executes the trajectory to predefined position (where object has to be placed) and opens the gripper
14) ***move looking down*** - the robot moves to the predefined configuration with the gripper tilted down

4 Results

The experiments with the implemented system were carried out in three various scenarios. In the first one, we tested the programming robot's moves and whole trajectory with the use of the pointer and voice commands. For this purpose, commands from 1 to 9 described in Sect. 3.6 were used. The aim was to compare our programming method with the programming methods available in the teach pendant. During this test, the operator was moving the pointer to various places and used a sequence of proper voice commands to program the robot. The three example sequences are described in Table 1. The example programming process is illustrated in Fig. 1. When the program is saved it can be executed after command *"execute program"*.

For the proposed method of robot programming, we measured the time required to program described trajectories. We compared the results with the time of programming the same move sequences in the traditional way only from the teach pendant. Also, the time for programming by a demonstration was measured. In this way, some of the commands (like gripper opening/closing or init pose configuration) were programmed from the panel, but waypoints were saved after placing the arm manually in the target pose (teaching by doing). The results are presented in Table 2 (seq 1–3).

The measured times for our method depend on the performance of the voice recognition system. First of all, the audio is recorded in the loop, so the operator has to start his utterance at the right moment. Secondly, the similar sound of some words (e.g. *"save"* and *"safe"*) can cause recognition mistakes, so there are many situations when the command has to be repeated 2–3 times. To obtain a successful speech-to-text conversion, the operator has to speak quite loud, clear, and with the correct accent. The approximated recognition time of single record is 6 s (5 s of recording and about 1 s for response from recognizer server). It can be seen that programming trajectories with a pointer and voice commands take more time, but it is easier than the classic ways of programming,

Table 1. Sequences of commands used during experiments.

Seq no.	Commands
1	gtic → sic → cg → sgc → src → gt → src → og → sgc → st
2	gtic → sic → og → sgc → src → gt → src → gt → src → cg → sgc → st
3	gtic → sic → sgc → src → gt → src → gt → src → gt → src → cg → src → sgc → gtic → sic → st
4	*grasp* (with pointer) → *release* → *move looking down*
5	*grasp sponge* (with object detection) → *release* → *move looking down*

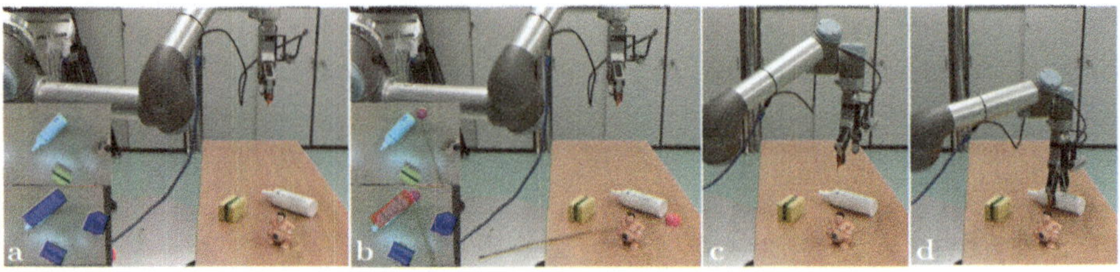

Fig. 4. Example of grasping with the pointer and voice commands: the robot's camera is focused on the objects (a), when the pointer appears in the image the selected object is marked in red (b), after voice command "grasp" the trajectory is planned and executed (c), when the goal is reached the object is grasped (d)

Table 2. Comparison of programming time of example move sequences for 3 programming methods

Seq.	Commands no.	Programming time [s]		
		Pointer + voice	Teach pendant	TbD
1	10	195	162	100
2	12	220	184	113
3	16	270	235	155
4	3	62	–	–
5	3	67	–	–

which require good teach pendant operating skills. In the proposed method, the operator should only know a set of available voice commands.

In the second scenario, the operator used the pointer and voice commands to program the robot to complete the simple pick-and-place task. For this experiment, we placed 3–4 objects on the table in front of the robot. The first stage was scene processing aimed at obtaining objects' models. This process is started by operator's command: *"inspection"* or *"scanning"*. When the scanning is done, the command *"move looking down"* lets the robot arm move to the configuration where the table is observed (by the wrist-mounted camera) from the top. Next, the operator places the pointer close to the selected object and says command *"grasp"*. When the trajectory is planned and executed, the gripper closes. If the grasp succeeds, the operator's command *"put down"* or *"release"* starts placing the object in a predefined position on the table. After that, the operator can move the robot to the home position and repeat the process for another object. The example of grasping the bottle is presented in Fig. 4.

In this scenario, commands sequence no. 4 from Table 1 was used. We measured time from saying the first command to the end of the trajectory, results are presented in Table 2. In this sequence, we skip the scanning process which is time-consuming (takes about 90 s for robot velocity limited to 40%), but it is performed only once for the scene independently from the number of objects

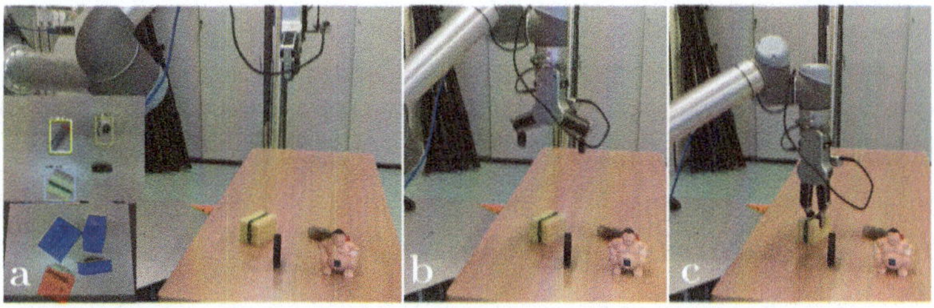

Fig. 5. Example of grasping with the object detection and voice commands: the robot's camera is focused on the objects which are detected by the CNN, after command "grasp sponge" the selected object is marked in red (a), the trajectory is planned and executed (b), when the goal is reached the object is grasped (c)

Table 3. Time measurement results of the pointer detection process and the system response to voice commands

Measurement category	Measurement description	Mean time [s]	Meas. no.
Pointer detection performance	2D pointer detection	0,008	1000
	3D pointer pose estimation	0,028	1000
Response times	From voice command to the end of detected object's pose estimation	7,6	150
	From command "grasp" (with pointer) to the robot movement beginning	7,2	10
	From command "grasp + object name" to the robot movement beginning	14,3	10

on the table. Time performance of important parts of the system is described in Table 3. Also, the pointer detection success rate was checked. For a test consisting of 1000 image frames, it was detected 925 times (92,5% efficiency). The failures appeared when the pointer was partially covered or moved fast.

In the last scenario, we used voice commands and the object detection method in the same pick-and-place task. The test starts with the same scanning process as before. The operator selects an object with the voice command, for example, *"grasp sponge"*. When the command is recognized, the trajectory is planned and executed. After picking the object user is able to give the "place" command and repeat the process for another object. The example of the described experiment is presented in Fig. 5. In this case, we used sequence no. 5 (Table 1) and measured the same time as in the second scenario (the scanning time was skipped). The time performance is presented in Table 3.

Classic methods of programming robots do not allow programming in the context of operating selected objects. Using the pointer for object selection is a bit faster than detecting the object mentioned in the voice command. This is due to the time of selected object pose estimation being longer than pointer pose estimation. The reason is that the first one consists of double inference time - first for checking labels of available items, second for getting selected object pose. However, the advantage of this intuitive and natural interface outweighs the slight increase in time.

5 Conclusions and Future Work

In this paper, we show that the visual pointer, voice commands connected with object detection, 3D perception, and grasp planning enables the operator to program the robot at a high abstraction level. The robot programmed once can save the trajectory and repeat the movement. A significant improvement of the system lies in the neural-based object detector, which simplifies objects selection. Another advantage of the proposed system is that trajectory planning is done in the 3D space with collision avoidance (using MoveIt).

In the future, we are going to extend the presented system and program the robot by observing an operator and mimicking the observed actions.

References

1. Azargoshasb, S., Korayem, A.H., Tabibian, S.: A voice command detection system for controlling movement of scout robot. In: 2018 6th RSI International Conference on Robotics and Mechatronics (IcRoM), pp. 326–330 (2018)
2. Chatterjee, A., Pulasinghe, K., Watanabe, K., Izumi, K.: A particle-swarm-optimized fuzzy-neural network for voice-controlled robot systems. IEEE Trans. Ind. Electron. **52**(6), 1478–1489 (2005)
3. Du, G., Zhang, P.: A markerless human-robot interface using particle filter and kalman filter for dual robots. IEEE Trans. Ind. Electron. **62**(4), 2257–2264 (2015)
4. Gualtieri, M., Kuczynski, J., Shultz, A.M., Ten Pas, A., Platt, R., Yanco, H.: Open world assistive grasping using laser selection. In: 2017 IEEE International Conference on Robotics and Automation (ICRA), pp. 4052–4057 (2017)
5. He, K., Gkioxari, G., Dollár, P., Girshick, R.: Mask r-cnn. In: 2017 IEEE International Conference on Computer Vision (ICCV), pp. 2980–2988 (2017)
6. Kemp, C.C., Anderson, C.D., Nguyen, H., Trevor, A.J., Xu, Z.: A point-and-click interface for the real world: Laser designation of objects for mobile manipulation. In: 2008 3rd ACM/IEEE International Conference on Human-Robot Interaction (HRI), pp. 241–248 (2008)
7. Kulecki, B., Młodzikowski, K., Staszak, R., Belter, D.: Practical aspects of detection and grasping objects by a mobile manipulating robot. Ind. Robot **48**(5), 688–699 (2021)
8. Lin, T.-Y., et al.: Microsoft COCO: common objects in context. In: Fleet, D., Pajdla, T., Schiele, B., Tuytelaars, T. (eds.) ECCV 2014. LNCS, vol. 8693, pp. 740–755. Springer, Cham (2014). https://doi.org/10.1007/978-3-319-10602-1_48
9. Liu, W., et al.: SSD: single shot multibox detector. In: Leibe, B., Matas, J., Sebe, N., Welling, M. (eds.) ECCV 2016. LNCS, vol. 9905, pp. 21–37. Springer, Cham (2016). https://doi.org/10.1007/978-3-319-46448-0_2
10. Lv, X., Zhang, M., Li, H.: Robot control based on voice command. In: 2008 IEEE International Conference on Automation and Logistics, pp. 2490–2494 (2008)

11. de la Puente, P., Fischinger, D., Bajones, M., Wolf, D., Vincze, M.: Grasping objects from the floor in assistive robotics: real world implications and lessons learned. IEEE Access **7**, 123725–123735 (2019)
12. Putthapipat, P., Woralert, C., Sirinimnuankul, P.: Speech recognition gateway for home automation on open platform. In: 2018 International Conference on Electronics, Information, and Communication (ICEIC), pp. 1–4 (2018)
13. Redmon, J., Farhadi, A.: Yolov3: an incremental improvement. arXiv (2018)
14. Sharan, S., Nguyen, T.Q., Nauth, P., Araujo, R.: Implementation and testing of voice control in a mobile robot for navigation. In: 2019 IEEE/ASME International Conference on Advanced Intelligent Mechatronics (AIM), pp. 145–150 (2019)
15. Szegedy, C., Reed, S., Erhan, D., Anguelov, D., Ioffe, S.: Scalable, high-quality object detection (2015)
16. Theis, C., Iossifidis, I., Steinhage, A.: Image processing methods for interactive robot control. In: Proceedings 10th IEEE International Workshop on Robot and Human Interactive Communication, ROMAN, pp. 424–429 (2001)

Seamless Multi-platform Tracing: Shadow Builder

Alexandre Malki(✉), Tomasz Kołcon(✉), and Mateusz Maciaś

Security and Defence Systems Division, Łukasiewicz Research Network - Industrial Research Institute for Automation and Measurements PIAP, Al. Jerozolimskie 202, 02-486 Warsaw, Poland
{alexandre.malki,tomasz.kolcon,mateusz.macias}@piap.lukasiewicz.gov.pl

Abstract. Tracing software application is a mandatory step toward producing quality code and software. Many of the tracing tools exist, with their set of advantages and limitations. This variety of tools forces developers to introducing readability over-head and additional software glues. This paper presents a solution for reducing code branching to obtain a multi-platform, cleaner code base along with a comparison to the already existing tools.

Keywords: Tracing · Multi-platforms

1 Introduction

Developing and working on stabilizing an application from the scratch to the final executing binary is a long and challenging task. Developers may come across stability and performances issues. In addition, some specified Quality of Services (QoS), essential to prove the relevance of a software solution, might be difficult to quantify. This was the case during the OFERA project [1], in which the task was to adapt the ROS2 (Robot Operating System 2) framework running on a Linux x86 platform to microcontroller. Depending on which the QoS to measure, one tracing tool was more profitable than another.

A set of selection criteria were established to narrow down the list of tools:

- Compatibility with the targeted platform.
- Tool's tracing purposes (e.g. network, execution, etc.)
- Popularity of a tool.
- Performances of the tracing tool
- Developer's familiarity with the tool.

Each of these criteria is difficult or impossible to meet by the existing tracing tools. Therefore, it was necessary to create a new tool. This paper provides a state of the art of the existing tools. The general architecture of the created tool will be discussed. Then the concepts and implementation of a tool called Shadow Builder (SB) [2] are discussed in detail. Finally, an example of using the tool to show its potential will be presented.

R. Szewczyk et al. (Eds.): AUTOMATION 2022, AISC 1427, pp. 244–250, 2022.
https://doi.org/10.1007/978-3-031-03502-9_25

2 Tracing State of the Art

There are many types tracing tools available for every type of application, platform, operating systems etc. In regards to the micro-ROS (OFERA) project [1], different QoS needed to be measured and traced on resource constrained microcontroller. There are many popular tracers and formats: LTTng (Linux Tracing Tool) [3], perf [4], kernel probes [5], eBPF (Berkeley Packet Filter) [6], Common Trace Format (CTF) [7], PerfView [8]. Within the above mentioned tracers, common features are listed in the Table 1.

Table 1. Tool format criteria selection

Tool	Platform(s)	Goal(s)	Popularity
LTTng	Linux	Tracing tools	++
perf	Linux	Profiling	+++
eBPF	Linux	Network	+++
PerfView	Windows	Event tracing	++
Tracealyzer	FreeRTOS	RTOS profiling	+

Among the tools listed above, some provide hardware or low-level kernel tracing information. After some code review, the Table 2 shows porting effort for tools providing low-level details information.

Table 2. Porting effort criteria

Tool	Porting difficulty
perf	Difficult
eBPF	Difficult
PerfViewer	Difficult
Tracealyzer	Difficult

All tools listed in the Table 2 are tied to a specific kernel. In the case of *eBPF, perf, PerfView, Tracealyzer*, the effort to port tracing facilities from one OS is enormous. For instance, the *eBPF* tools make use of a *Just In Time* compiler available in the kernel. Additionally, the tools *perf, PerfView* and *Tracealyzer* tools are also relying on the kernel instrumentation format and mechanisms as well. Porting is possible however it would be long or maybe an impossible task when targeting microcontroller.

Kernel and hardware tracing are not the only tracing methods available. User space application tracing is also needed in some cases. These methods often introduce additional overhead code. The most popular one is called *LTTng*. For instance, to trace and monitor a simple string variable using *LTTng*, a considerable amount of lines are added.

Additional tools - the data collectors - aim to converge the multitude of data produced by the tracers. OpenTracing [9], Fluentd [10] are examples of data collector/filters.

They answer one of the common recurring issues which is the multitude of trace formats. The trace collectors (also called data collectors or logging collectors) are useful to filter specific trace patterns. Also they provide a way to translate the trace from a specific format to another. Despite the improved flexibility, data collectors do not address all the tracers' issues.

In the next section, the paper presents an architecture developed to overcome the issues listed above.

Many tools offer different interesting tracing solutions for one specific platform. However, these tools are inherently suffering from being linked to one OS and/or complexify the source code base. Therefore, within the frame of the OFERA project [1] the creation of a new tool was necessary. It had to address the key requirements as listed below:

- Re-use of the available tracing tool.
- Multi-platform support.
- Different trace formats.
- Keeping the code readable and clean.

To solve these key points, the SB shall provide ways to:

- Unify existing tracing tools.
- Be platform independent.
- Reduce developing resources.
- Improve code base clarity.
- Not add execution overhead.

3 Architecture and Implementation

This section introduces the global architecture and its implementation called the Shadow Builder. The high-level block diagram from Fig. 1 shows the SB and its components.

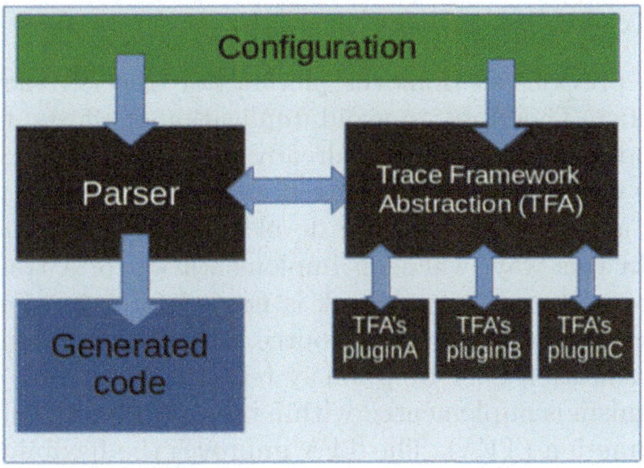

Fig. 1. Shadow Builder - Overall description

3.1 Code Platform Independent

Once the tracing tool(s) are selected, the execution scope might be limited to the platform supported by the tracing tool. Therefore, it is necessary to provide a way to limit this dependency. The solutions provided to clear this issue are:

- Non-intrusive code syntax.
- Configurable code preprocessing.

A non-intrusive code syntax is a way of removing unnecessary code reserved for tracing only. Using such syntax offers a clarification on what is and what is not relevant for proper execution of the functions.

The idea behind configurable preprocessing is to provide a way to generate from the code base when a tracing beacon is parsed. In this way, the original code written by the developers is untouched regardless of the platforms it's executed on.

In practice, the SB implements a unique syntax wrote in code commentaries. The format of the SB's unique syntax is named the Tag, and format in C/C++ is the following:

```
/** XX::YY::Function */ .
/** XX::YY::Function(X1,...,Xn) */
```

A *tag's* string is constructed of 3 parts. According to the example above, XX and YY can represent respectively the type of tracing (i.e. Benchmarking etc.) and the tracing category used (i.e. Timer, User etc.). The last part of the *tag*, is the function that the plugin (here module YY) is going to be replaced by.

The *tag* is browsed by a parser. The parser is a back-end tool that indicates to the Trace Framework Abstraction module, detailed later, that a *tag* was found in the source code. It should be noted that the parser is modular and while the current version implements a parser for C/C++, it is possible to adapt it to any other programming language.

3.2 Tracing Tool Abstraction

As mentioned in previous sections the plethora of tracers, tracing formats and data collectors exist. Therefore, to avoid duplication of efforts, the new tool had to build upon and leverage what was already available.

To do so, the tool has to give the freedom to specify what tracer should be used. Therefore, a framework must be developed in order to interface the tool with the code. In this way, when an implementation of a tracer is interfaced with the framework, no additional work is needed for future projects to re-use it. Moreover, because the tool is open-source, the whole community can benefit from this implementation once its publicly released.

Such a mechanism is implemented within the SB under the name of the Trace Framework Abstraction (TFA). The TFA improves the flexibility, increase code re-usability and give a common interface to use different tracers. This is achieved by relying on the use of plugins. Each plugin is in charge of decoupling the code base from the actual tracer's function. Decoupling is done by a set of rules and mechanisms which each plugin has to comply with. The minimal information that the plugin needs to provide is:

- The plugin's name.
- The platform it supports.
- The list of functions it will respond to and the corresponding callbacks in charge of replacing the Tag with a specific piece of code.

Each callback that the plugin provides to the TFA is tied to a *tag*. The TFA is in charge of executing the plugin's callback when the corresponding Tag is found and the configuration requirements are met.

3.3 Reduced Overhead

Aiming for low overhead is one of the corner stones for the SB. Indeed, the SB aims to reduce the additional code needed to write the application. However, from the tracing tool designer's perspective, the metric used for quantifying overhead is the execution. Though minimizing overhead execution is crucial, the idea of a cleaner code is often left aside and therefore some basic function can become crowded with tracing related function call.

Readability can be increased by introducing a non-intrusive code syntax. As for the execution overhead, it is as low as ensured by the tracer tool chosen.

3.4 Configurability

One of the requirements was also to provide a multi-platform support. The TFA provides a plugin that supports different platforms. However, one may want to get only timer information for ARM baremetal platform. Thus, the SB needs to be specified with what should be generated and what should not. Therefore, before generating a specific code, the SB is configured via a file, with the following parameters:

- An operating system targeted.
- Trace functions expected to be used (Timer, variable monitoring ...).
- Plugins available to be used.

This configuration can limit the number of traces provided by activating/deactivating and/or limiting the scope to selected plugins.

4 Concrete Workflow

In this section, a real life scenario is demonstrated. The scenario encompasses observation of the value of a variable after the execution of a function.

The code has the capability to run on a baremetal microcontroller (STM32F4) and a fully fledged x86 computer. Figure 2 is depicting the global code generation flow.

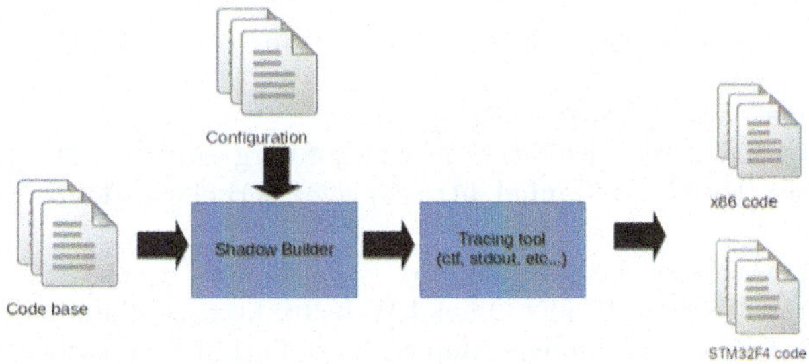

Fig. 2. Code generation flow.

In the case that plugins already exist, to use the SB the developer shall proceed as follows:

- Find a plugin that meets the platform requirement.
- Put the corresponding Tag in the code.
- Put in the configuration file of which plugins to use and which platform to run on.
- Execute the SB.
- Compile the binary with the code generated for each platform.

In the case that no plugins exist, then two additional tasks have to be added:

- Create a plugin that meets the requirement for the platform.
- Document the plugin to increase re-usability.

Even though the developer has to spend additional times to understand and write a plugin, a team of developers can work on this task in parallel. The development of plugins only has to be done once and re-used as many time as needed. Additionally, plugins can be made publicly available to leverage the open source community.

An example of use can be found in the project's repository on GitHub [11].

5 Conclusions

In this paper, the architectural design and the implementation of a seamless tracing tool were presented. This seamless tracing tool called the SB brings a solution to the inherent problems that the tracers suffer from. But instead of acting as a yet another tracer, the SB rather completes the tracing processes and provides a way to use different tracing tools seamlessly.

Acknowledgments. The project leading to this results has received funding from the European Union's Horizon 2020 research and innovation programme under grant agreement No. 780785.

References

1. Ofera Project Official Website. http://www.ofera.eu/
2. Shadow Builder Repository. https://github.com/micro-ROS/benchmarking_shadow-builder
3. LTTng Official Website. https://lttng.org/
4. GNU Gperf Official Website. https://www.gnu.org/software/gperf/
5. Kernel Probes (Kprobes) Manual. https://www.kernel.org/doc/html/latest/trace/kprobes.html
6. eBPF Official Website. https://ebpf.io/
7. Common Trace Format (CTF) Official Website. https://diamon.org/ctf/
8. Windows PerfView Performance Analysis Tool Github Repository. https://github.com/microsoft/perfview
9. Open Tracing Website. https://opentracing.io/
10. Fluentd Website. https://www.fluentd.org/
11. Shadow Builder Repository - Example. https://github.com/micro-ROS/benchmarking_shadow-builder/tree/master/examples/example_loop_main

Vision Based Navigation Securing the UAV Mission Reliability

Tomasz Pogorzelski[1] and Teresa Zielińska[2(✉)]

[1] Łukasiewicz Research Network – Institute of Aviation, Warsaw, Poland
[2] Institute of Aeronautics and Applied Mechanics, Warsaw University of Technology, Warsaw, Poland
teresaz@meil.pw.edu.pl

Abstract. Redundant location is essential for the reliable mission completion by un-manned aerial vehicles. It plays a key role, e.g., in cities where radio signals are particularly susceptible to interference, reflections and where the transmission of satellite location signals may be temporarily interrupted. This article re-views the publications on this topic. The commercial versions of visual navigation systems are also addressed, is worth emphasizing that they are available in both closed and open versions (both in terms of hardware and software). Basis on the state of the art the concept of redundant localization system is presented, the hardware and software configuration are briefly described. The method of generating data with the use of simulators was also discussed. This improves system testing. At the end, the preliminary results of the research were shown as a graphical visualization of the information collected by the camera and the corresponding maps were also shown. Finally, conclusions and directions for further work are presented.

Keywords: Vision-based navigation · UAV · GPS-denied environment · Absolute localization

1 Introduction

Unmanned aviation is one of the faster growing areas of engineering research. As a result of friendly legal regulations and technological solutions supporting aerial operations, UAVs are used in many applications, such as various types of monitoring, or energy inspection of buildings [1], detection of events and anomalies, or acquisition of precise images for photogrammetry and remote sensing [2]. The need for reliable navigation is also associated with the development of Unmanned Aerial Vehicle (UAV) applications for transport purposes.

Currently, UAV navigation is based on GNSS (Global Navigation Satellite Systems) [3], which, in the event of a loss of communication or lack of access to the service, makes it impossible to control UAVs efficiently. In view of this fact, there is a need to ensure flight safety and uninterrupted navigation by introducing redundancy. This need also arises from the vulnerability of the traditional satellite navigation system to jamming/spoofing. Due to the expected universality of the solution, navigation methods that do not require additional external infrastructure are preferred.

R. Szewczyk et al. (Eds.): AUTOMATION 2022, AISC 1427, pp. 251–263, 2022.
https://doi.org/10.1007/978-3-031-03502-9_26

One possible solution to ensure continuous navigation for UAVs without the need for additional external infrastructure is a location system using image analysis acquired from on-board cameras. In this case, information is processed inside the UAV, making such the system immune to external jamming and spoofing, which significantly increases the security and reliability of operations. This is critical for Beyond Visual Line of Sight (BVLOS) operations for automated missions.

2 State of the Art

Scientific publications have been reviewed to identify similar solutions. Publication [4] presents an extensive literature review of vision-based UAV navigation methods and the division of navigation depending on the availability of maps of the surroundings. As shown in the Fig. 1, map-based systems are divided into those that use pre-built maps and those that build such maps. Systems that use maps can determine location either globally or incrementally. Map building systems use SLAM (Simultaneous Location and Mapping) methods, mapless systems perform location based on the location of environmental features or optical measurements.

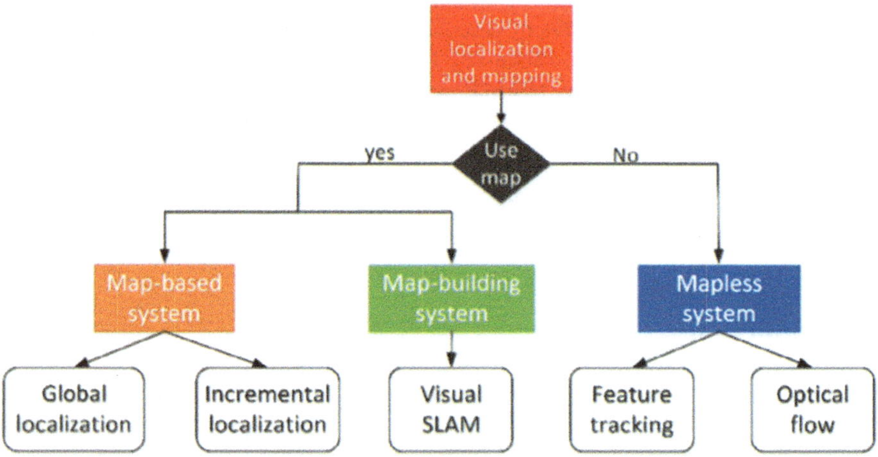

Fig. 1. Classification of vision location systems. Source: [4]

The publication [5] describes the navigation method using photographs of the terrain. The authors used here a neural network to analyze the aerial photographs. The deep neural network looks for some characteristic features in the on-line images of the terrain. Then, the location of objects with these features is determined in the known off-line maps in that way delivering its geographic coordinates. Unfortunately, the authors did not provide the exact architecture of the applied neural network, but they only discussed the test results obtained for three cities in Pakistan. A total of 2052 aerial photographs were used covering an area of 2 km^2 (Fig. 2).

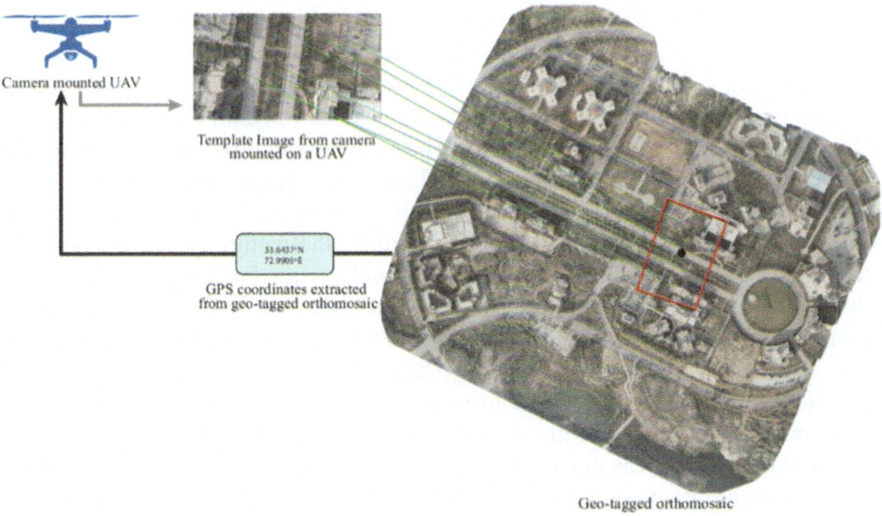

Fig. 2. Visualization of the operation of the absolute location system. Source: [5]

Publication [6] uses an optical flow algorithm based on translation between two image frames using appropriate filters to obtain more accurate readouts. Paper [7] presents an object tracking method involving edge detection. The method uses images from a single camera. In order to improve the accuracy and reliability of the edge identification, a Kalman filter was used, which additionally allows the estimation of the displacement and revolute velocity of flying objects (Fig. 3).

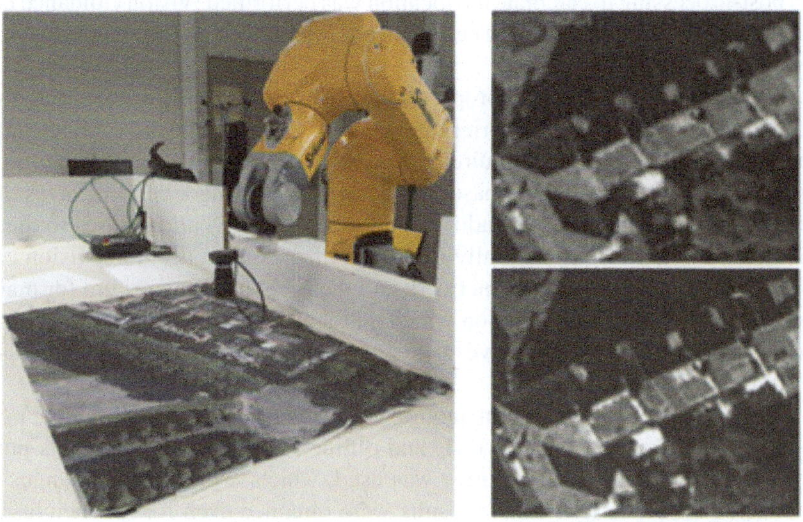

Fig. 3. Simulation of a UAV flight with the camera pointing downwards. Source: [8]

System in [8] uses a terrain map stored in internal memory and a model of possible trajectories. Using a graph approach, the shortest route between two points is designed, which is then implemented using visual feedback. Validation of the correct operation of elaborated navigation method is carried out using a 6° of freedom robotic arm equipped with a camera observing the printed map. The Extended Photometric Visual Servoing (Extended PVS) algorithm was proved by authors to be more accurate than the Oriented FAST and Rotated BRIEF (ORB) based algorithm. Future work will be devoted to more advanced tests using flat and spatial terrain models, the pathfinding and following in more complex scenarios will be tested using not only vision but also the other sensors.

In [9], authors used the Building Ratio Map feature for circular mark buildings by a segmentation neural network. The ratio of the building area to the circular area is a feature. This feature is rotation invariant, so it has the advantage of estimating the position candidate group. Additionally, the starting position is not required. This works more robustly in terms of the segmentation accuracy than other shape matching algorithms. The authors use the maps available on the Internet to train the network in order to ensure the universal solution for various areas. Similarly, [10] presents a method for localizing an Unmanned Aerial Vehicle (UAV) using aerial images. For the purposes of the work, the following algorithms were considered: Sum of Squared Differences (SSD), Sum of Conditional Variance (SCV) and Normalized Cross Correlation (NCC). However, these methods are very sensitive to changes in lighting, which disqualifies their use. Ultimately, the authors used mutual information (MI), which is more resilient to global changes and local lighting at the cost of greater computational complexity. The MI is based on the amount of information provided by the current and reference (georeferenced) images. To check the behavior for bad weather conditions, an artificial fog was applied on orthophoto maps. The introduction of this disturbance did not deteriorate the accuracy of the result. To validating the developed solution, a 4-min flight mission of the platform was carried out and post-processing to calculate the location was performed. Vision Guidance needed the last known position at the start. The calculated trajectory was very close to that given by GNSS.

In [9] publication, the authors proposed the BRM (Building Ratio Map) algorithm. It consists in comparing the numerical map to the image from the on-board camera with no known start position is required. The flight took place at a constant altitude of 150 m. A semantic neural network based on UNet is used to mark building areas that are divided into circles with different radii. Then the ratio of the building area to the circle is used as a feature that is rotationally invariant. Images from the stereovision camera are processed by a semantic neural network and then sent to the algorithm for matching the proportions of buildings, based on which a global location is generated. It has been proven that the proposed method gives better accuracy than conventional methods, e.g., the VINSmono algorithm.

In paper [11], the authors present another system for determining the global position of the UAV using a stereo camera, IMU and offline satellite images. For this purpose, the Mutual Information (MI) technique was used, which is resistant to changes in the appearance of the scenario. Positive results were obtained even for low heights of the order of 36 m above the ground. SURF descriptors were used to calculate stereo image triangulation. Several passages were made at different times of the day with different light intensity. The highest accuracy was obtained in the conditions with the best lighting, and it was 3 m for the position and 3° for the course.

The authors of article [12] integrated the IMU and a camera with a 2D map loaded into it for the purposes of obtaining a visual navigation system. An additional distinguishing feature of their approach is the use of a rigidly mounted camera without a gimbal. The publication presents a comparative table of eight visual navigation methods, from which the authors obtained the most accurate results.

2.1 Motivation and Objectives

Based on a review of the state of the art, it was concluded that there is a need to develop efficient vision-based navigation methods. The methods should be resilient to interferences with performing well in the real time. The methods described in the literature are in many cases in the testing phase and there is room for improvement or extension of their areas of application. This conclusion leads to the purpose of our paper.

Fig. 4. SUDIL platform. Source: [13]

The immediate objective of the research is to develop a navigation system such that it can be used in the Drone Destruction System of the Institute of Aviation (SUDIL) [13]. The SUDIL system neutralizes unwanted other drones using a launching net mounted on a multi-rotor platform, which is done without the need for operator intervention (Fig. 4). When the drone intruder becomes entangled in the net, it should be towed to a safe place or lowered to the ground, secured with a parachute. The interception procedure is as follows:

1. Intruder detection by ground radar system
2. Launch of the SUDIL platform and flight to a designated location
3. Automatic homing of SUDIL to the intruder
4. Interception of the intruder using the net
5. Towing the intruder to a safe place
6. SUDIL landing on the launch platform

To improve performance, the SUDIL platform can be equipped with a short-range radio jamming device for local jamming when capturing an intruder. In the event of a terrorist attack, there is a real risk of losing communication and the satellite signal needed for positioning. Due to the possibility of automatic operation, communication with the operator or the Ground Control Station (GCS) is not necessary. However, jamming the satellite navigation signal of the SUDIL system prevents it from continuing to operate, which translates into a real need for redundancy in the operation of the navigation system.

3 Approach and Methodology

This paper focuses on describing the basic version of the navigation system, which will be enhanced in future versions. This version allows visual geolocation of the UAV based on the last known GPS position and using camera images that are compared with an off-line map - according to the "map-based system" principle (Fig. 1). Currently, inertial information (AHRS) is not taken into account, because in the first version of the system, only flights at low speeds will be performed, and vertical deviations will be corrected by the gimbal mechanism.

3.1 World Simulation

An AirSim based simulation environment was elaborated prior to the acquisition of specialized apparatus. AirSim is an unmanned vehicle simulator available as an open source. The simulator was built using Epic Games' Unreal Engine 4, a dynamic environmental imaging system used in computer games development. Using Blender software, a selected three-dimensional area was imported from Google Earth data. Blender is an open-source software for creating and rendering images (rendering involves analyzing a model of a scene and creating its 3D animation). After connecting to the simulator, the PixHawk autopilot in HIL (Hardware in the Loop) mode, it is possible to simulate both: - the movement of the UAV and - the video observation obtained using the stabilized camera pointing downwards (Figs. 5 and 6).

Fig. 5. Visualizing the world in AirSim. Source: [14]

The second way to get a real-world simulation is to use Google Earth Studio. It allows to render an image from a simulated camera by trajectory and direction of view which can be designed. Thanks to 3D rendering of buildings and terrain topography and the adjustment of the time of day or camera FOV (Field of View), it is possible to obtain many recordings with a similar appearance to the real one. In this way, several recordings were made, on which the operation of the described solution is currently being tested.

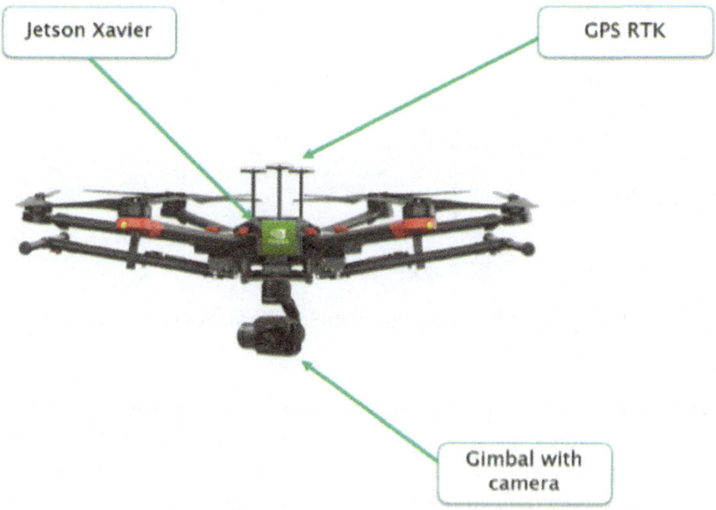

Fig. 6. Vision navigation test platform. Source: developed by the author

3.2 Hardware Setup

The system currently consists of a Matrice 600 Pro multi-rotor platform used as flying device, equipped with high-performance NVIDIA Jetson Xavier AGX computing unit, and a 3-axis gimbal providing image stabilization for downward-facing sensors, and a visible light camera with a wide-angle lens.

3.3 Vision-Based Navigation System

The diagram from Fig. 7 illustrates the vision-based navigation system concept developed as part of our research. When the satellite location is functioning properly, the PixHawk autopilot uses this data to determine its geographical location (green rectangles in the drawing). The Data Extraction module continuously receives the data from the PixHawk autopilot using the MAVLink protocol (Micro Air Vehicle Link – a communication protocol used in unmanned vehicles). The status of these data specifying the location is checked.

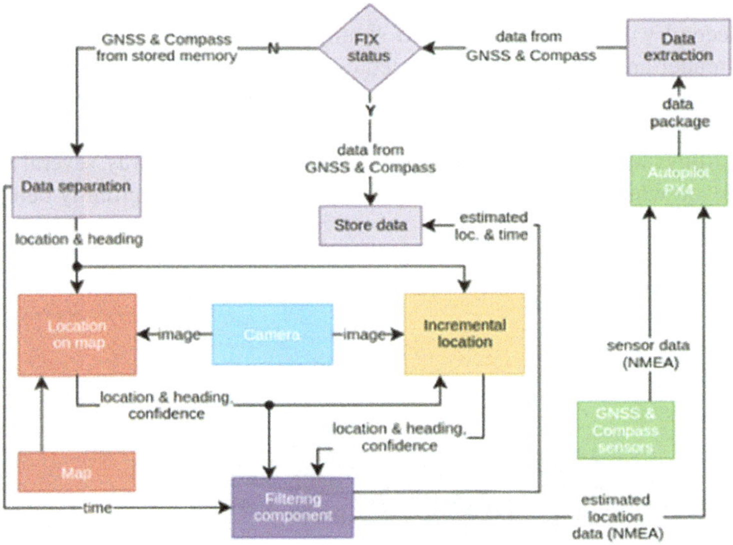

Fig. 7. Conceptual diagram of the vision navigation system. Source: developed by the author

It takes place in the decision block marked as Fix status in the diagram. If the status is correct (data complete), these data transmitted by the autopilot from the location sensors are stored in repository. If the status is incorrect (received data are incomplete or corrupted) the last correct data stored in memory are transferred to the Location on map and Incremental location modules, while the timestamp at which these data were recorded is transferred to the Filtering component. The Location on map process evaluates the drone position by comparing the actual image of the ground with the known map (stored in the memory). Concurrently the Incremental location module estimates the drone position by evaluating the displacement of feature points seen in the actual image towards their last known location. The location determined in the Location on map module is transferred to the Filtering component and to the Incremental location to update the reference position. Both modules use the OpenCV function to determine the confidence coefficients of the result. In the next step, the Filtering component determines the final estimate of the position taking into account the calculated confidence coefficients. With respect to the last time with known data, the Filtering component writes the finally estimated position into NMEA-compliant data packet that is sent to the autopilot. The location determined using the video data is also deposited in the data buffer (Store data block) for reuse in the case that data from satellite systems (GNSS & Compass sensors data) are not available. The above navigation system concept has been implemented in a basic version and is subject to testing, as described in the following sections of the paper.

3.4 Incremental Location Block

The Incremental location block calculates the drone's displacement at an interval corresponding to two consecutive frames of the camera image. For this purpose, the feature-based Lucas-Kanade (LK) method [15] with pyramid representation of images was used. This algorithm is based on analyzing the similarities between consecutive image frames. It is made by an iterative search for a set of transformation parameters that will best match the image to the template. Due to the low computational effort, the Incremental location block module is executed for every frame, unlike the Location on map module. On the basis of the characteristic points identified, the translation of those points is estimated. Currently this algorithm does not take into account changes in image size or perspective (viewpoint), so the viewpoint is delivered by the Location on map module.

3.5 Location on Map Module

The Location on map module consists of a map generation subsystem and a location subsystem for finding positions on an offline map. The global offline map is constructed using a geoinformation software. It allows to download orthophotos in geoTIFF format (containing geo-referenced information). Using this software, an area of 9 km^2 of lowland and rural areas was delimited. Currently, projection distortions are not taken into account. The map generation subsystem creates a map of the area where the drone is most likely to be located (region of interest (ROI)) based on the last known GNSS position. For this purpose, is used the GDAL library converting the raster and vector geospatial data to the required formats. The task of the localization subsystem is to compare the image from the camera mounted on the UAV with the ROI map to determine the geographical location of the drone. To this date, typical algorithms have been implemented: Scale-invariant-feature-transform (SIFT), Oriented FAST and Rotated BRIEF (ORB). The SIFT algorithm detects repetitive sequences in object tracking, as well as merges images into panoramas. The basis of SIFT is the determination of points that are invariant with the scaling and rotation and partially invariant due to changes in illumination and viewpoint (changes in perspective). The ORB algorithm performs as well as SIFT in the task of feature detection, while being almost two times faster. In addition, SuperGlue

[16] algorithm using deep neural networks was used to match two sets of local features by finding similar points. After the focal points have been extracted, homography (isomorphism of projective spaces) and RANSAC (iterative method to estimate parameters of a mathematical model from a set of observed data that contains outliers) [17] are used to calculate the perspective quadrilateral corresponding to the camera's field of view. RANSAC is capable of interpreting data containing a significant number of gross errors, and therefore is suitable for automated image analysis The point of geometric center of quadrilateral are converted using the GDAL library to geographic coordinates expressed in standard WGS-84 format.

To summarize - in the proposed method the last successfully received GPS coordinates are taken into account and actual localization using the local maps is estimated as it was described above. In the next steps, it is planned to train the neural network using our data generated by the simulation software and data from actual flights. The use of a modern approach such as video image processing and neural networks contributes to the universality of the solution.

3.6 System Response

In order to validate the information generated by the system, a program has been written to graphically visualize the results obtained, which makes it easy to check if the position of the vehicle is correct. On the basis of the video recording and the logs of the actual UAV flight, the motion trajectory is drawn on the map. This makes it possible to compare the obtained calculation results of the Location on map and Incremental location modules with true location.

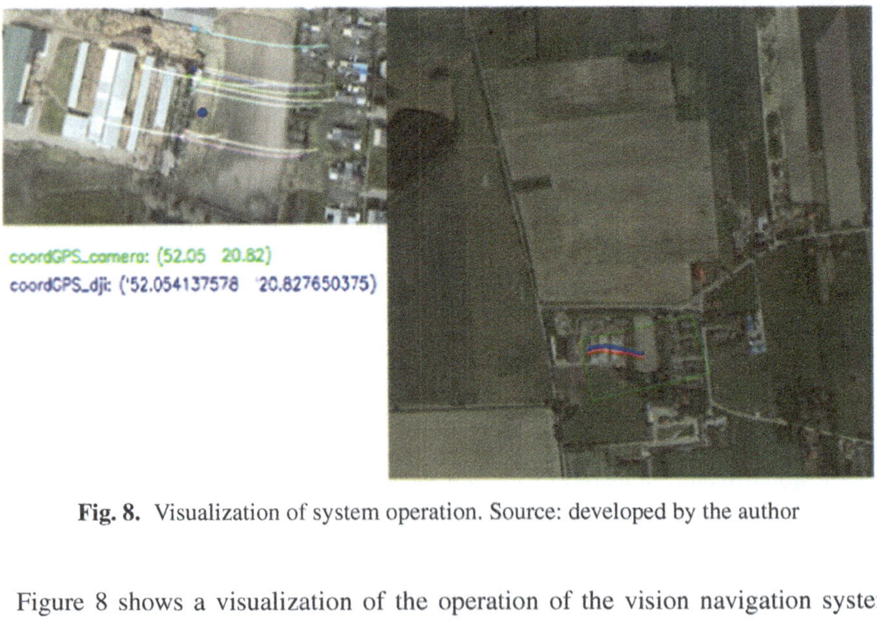

Fig. 8. Visualization of system operation. Source: developed by the author

Figure 8 shows a visualization of the operation of the vision navigation system. On the left the image from the camera is displayed with the drone displacement lines, on the right the image from the map with the visualization of the image perspective (green), the vision-based trajectory (red) and the GNSS based trajectory (blue) are given. The tests were carried out using available hardware resources. It will be followed by comparative testing conducted on the target equipment. This result obtained from a UAV flight test performed in a rural area - 20 km west of Warsaw. The video image (left) shows the displacement trajectories of the feature points obtained using the optical flow. The effect of the Location on map module is the visualization of the perspective (camera viewpoint) in the form of a green quadrangle shown on the map on the right. Its center is an initial estimate of the location. The Filtering component module, using a Kalman filter, generates a smoothed motion trajectory of the drone based on Location on map and Incremental location data, which is then drawn in red on the map. In addition, using the open source GDAL library for reading geospatial data, the drone geographical coordinates are obtained, which are then sent to the autopilot.

3.7 Specification of the Vision System

Two companies have now been identified that specialize in developing navigation systems for UAVs based on camera image processing. The first company from Israel called Sightec, on its official website [18] presents the NavSight vision navigation system, which enables real-time location using computer vision (CV) and artificial intelligence (AI). The solution has been filed for patent protection [19]. The other company, Asio Technologies also from Israel, has developed a finished module with a camera called NavGuard [20]. It is a stand-alone and compact module that can generate the current location of a UAV platform based on maps available in the device's memory. The above systems, due to their commercial nature, are closed, both in terms of software and hardware (Fig. 9).

Fig. 9. PX4FLOW module. Source: [21]

One of the open-source solutions is the PX4FLOW [21]. It is an intelligent camera module with an optical flow (OF) algorithm implemented. Using image analysis, the algorithm calculates the edge displacement and sends this information to the PixHawk autopilot via an I^2C serial interface. The PX4 autopilot software, when properly configured, takes these corrections into account when determining its position. Unlike previous proposals, PX4FLOW is based on displacement only, which precludes the determination of an absolute position. It is mainly used for indoor navigation.

4 Conclusions

The aim of presented research is to develop methods and implement a vision navigation system. The result of the work will be a set of algorithms that will allow to determine drone's geographical location (with an acceptable tolerance) basing on images from the camera in combination with a priori known map of the area of the planned mission. A vision navigation hardware is envisaged, consisting of dedicated instruments such as cameras and laser sensors and a high computing power vision processing system, which will be installed on board unmanned vehicles and will operate autonomously.

In this paper the results of the redundant navigation testing were presented. The functional structure of such system which, besides of data from navigation satellites, uses the maps and visual information was elaborated. The system was developed it its basic version, and successfully tested using available software and hardware.

The system together with its input interfaces, allows real-time testing using conventional algorithms but, in the future, it will be used to test the neural networks approach too for selecting the best option. The test rig allows neural networks to be trained for a dedicated flight area using both, simulation, and real data. The use of the NMEA data format ensures compatibility with popular UAV autopilots, including the PixHawk autopilot. The research will be followed by hardware integration in its final version and by in-depth testing of the navigation method using the hardware in the loop option, and the real environment as well.

Acknowledgements. The presented works were carried out as part of the doctoral dissertation carried out by Tomasz Pogorzelski, registered at the Doctoral School of the Warsaw University of Technology. The works were supervised by T. Zielińska. The test rig is located at the Centre for Unmanned Aerial Vehicle Technology, Research Network - Łukasiewicz Institute of Aviation.

References

1. Krawczyk, J.M., Mazur, A.M., Sasin, T., Stokłosa, A.W.: Infrared building inspection with unmanned aerial vehicles. Trans. Inst. Aviat. **3**(240), 32–48 (2015)
2. Kacprzak, M., et al.: Budowa i zastosowanie platformy wielosensorowej w badaniu wybranych parametrów środowiska. Prace Instytutu Lotnictwa **1**(234), 126–142 (2014)
3. Fellner, A., Konieczka, R.: Rotorcraft in the performance based navigation international civil aviation organization implementation. Trans. Aerosp. Res. **1**, 53–64 (2019)
4. Lu, Y., Zhucun, X., Gui-Song, X.: A survey on vision-based UAV navigation. Geo-spatial Inf. Sci. **12**(1), 21–32 (2018)
5. Mughal, M.H., Khokhar, M.J., Shahzad, M.: Assisting UAV localization via deep contextual image matching. J. Appl. Earth Observations Remote Sens. IEEE J. Sel. Top. Appl. Earth Observations Remote Sens. **26**(1), 2445–2457 (2021)
6. Shan, M., Wang, F., Lin, F., Zhi, G., Tang, Y.Z., Chen, B.M.: Googlemap aided visual navigation for UAVs in GPS-denied environment. In: 2015 IEEE International Conference on Robotics and Biomimetics (ROBIO), pp. 114–119, 6–9 December 2015

7. Anthea, C., Emmanuel, Z., Christine, E., Vincent, D.: Vision-based navigation for autonomous space rendezvous with non-cooperative targets. In: International Conference on Information, Intelligence, Systems and Applications (IISA), pp. 1–8, 15–17 July 2020
8. Rodríguez Martínez, E.A., Caron, G., Pégard, C., Alabazares, D.L.: Photometric path planning for vision-based navigation. In: 2020 IEEE International Conference on Robotics and Automation (ICRA), pp. 9007–9013, 15 August 2020
9. Choi, J., Myung, H.: BRM localization: UAV localization in GNSS-denied environments based on matching of numerical map and UAV images. In: IEEE/RSJ International Conference on Intelligent Robots and Systems, pp. 4537–4544 (2020)
10. Bertrand, D., Yol, A., Dame, A., Dartois, J.E., Marchand, E.: Vision-based absolute localization for unmanned aerial vehicles. In: 2014 IEEE/RSJ International Conference on Intelligent Robots and Systems, pp. 3429–3434, 14–18 August 2014
11. Patel, B., Barfoot, T.D., Schoellig, A.P.: Visual localization with google earth images for robust global pose estimation of UAVs. In: IEEE International Conference on Robotics and Automation (ICRA), pp. 6491–6497 (2020)
12. Mao, J., Zhang, L., He, X., Qu, H., Hu, X.: Precise Visual-Inertial Localization for UAV with the Aid of a 2D Georeferenced Map. arXiv (2021)
13. Łukasiewicz Research Network – Institute of Aviation. www.ilot.lukasiewicz.gov.pl, https://ilot.lukasiewicz.gov.pl/oferta-pdf/technologie-bezzalogowe.pdf. Accessed 18 Nov 2021
14. Microsoft: GitHub AirSim. https://microsoft.github.io/AirSim/. Accessed 18 Nov 2021
15. Bouguet, J.-Y.: Pyramidal implementation of the Lucas Kanade feature tracker, Intel Corporation, Microprocessor Research Labs (2000)
16. Sarlin, P.-E., DeTone, D., Malisiewicz, T., Rabinovich, A.: SuperGlue inference and evaluation demo script. In: CVPR 2020 (2019)
17. Fischler, M.A., Bolles, R.C.: Random sample consensus: a paradigm for model fitting with applications to image analysis and automated cartography. Commun. ACM **24**(6), 381–395 (1981)
18. www.sightec.com. Accessed 18 Nov 2021
19. Sightec: Image comparison system and method. US Patent 10699157B2 (2017)
20. www.asiotech.com, https://asiotech.com/navguard/. Accessed 18 Nov 2021
21. www.docs.px4.io, https://docs.px4.io/v1.12/en/sensor/px4flow.html. Accessed 18 Nov 2021

Smart Warehouse as an Example of Micro-ROS Application

Tomasz Kołcon(✉), Alexandre Malki, and Mateusz Maciaś

Security and Defence Systems Division, Łukasiewicz Research Network - Industrial Research Institute for Automation and Measurements PIAP, Al. Jerozolimskie 202, 02-486 Warsaw, Poland

{tomasz.kolcon,alexandre.malki,mateusz.macias}@piap.lukasiewicz.gov.pl

Abstract. The use of ROS/ROS2 in real life is becoming more and more popular. One of the ROS2 extensions is micro-ROS. Smart warehouse is a use case that aims to demonstrate how micro-ROS can be used in a low power wireless network environment. In the described case, the mobile base platform moves around a laboratory simulating a warehouse performing scheduled tasks. It shows how, thanks to micro-ROS, it is possible to easily connect basic elements such as sensors and actuators, which are based on microcontrollers, to a large system based on ROS/ROS2.

Keywords: ROS · ROS2 · Micro-ROS · Robot · Sensor · Actuator · Microcontroller · IoT

1 Introduction

Information management in warehouses has an increasingly important role in the supply chain of goods, inventory control, and storage conditions control [1,2]. There is a need for a constant exchange of information. IoT devices are used more and more often for such purposes. Unfortunately, the number of sensors, their diversity in terms of types and protocols used, make such systems very complicated [3–5]. It is therefore necessary to try to find a solution that unifies and simplifies it. ROS can be used for warehouse management [6]. On the other hand, to solve the problem of supporting the aforementioned IoT devices, it is best to use standardized solution like micro-ROS.

Micro-ROS is an extension of ROS2 and enables seamleass integration of low resource devices (IoT - Internet of Things). It was developed as part of the OFERA project [7]. Detailed description of the use case can be found in the OFERA deliverables: [8, D6.8] and [9, D6.13]

It should be noted that there are solutions similar to micro-ROS, but none of them works in such a comprehensive way [10].

The purpose of the use case is to demonstrate that micro-ROS can successfully operate in a dynamic and heterogeneous environment. Integration of micro-ROS enabled devices is easy and without extra effort, almost like in standard ROS2.

R. Szewczyk et al. (Eds.): AUTOMATION 2022, AISC 1427, pp. 264–272, 2022.
https://doi.org/10.1007/978-3-031-03502-9_27

The mobile base platform moves around the warehouse and in the meantime collects data from sensors such as humidity and temperature sensors and lidar according to planned tasks. After reaching a certain place, mobile platform can use an actuator to perform an action, e.g. open the gate. Communication with sensors and actuators is carried out using the 6LoWPAN [11] and the micro-ROS stack.

In this use case, two test missions were performed, but only one selected will be described. In addition, the mobile base platform as well as sensors and actuators will be described. Finally, conclusions will be presented.

2 Smart Warehouse Use Case Architecture

One of the laboratories located in Łukasiewicz Research Network - Industrial Research Institute for Automation and Measurements PIAP has been arranged as a warehouse. Micro-ROS enabled sensors and actuators have been installed there. An autonomous mobile robot platform moves inside and outside the warehouse. The autonomy of the mobile base platform is based on ROS. The mobile base platform's communication with sensors is based on ROS2 and micro-ROS.

Schematic diagram of the test system is shown in Fig. 1.

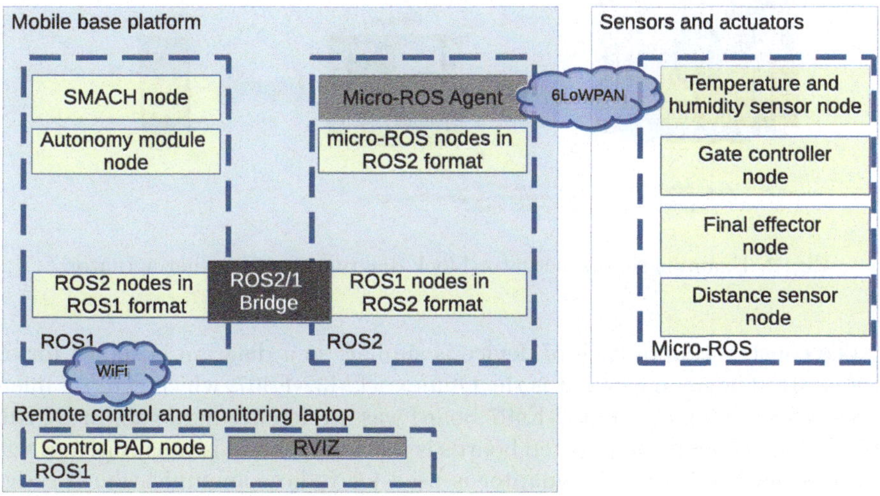

Fig. 1. Schematic diagram of the test system architecture.

Smart warehouse use case uses a mixture of ROS, ROS2 and micro-ROS systems. The combination of many different devices and their location creates a distributed environment.

Warehouse, sensors and actuators as well as mobile base platforms will be described below.

2.1 Micro-ROS Enabled Sensors and Actuators

For use case purposes, 2 sensors and 2 actuators were created. Each of these devices performs a different task and has been selected to show the wide range of micro-ROS applications. Of course, in other cases there may be more (more research is needed to determine the maximum number), but for this demonstration, the number of devices used is sufficient.

List of the sensors and actuators:

- Temperature and humidity sensor.
- Distance sensor.
- Door/gate opener - actuator.
- Final effector (light) - actuator.

Example of actuator is shown in the Fig. 2.

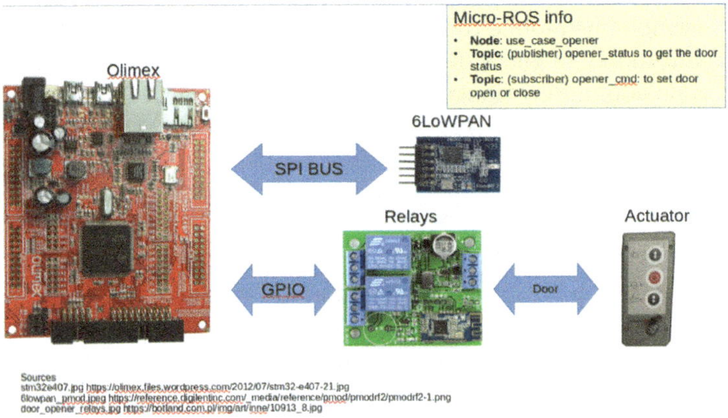

Fig. 2. Example sensor/actuator block diagram - gate opener actuator.

The construction of each of device is similar, so a diagram is shown for one device only. The main element is the Olimex STM32-E407, which runs on micro-ROS software. Olimex STM32-E407 board was a reference hardware in OFERA project, but number of supported boards is much larger and is constantly growing [12]. The 6LoWPAN to SPI adapter is used for radio communication. In Fig. 2 is shown a door opener which has a relay module. It controls the buttons in the warehouse gate driver. The second actuator (final effector) is very similar, but instead of the relay module it has a circuit with an additional converter that is used to power the warning light. The other two sensors differ in that they have a temperature and humidity sensor and an optical distance sensor, respectively.

Sensors/actuators firmware build procedure is available on smart warehouse Github repository [13]. In this case, NuttX RTOS (Real-time Operating System) is used. The micro-ROS API based application is as simple as the standard ROS2 node.

2.2 Smart Warehouse Equipment

From the use case point of view, the most important equipment in the hall are sensors and actuators based on micro-ROS. Temperature and humidity sensor is installed on the opposite side of the hall from the gate. Other devices are located in the vicinity of the gate (Fig. 5).

The modified gate control panel with additional IoT interface is shown in the Fig. 3.

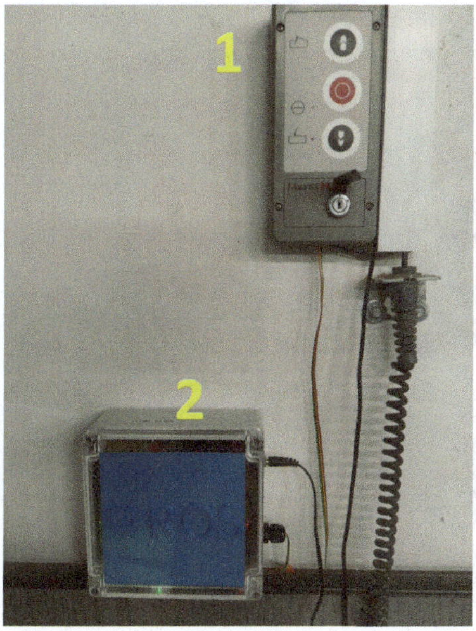

Fig. 3. Gate controller (1 - gate control panel, 2 - IoT interface).

It allows remotely controlling the gate and read the status of the gate opening. Next is a distance sensor, which in planned scenarios is used to check whether the road is empty and allows the robot mobile platform to travel. The last is the final effector device. It is a module equipped with warning light. This device is used in scenarios to confirm the execution of a sequence of actions.

The temperature and humidity sensor is placed far away from other three devices. Such locations of these devices is intended to simulate the situation of loss of radio communication in a situation when the robot is on the opposite side of the hall.

Artificial obstacles simulating storage shelves have been placed in the hall. Warehouse hall is also equipped with UWB (Ultra Wide Band) anchors to provide indoor localisation capabilities.

2.3 Mobile Base Platform

As the mobile base platform was used the PIAP Scout robot equipped with:

- Autonomy module (PC computer/lidar/indoor positioning/outdoor positioning)
- Communication mediums (6LoWPAN/WiFi),
- Other compoments (Wireless emergency stop).

Robot with its payloads is shown in the Fig. 4:

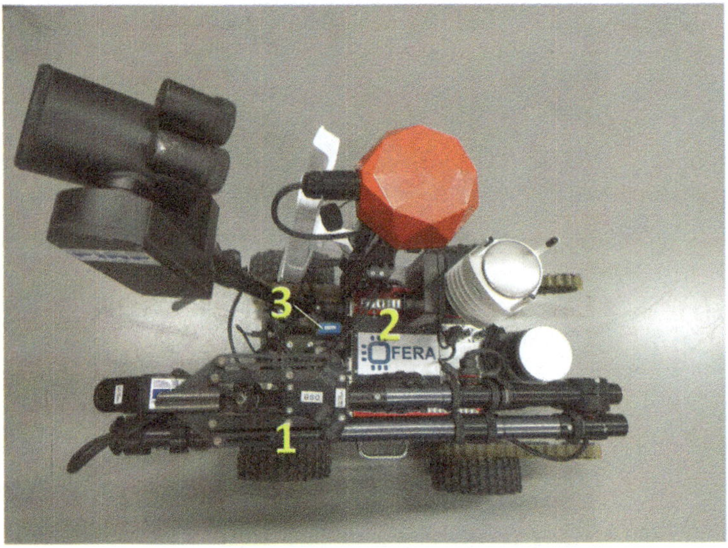

Fig. 4. Top view of the mobile base platform (1 - PIAP Scout robot, 2 - autonomy module, 3 - 6LoWPAN communication module).

Mobile base platform architecture is shown in Fig. 1 as a "Robot" block. This is a very interesting case due to the unique example of ROS, ROS2 and micro-ROS cooperation. Autonomy related hardware components work under the control of ROS. Communication is based on ROS2 and micro-ROS. By using the ROS2 bridge, all communication between the sensors and the ROS-based autonomy module is transparent. All topics from/to sensors in their original form can be used on the mobile base platform. Low level communication with sensors is done by radio using the 6LoWPAN adapter. Comparing to the standard ROS2, it is also necessary to run the micro-ROS agent. To run micro-ROS agent, which is a special ROS2 node that enables interaction with ROS2:

```
docker run -it --net = host --rm microros / micro-ros-agent: \
foxy udp4 --port 9999
```

3 Mission Example

As part of the OFERA project, during which the micro-ROS was developed, two simulation scenarios were created in the smart warehouse use case. Due to the fact that they are quite similar, only one will be described.

The mission steps are shown on the state machine in the Fig. 5 and in the Fig. 6. In this case, SMACH is used. SMACH is a library for task-level execution and coordination, and provides several types of "state containers" [14].

First, the mobile platform reads the values of humidity and temperature from the first sensor. At this point, there is no communication with other devices because there is no radio range. Then, using the autonomy module, it moves to the next point, near the gate. During the approach, it communicates with the gate. After reaching it, the mobile base platform sends command to open it. After opening the gate, the mobile base platform moves outside the warehouse and checks if the path is clear by reading value from the distance sensor. Finally, the robot approaches the final effector to turn it on. Lighting the warning light means that the mission has been successfully completed and we have communicated with all micro-ROS enabled devices in the warehouse and surrounding area.

The complete mission was video documented and available on YouTube on the Łukasiewicz-PIAP channel [15].

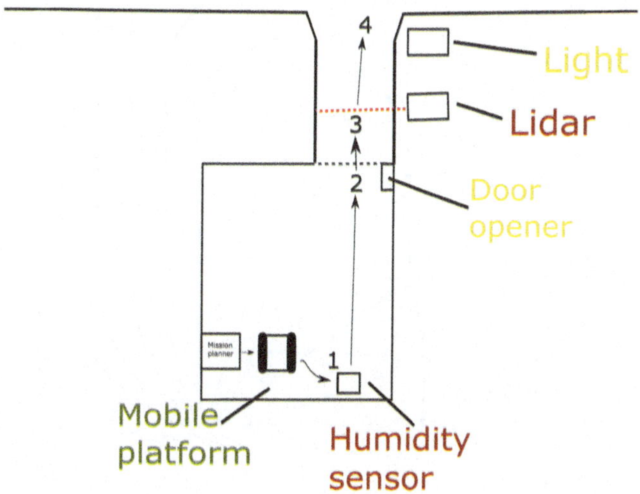

Fig. 5. Mission example.

It should be emphasized that there is no constant communication with the devices due to the limited radio range. Communication is established when approaching the sensor/actuator. Similar scenarios can be realized in a real smart warehouse, where there are many more different devices that communicate in the standardized way that micro-ROS enables.

4 Data Visualisation

Micro-ROS can be integrated with the FIWARE Context Broker [16]. The FIWARE context broker enables data visualization with micro-ROS enabled sensors and can control actuators. It is worth noting that this can be done remotely via Internet. It is needed to run the IoT agent on the mobile base platform. API for the IoT agent are ROS2 topics. An example of integrating a smart warehouse use case with it is shown in Fig. 7.

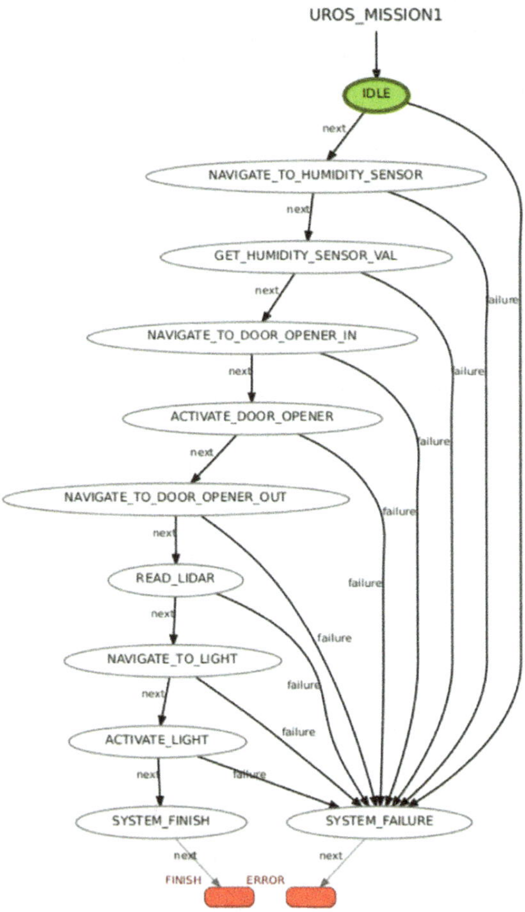

Fig. 6. Mission example state machine.

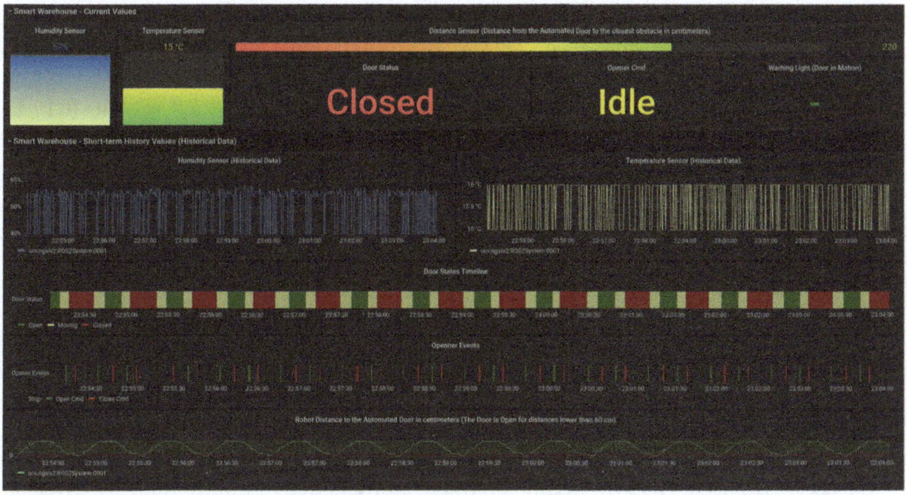

Fig. 7. Example dashboard.

5 Conclusions

The purpose of the smart warehouse use case was to demonstrate the usefulness of micro-ROS in a dynamic, distributed system of heterogeneous devices. Although presented demo is only a mockup, all solutions can be used in real smart warehouse. Obviously, this is just one example of a micro-ROS application.

Many sensors or actuators can be installed in a place such as warehouse. From now on, in order to integrate these devices into ROS/ROS2 it is no longer necessary to use computer with operating system (e.g. Ubuntu) with ROS/ROS2 installed on each of them or use a computer with ROS/ROS2 to which individual sensors are connected in a custom and non-standard way. Instead, thanks to micro-ROS, standard microcontrollers can be integrated seamlessly, almost as easy as in ROS/ROS2. The idea behind micro-ROS is to make it the standard for such applications.

Finally, it should be mentioned that the micro-ROS community is constantly growing and new companies are joining, including world leaders.

Acknowledgments. The project leading to this results has received funding from the European Union's Horizon 2020 research and innovation programme under grant agreement No. 780785.

References

1. Yerpude, S., Singhal, T.: Smart warehouse with internet of things supported inventory management system. Int. J. Pure Appl. Math. **118**, 1–15 (2018)

2. Žunić, E., Delalić, S., Hodžić, K., Beširević, A., Hindija, H.: Smart warehouse management system concept with implementation (2018). https://doi.org/10.1109/NEUREL.2018.8587004
3. Ding, W.: Study of smart warehouse management system based on the IoT. In: Du, Z. (ed.) Intelligence Computation and Evolutionary Computation. Advances in Intelligent Systems and Computing, vol. 180, pp. 203–207. Springer, Heidelberg (2013). https://doi.org/10.1007/978-3-642-31656-2_30
4. Zhen, L., Li, H.: A literature review of smart warehouse operations management. Front. Eng. Manag. (2022). https://doi.org/10.1007/s42524-021-0178-9
5. Detzner, P., Pose, T., Fumagalli, L., Matteucci, M.: Towards a plug and play architecture for a materialflow handling system. In: 2019 IEEE Conference on Open Systems (ICOS), pp. 28–33 (2019). https://doi.org/10.1109/ICOS47562.2019.8975705
6. AWS RoboMaker Small Warehouse World Github repository. https://github.com/aws-robotics/aws-robomaker-small-warehouse-world
7. Ofera Project Official Website. http://www.ofera.eu/
8. Kołcon, T., Malki, A., Maciaś, M.: Smart warehouse micro-ROS use-case release - Revised (2020). http://ofera.eu/storage/deliverables/M36/OFERA_65_D6.8_Smart_warehouse_micro-ROS_use-case_release_Revised.pdf
9. Kołcon, T., Malki, A., Maciaś, M.: Smart warehouse micro-ROS use-case release - Extension (2021). http://ofera.eu/storage/deliverables/M48/OFERA_98_D6.13_Smart_warehouse_micro-ROS_use-case_release_Extension.pdf
10. Micro-ROS Comparison to Related Approaches. https://micro.ros.org/docs/overview/comparison/
11. IPv6 over Low-Power Wireless Personal Area Networks. https://en.wikipedia.org/wiki/6LoWPAN
12. Micro-ROS Supported Hardware. https://micro.ros.org/docs/overview/hardware/
13. Łukasiewcz-PIAP: Smart warehouse Github repository. https://github.com/micro-ROS/micro-ROS_warehouse_demo
14. SMACH module. http://wiki.ros.org/smach
15. Łukasiewcz-PIAP: The OFERA project – smart warehouse use case - mission 1. https://youtu.be/XsUw1w_f8ks
16. FIWARE: Joint use-case with FIWARE Context Broker (2021). https://github.com/FcoMelendez/micro-ROS_jointUseCase/blob/master/pdfs/OFERA_99_D6.14_Joint_use-case_with_FIWARE_context_broker.pdf

Prototype of the Arm-Z Modular Solar Tracker

Ela Zawidzka[1], Jacek Szklarski[1], Wojciech Kiński[2], and Machi Zawidzki[1(✉)]

[1] Institute of Fundamental Technological Research, Polish Academy of Sciences, Warsaw, Poland
zawidzka@ippt.pan.pl

[2] Industrial Research Institute for Automation and Measurements PIAP, Warsaw, Poland

Abstract. Arm-Z - a hyper-redundant manipulator based on linearly joined sequence of congruent units is presented. Each unit has 1-DOF (one degree of freedom) only, namely a twist relative to the previous unit in the sequence. Arm-Z has a potential of being economical and robust. The control of Arm-Z, however, is not intuitive and difficult. This paper presents the preliminary results of designing a prototype of a modular Sun-tracking device comprised of four congruent units with possible application as a solar energy harvester or a Sun-shade.

Keywords: Extremely modular system · Arm-Z · Hyper-redundant manipulator · Rapid prototyping · Low-tech

1 Introduction

Complex three-dimensional tubular forms can be constructed with relatively simple congruent units, as described in [1]. Ref. [2] introduced an analogous parametric design system composed of one type of modular unit. Arm-Z is a concept of robotic manipulator introduced in [3] based on that idea. Biological snakes are extremely well adapted for different environments. This is mostly the result of the high redundancy of the snake mechanisms. In many instances of irregular environments the bio-inspired robots outperform conventional wheeled, legged or tracked robots. The snake-resembling robots are researched already for a few decades. This type of locomotion has been studied already in the 1940s [4], and a half century later, its rigorous mathematical model has been developed. In the late 90's, a trunk-like locomotors and manipulators have been introduced in [5]. Alike biological snakes, in various environments the characteristic type of motion gives this type of manipulators certain advantage over conventional robotic manipulators. They can operate in geometrically complicated environments which are not accessible by other ways. Depending on the required task, various working heads can be installed on such manipulators, e.g. for: welding, cleaning, monitoring, etc. as shown in Fig. 1.

R. Szewczyk et al. (Eds.): AUTOMATION 2022, AISC 1427, pp. 273–282, 2022.
https://doi.org/10.1007/978-3-031-03502-9_28

Fig. 1. Oliver Crispin Robotics Ltd. (http://www.ocrobotics.com). Snake-arm robots, series II, X125 System. On the left: unsleeved, integrated manipulator with inspection camera and light tool. On the right: sleeved and integrated manipulator with a laser cutting head.

A conventional industrial manipulator has low number of degrees of freedom, whereas a trunk-like or snake-line manipulator has redundant (considerably large) number of degrees of freedom. Arm-Z manipulator has as many degrees of freedom as the number of units less of one. This redundancy allows the Arm-Z to perform complicated spatial movements, but also may improve the robustness and fault tolerance of the system. Therefore Arm-Z belongs to the family of so-called hyper-redundant-manipulators (HRM [6]). Since the inverse kinematic problem of a typical industrial manipulator can be solved easily [7], its control is straightforward. On the other hand, because HRMs are highly non-linear systems, their control is by no means straightforward, and requires application of artificial intelligence methods [8–10]. For more information on this type of manipulators see [11].

2 The Geometrical Parameters of Arm-Z

The Arm-Z units are geometrical objects which are analogous to sectors of circular tori. Each unit is defined by the following parameters: size r, offset d, and ζ, that is the angle between upper (**T**) and lower (**B**) faces of the unit (see Fig. 2).

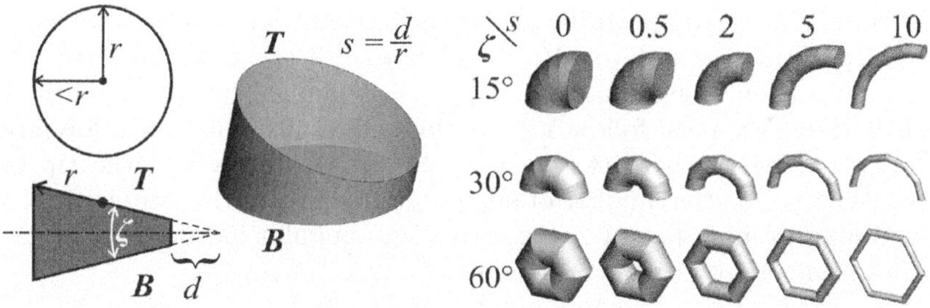

Fig. 2. On the left: visualization of the Arm-Z unit defined by three parameters: r, d and ζ. On the right: simple examples of assemblies of units for various values of ζ and s (slenderness), which is an additional parameter, a d to r ratio.

The form of Arm-Z depends on: the number of units, the geometric parameters of each unit, and the relative twists between the modules. As an example, two Arm-Z units at 6 successive rotations from 0 to 180° in 30° steps are shown in Fig. 3.

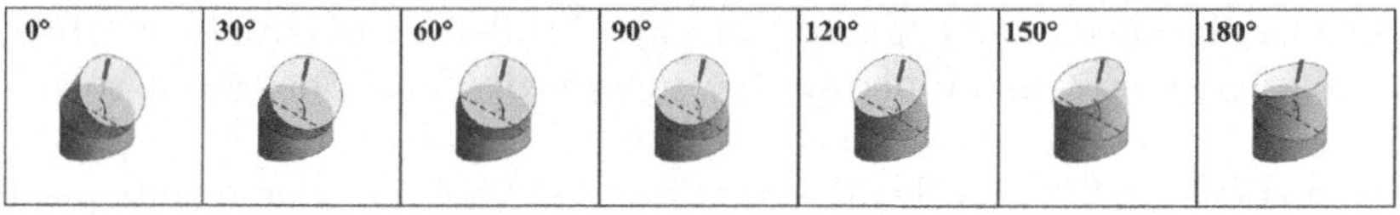

Fig. 3. The relative rotation of an Arm-Z unit. The bottom (fixed) unit is shown in dark grey. Red line indicates the axis of rotation (perpendicular to the top face of the bottom unit) for the subsequent unit.

3 The Sun-Tracking Task

The position of the Arm-Z in the coordinate system and explanation of its tracking parameters are shown on the left of Fig. 4. The goal for Arm-Z is to follow the Sun direction during summertime (from summer solstice to September equinox) between 10 a.m. and 6 p.m., as illustrated on the right in Fig. 4.

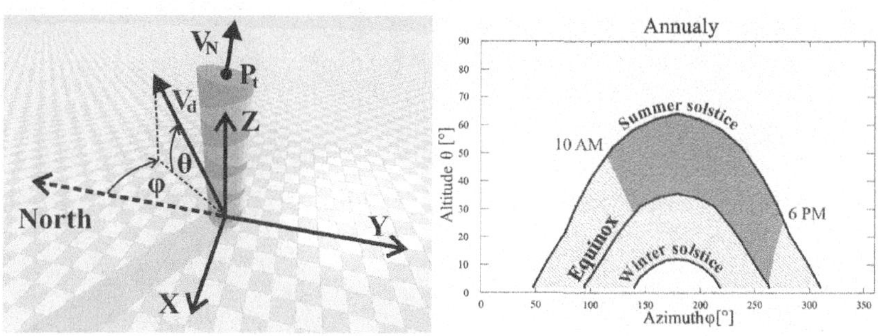

Fig. 4. On the left: the coordinate system, notation and the control parameters. On the right: the possible sun location during a year and for the given periods during summertime.

$\mathbf{V_N}$ is the normal vector to the face of the last unit, i.e. it is a vector directing from the tip of the manipulator. $\mathbf{V_d}$ is the desired direction of the normal vector (here, the direction along sunlight). $\mathbf{V_d}(\phi, \theta)$ is defined by the azimuth ϕ and altitude θ. Here, $\mathbf{V_N}$ must follow $\mathbf{V_d}$, i.e. this is formally a minimization problem of the angle between these two vectors. $\mathbf{P_t}$ is the position of the tip of the manipulator. N_{dof} is the number of degrees of freedom of the system, it is equal to the number of units less of 1. Figure 9 shows angular inaccuracies for Arm-Z of various numbers of degrees of freedom: 1, 2, 3, 4, 6 and 7.

As Fig. 9 indicates, a four-module Arm-Z manipulator already covers almost all required directions. The last sub-figure indicates that an Arm-Z manipulator with seven degrees of freedom covers most directions, including almost the entire hemisphere below the manipulator).

3.1 The Optimization of the Tracking Motion

Bullet Physics software development kit [12] is a widely used computational tool for robotics, machine learning etc., in particular for: real-time collision detection and multi-physics simulation for virtual reality. For realistic simulations, the manipulator has been implemented in this simulation environment. In order to control the Arm-Z manipulator, the inverse dynamics has been solved, for comparison, by two numerical methods: the classical Recursive Newton-Euler algorithm (RNEA), [13] and Dual Annealing (DA) [14,15]. The results produced by the latter outperformed the former in terms of the mean errors. This most likely due to the fact, that the system is highly non-linear and even small changes in position of the units at the beginning of the manipulator significantly influence the position of the tip at the end.

In order to control the manipulator continuously, the twist velocities and spins of three units have been optimized with DA method. Large deviations from the previous state have been penalized. As a result, the motion of the manipulator is relatively smooth. Figure 5 shows the resulting angular velocities of the individual units on the left. The angular errors throughout the tracking period are shown on the right in the same figure.

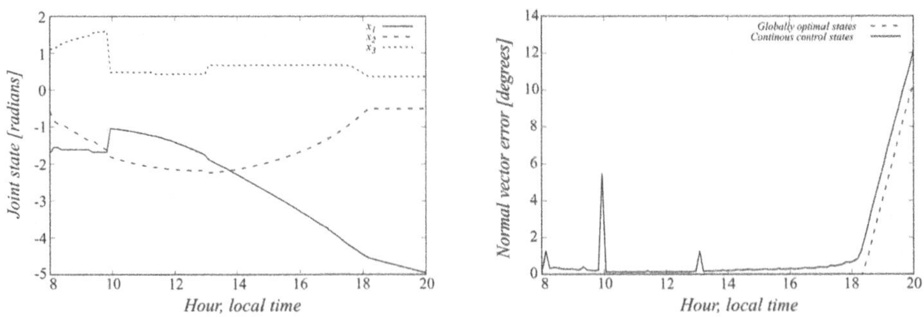

Fig. 5. On the left: the joint states x_1, x_2, x_3 for a 4-unit Arm-Z following the Sun on July 1st between 8:00 and 20:00 h. On the right: the error, i.e. the angle between $\mathbf{V_d}$ and $\mathbf{V_N}$ for globally optimal states and for the continuous control scenario.

As the right sub-figure of Fig. 5 indicates, there are error peaks in the continuous control scenario. Moreover, the errors continuously grow after 18:00 h since the tip's normal vector $\mathbf{V_N}$ cannot follow the Sun for low latitudes, especially during sunrise and sunset (compare with Fig. 9).

The concept of low-tech prototype is based on passive units, in other words, they do not have individual motors and control units. Only the first (base) unit is driven. Next the drive and its direction are transferred to the subsequent units solely by gears and switches. This means that each unit will have a fixed direction and gear set for the entire duration of the tracking action.

The mean tracking error for given time range t_1 - t_2 is defined as follows:

$$T_{\text{err}} = \int_{t_1}^{t_2} v_s(t) \cdot v_d(t) \mathrm{d}t,$$

where v_d is the vector perpendicular to the surface of the last module and v_s points towards the sun at the time t.

The goal is to find module velocity multipliers d_1 and d_2 together with the angular velocity of the base module v_0 which minimize T_{err}. It was done by means of brute-force search for the 3 parameters, d_1, d_2 and v_0 in the range which is defined by possible mechanical constraints. Tables 1 and 2 show the optimization results. Table 1 indicates that the minimal error is for $d_1 = -0.5$, $d_2 = -1.0$ and the corresponding base velocity is -5.95 deg/h (Table 2).

Table 1. Minimum mean tracking error for various velocity multipliers and an optimal base angular rotation velocity.

d_2	d_1									
	−1.50	−1.25	−1.00	−0.75	−0.50	0.50	0.75	1.00	1.25	1.50
−1.50	12.24	10.66	8.70	6.90	6.19	14.32	17.96	21.08	23.68	25.73
−1.25	13.11	11.28	8.95	6.76	6.03	13.35	16.44	19.12	21.39	23.34
−1.00	14.39	12.22	9.46	6.68	**6.00**	12.45	15.03	17.25	19.13	20.73
−0.75	16.29	13.61	10.25	6.84	6.03	11.60	13.75	15.56	17.07	18.36
−0.50	19.11	15.73	11.45	7.34	6.25	10.84	12.61	14.07	15.25	16.26
0.50	26.43	28.71	25.03	18.41	10.92	9.01	9.84	10.61	11.19	11.72
0.75	24.25	26.32	27.77	21.90	13.41	8.86	9.76	10.51	11.09	11.60
1.00	22.65	24.43	27.26	24.78	16.12	8.84	9.78	10.51	11.10	11.59
1.25	21.41	22.98	25.48	27.08	18.77	8.90	9.82	10.56	11.20	11.66
1.50	20.46	21.84	24.06	28.00	21.18	8.98	9.90	10.65	11.26	11.72

Table 2. Angular rotation velocities of the base module in [deg/h] which give minimum tracking errors for the corresponding velocity multiplication values.

d_2	d_1									
	−1.50	−1.25	−1.00	−0.75	−0.50	0.50	0.75	1.00	1.25	1.50
−1.50	−3.78	−4.33	−4.86	−5.13	−5.40	−6.22	−5.95	−5.68	−5.68	−5.68
−1.25	−4.06	−4.60	−5.40	−5.68	−5.68	−5.95	−5.40	−5.13	−4.60	−4.33
−1.00	−4.60	−5.40	−5.95	−6.22	**−5.95**	−5.68	−5.13	−4.60	−4.33	−3.78
−0.75	−5.40	−5.95	−6.76	−6.76	−6.49	−5.40	−4.86	−4.33	−4.06	−3.78
−0.50	−5.68	−6.76	−7.84	−7.57	−7.02	−5.13	−4.60	−4.06	−3.78	−3.51
0.50	3.78	3.24	−6.22	−8.38	−9.19	−4.33	−3.78	−3.51	−2.97	−2.70
0.75	3.51	4.06	−4.60	−7.30	−9.19	−4.33	−3.51	−3.24	−2.70	−2.44
1.00	2.97	3.78	4.33	−6.22	−8.65	−4.06	−3.51	−2.97	−2.44	−2.16
1.25	2.70	3.24	4.06	−5.40	−8.11	−3.78	−3.24	−2.70	−2.16	−1.89
1.50	2.44	2.97	3.78	4.33	−7.57	−3.78	−2.97	−2.44	−2.16	−1.89

4 The Prototype

The prototype of Arm-Z solar tracker is equipped with advanced gears which allow for: switching the spin of given module (right/left) about its axis and transferring of the motion to the subsequent module. The cogwheels with involute outline has been used for the gear train. In order to avoid slipping, they are embedded on the rods with square section. Moreover, in order to minimize the rolling resistance, 608ZZ bearings have been used. The gear train has been equipped with so called "reverse switch". A dedicated left/right lever switch has been placed in the side of the casing of each module as shown in Fig. 6.

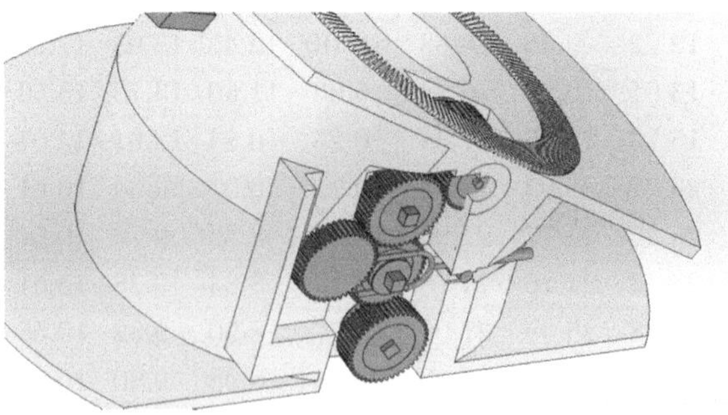

Fig. 6. The switch for changing the spin of the rotation of the module.

This corresponds to along/counter spin of the previous module. Each module is equipped with toothed ring as show in Fig. 7.

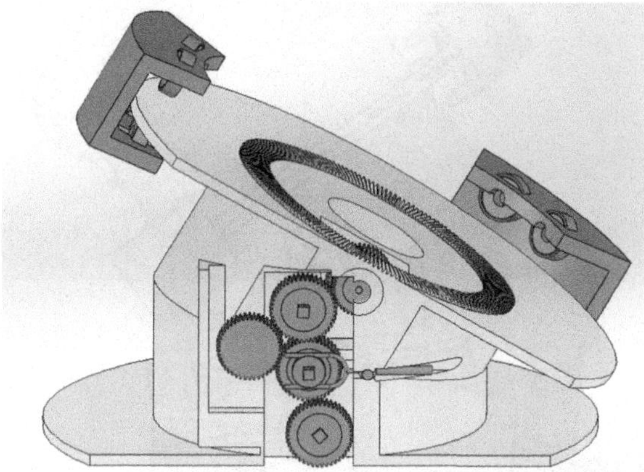

Fig. 7. The drive cogwheel (on the bottom) an corresponding tooth wheel (on the top).

In each module there is a sleeve which forms a continuous pipe through entire Arm-Z device. Its main purpose is for the wiring. In order to form a firm and smooth connections of the entire Arm-Z, the modules are attached to each other with special bracing equipped with bearings. The gear train can be freely modified inside each module. The gear ratio of each module can be adjusted by alteration of the number of teeth of the cogwheels of gear trains. The cogwheel sets are also modular and can be easily replaced. In this project, the design principle was to use as many elements as possible off-the-shelf, in other words, not designed or made to order but taken from existing stock or supplies, e.g.: commercially available bearings, standard rods, etc. Certain elements have been 3D-printed using FDM/FFF (Fused Deposition Modeling/Fused Filament Fabrication) technology and enhanced materials, e.g. with increased abrasion resistance by use of polyamide (PA) filaments with carbon fiber. The cogwheels have been fabricated in two versions: 3D printed and laser-cut from aluminum sheet. The CAD visualization of the prototype of a four-unit Arm-Z solar tracker is shown in Fig. 8.

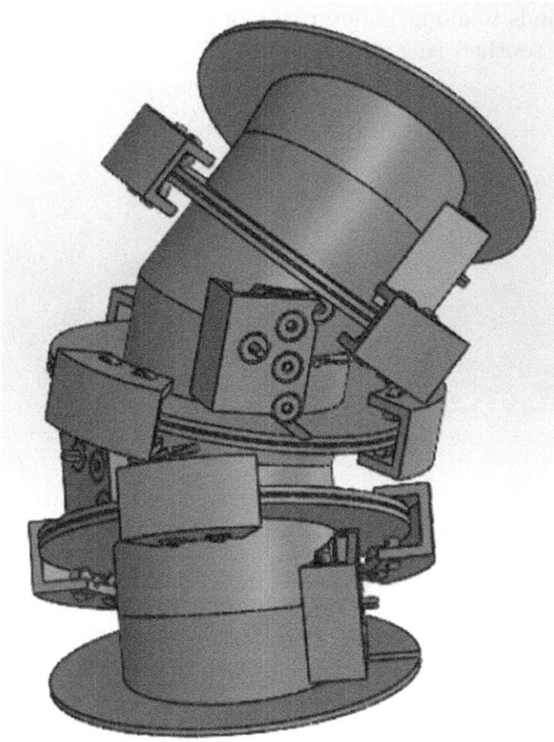

Fig. 8. The CAD visualization of the prototype of a 4-unit Arm-Z solar tracker.

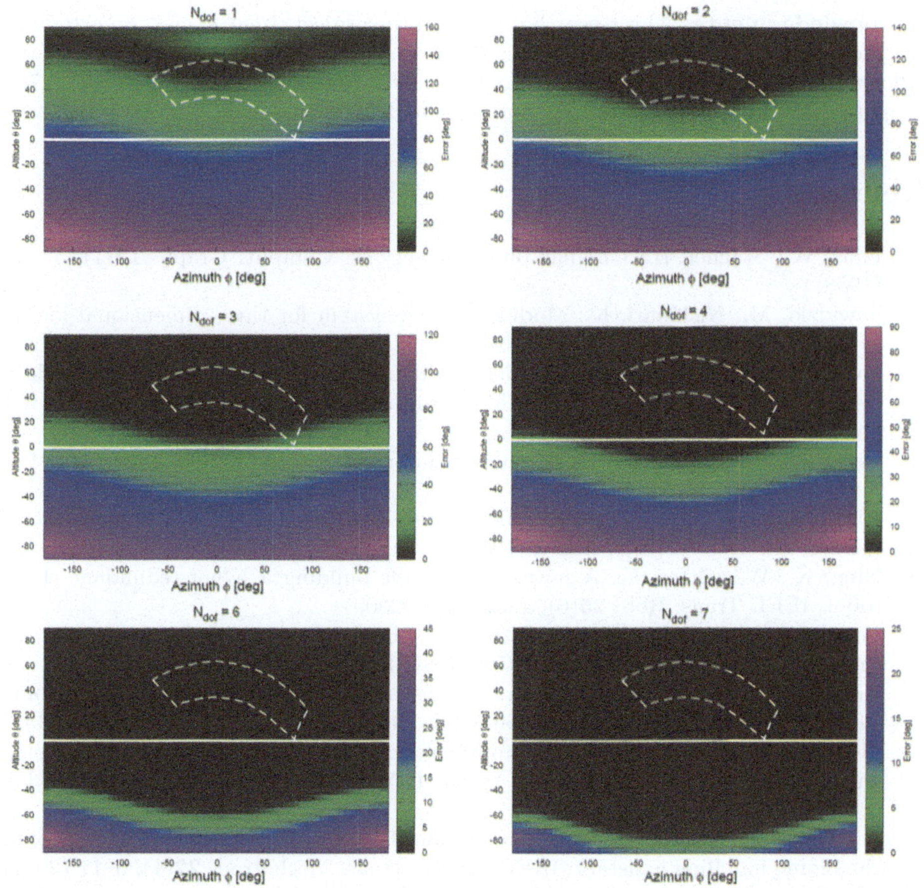

Fig. 9. The angular inaccuracies for Arm-Z manipulator of various numbers of degrees of freedom: 1, 2, 3, 4, 6 and 7. White line indicates the horizon. The tracking period from Fig. 4 is superimposed as the area surrounded by the white dashed line. Very small errors are shown in black.

5 Conclusions

- The direction of the Sunlight can be effectively tracked with three rotating Arm-Z modules.
- Theoretical accuracy of tracking action is good for this task.
- The preliminary simulations and early prototypes confirm these findings.
- The operating physical prototype is currently being made in order to ultimately verify the theoretical results.
- The structure of the modules should also be optimized. For an example where the authors have effectively optimized other extremely modular systems see: [16].

Acknowledgements. This research is a part of the project titled *Arm-Z: an extremely modular hyperredundant low-cost manipulator – development of control methods and efficiency analysis* and funded by *OPUS 17* research grant No. 2019/33/B/ST8/02791 supported by the National Science Center, Poland.

References

1. Fuhs, W., Stachel, H.: Circular pipe-connections. Comput. Graph. **12**(1), 53–57 (1988)
2. Zawidzki, M., Nishinari, K.: Modular pipe-z system for three-dimensional knots. J. Geom. Graph. **17**(1), 81–87 (2013)
3. Zawidzki, M., Nagakura, T.: Arm-Z: a modular virtual manipulative. In: Schröcker, H-P. (ed.) Proceedings of the 16th International Conference on Geometry and Graphics, pp. 75–80 (2014)
4. Gray, J.: The mechanism of locomotion in snakes. J. Exp. Biol. **23**(2), 101–120 (1946)
5. Hirose, S.: Biologically Inspired Robots: Snake-Like Locomotors and Manipulators. Oxford University Press, Oxford (1993)
6. Ning, K., Wörgötter, F.: A novel concept for building a hyper-redundant chain robot. IEEE Trans. Rob. **25**(6), 1237–1248 (2009)
7. Murray, R.M., Li, Z., Shankar Sastry, S., Shankara Sastry, S.: A Mathematical Introduction to Robotic Manipulation. CRC Press, Boca Raton (1994)
8. Rolf, M., Steil, J.J.: Efficient exploratory learning of inverse kinematics on a bionic elephant trunk. IEEE Trans. Neural Netw. Learn. Syst. **25**(6), 1147–1160 (2014)
9. Melingui, A., et al.: Qualitative approach for forward kinematic modeling of a compact bionic handling assistant trunk. IFAC Proc. Volumes **47**(3), 9353–9358 (2014)
10. Falkenhahn, V., Hildebrandt, A., Neumann, R., Sawodny, O.: Dynamic control of the bionic handling assistant. IEEE/ASME Trans. Mechatron. **22**(1), 6–17 (2017)
11. Chirikjian, G.S., Burdick, J.W.: A hyper-redundant manipulator. IEEE Robot. Autom. Mag. **1**(4), 22–29 (1994)
12. Coumans, E., Bai, Y.: Pybullet, a python module for physics simulation for games, robotics and machine learning (2016–2021). http://pybullet.org
13. Luh, J.Y.S., Walker, M.W., Paul, R.P.C.: On-line computational scheme for mechanical manipulators. J. Dyn. Syst. Measur. Control **102**(2), 69–76 (1980). https://doi.org/10.1115/1.3149599. ISSN 0022-0434
14. Xiang, Y., Sun, D.Y., Fan, W., Gong, X.G.: Generalized simulated annealing algorithm and its application to the Thomson model. Phys. Lett. A **233**(3), 216–220 (1997). https://doi.org/10.1016/S0375-9601(97)00474-X. https://www.sciencedirect.com/science/article/pii/S037596019700474X. ISSN 0375-9601
15. Virtanen, et al.: SciPy 1.0: fundamental algorithms for scientific computing in python. Nat. Methods **17**, 261–272 (2020). https://doi.org/10.1038/s41592-019-0686-2
16. Zawidzki, M., Jankowski, Ł: Optimization of modular truss-z by minimum-mass design under equivalent stress constraint. Smart Struct. Syst. **21**(6), 715–725 (2018)

Measuring Techniques and Systems

Experimental Comparison of Selected Triangulation and TOF Optical Distance Sensors

Bogdan Kreczmer(✉) and Piotr Portasiak

Department of Cybernetics and Robotics, Faculty of Electronics, Photonics and Microsystems, Wrocław, Poland
bogdan.kreczmer@pwr.edu.pl

Abstract. In the paper, features of two optical distance sensors are compared and discussed. The considered sensors are the triangulation sensor GP2Y0A02YK0F and the time-of-flight sensor VL53L0X. To compare and illustrate their features, the results of conducted experiments were used. They are discussed and explanations for some of the observed phenomena are given.

Keywords: Optical sensor · Distance measurement · Triangulation · Time-of-flight

1 Introduction

Measuring distance to objects is an essential capability that mobile robots must have. It allows robots to determine location of obstacles, build an environment map, perform self-localization, and then plan a collision-free trajectory. Among the sensors that provide this capability, optical distance sensors have several important advantages. They are cheap and very compact in size. Examples of such sensors are GP2Y0A02YK0F produced by Sharp and VL53L0X produced by ST Microelectronics. In the first one, the triangulation method is exploited to determine distance. The second uses the time-of-flight (TOF) measurement method. In this paper, some features of these two sensors are compared and discussed by using results of experiments.

2 Optical Linear Scanner

In order to check properties of the sensors of interest, a linear scanner (see Fig. 1) was used. It is the result of the work presented in [17]. This scanner uses a module DryLin SLW-1040 provided by Igus as a linear drive. Its positioning accuracy in this application is 0.2 mm. The range of its movement is 45 cm. The main part of the scanner is its movable sensor head. It contains a set of active optical distance sensors. Their emitters are located along the same vertical axis.

R. Szewczyk et al. (Eds.): AUTOMATION 2022, AISC 1427, pp. 285–295, 2022.
https://doi.org/10.1007/978-3-031-03502-9_29

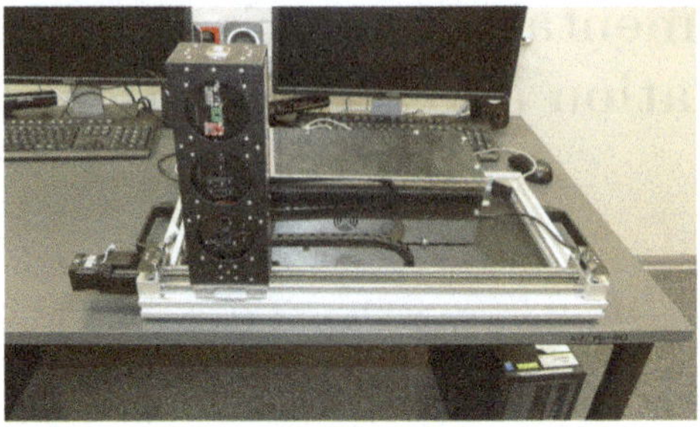

Fig. 1. The optical linear scanner

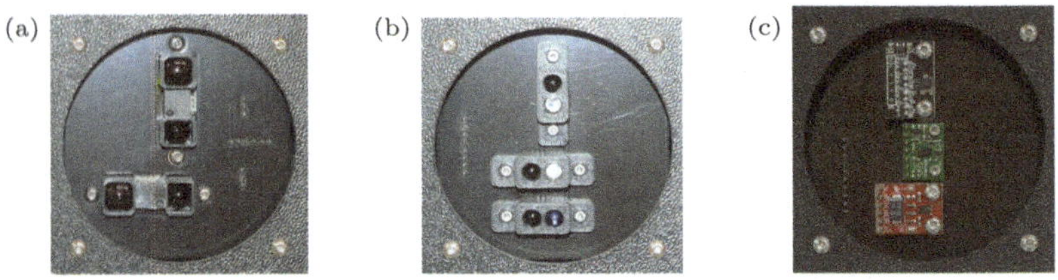

Fig. 2. The location of sensors mounted in the sensor head. a – triangulation distance sensors, b – ambient sensors, c – TOF and ambient sensors

In the lowest part, there are two distance sensors GP2Y0A02YK0F (see Fig. 2a). One of them was oriented horizontally, and the other vertically. In the middle part of the scanner head, there are mounted three sensors (see Fig. 2b). They measure intensity of reflected and ambient light. Each of them is a couple of an IR emitter and photo-transistor. In the top part of the scanner head, there are mounted three sensors (see Fig. 2c). Two of them (VL6180X and VL53L0X) measure distance exploiting a time-of-flight method. The third one (APDS-9960) measures ambient light and colors. Ambient light is also measured by VL6180X.

3 Distance Sensors

In the further part of this paper the focus is paid on the GP2Y0A02YK0F and VL53L0X sensors. They are discussed in more detail and the results of the experiments are presented. The GP2Y0A02YK0F sensor has been available on the market for many years. There are many of its applications in object detection for mobile robots [2,3,6,19] or their cooperative control [8]. There are also applications in control of unmanned aircrafts [16,20] or as a support for impaired and elderly people activity [4,15]. It uses the triangulation method. In the VL53L0X, the TOF measurement method is applied. This sensor is a relatively new one and has been available on the market from a few years. Its important advantage is that it is compact in size. Therefore, it can be used in mobile devices. Nevertheless, there are already examples of its implementation in obstacles detection [1,5,18]. Because this sensor has very small deadzone (about 7.5 mm [12]), it also has a lot of other applications. On the one hand it can be used for measuring the elevation of the hexapod robot body over the ground [14]. On the other hand, the elevation can be measured for a quadrotor [7] or distance to a wall for aerial vertical climbs [13].

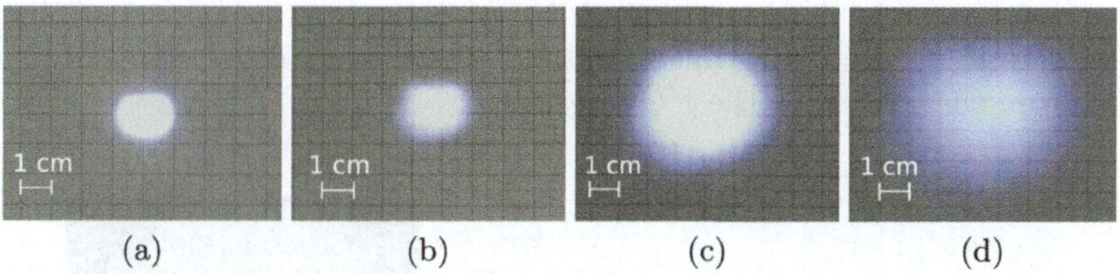

Fig. 3. A light spot formed by the GP2Y0A02YK0F sensor arranged vertically. The spot was created at the surface at a distance of a) 30 cm b) 40 cm c) 90 cm d) 140 cm

3.1 GP2Y0A02YK0F Sensor

For the GP2Y0A02YK0F sensor, its distance range is from 20 cm to 150 cm [10]. Due to using the triangulation method, the sensor is equipped with an emitter of coherent infrared light and a position sensitive detector (PSD). This detector is used for determining the position of the image of the light spot created by the emitter on the object of interest. As a source of light emission, an infrared LED is used. Moreover, the emitter is equipped with a lens which makes the light beam sufficiently coherent for the measured distance range. For this triangulation sensor, the light spot size of its beam is approximately 2 cm at a distance of 30 cm (see Fig. 3a). The spot of light is slightly deformed. It is better visible for the distance of 90 cm. Its size is approximately 3 × 4 cm (see Fig. 3c). For a distance of 140 cm, the size of the light spot increases up to 5 × 7 cm (see Fig. 3d). Moreover, its blur increases and is clearly visible. The presented pictures were taken by Microsoft LifeCam Studio web camera after removing its infrared filter. The size of the light spot and its intensity are greater than for short-range sensors e.g. GP2D120 [21] or GP2Y0A21YK0F [9]. The output quantity of measurement is voltage obtained from PSD detector. Due to the triangulation approach, this voltage U refers to the distance d according to the hyperbolic relation

$$d(U) = \frac{a}{U} + b, \tag{1}$$

where the coefficients a and b should be determined by using the calibration procedure. This relation is valid in the range in which the image of light spot created by the sensor emitter is correctly projected onto its PSD detector.

3.2 VL53L0X Sensor

The range of distance measuring of the VL53L0X sensor is up to 2 m. This sensor is equipped with lasers VCSEL (Vertical Cavity Surface-Emitting Laser) of Class 1 [11]. Its infrared light beam is not so coherent as it is for the GP2Y0A02YK0F sensor. Therefore, the light spots are much larger. Their images for the VL53L0X sensor obtained for a distance of 40 cm and 90 cm are shown in Fig. 4. At the distance of 40 cm, the diameter of the infrared light spot is about 19 cm. Its size increases up to 39 cm at a distance of 90 cm. The result of distance measurement is read from the sensor via the I^2C interface.

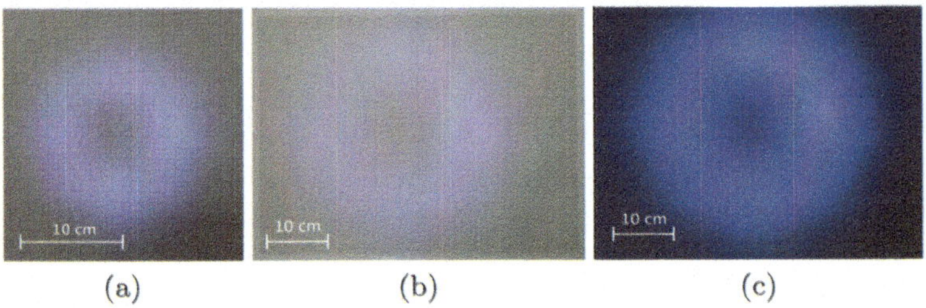

Fig. 4. A light spot created by the sensor VL53L0X at a distance of a) 40 cm b) 90 cm in a dimly lit room c) 90 cm in a dark room

4 Experiments

The main purpose of the experiments was to compare the results obtained for cylinder surface, edge effects and plain surfaces placed at an angle of 45° in relation to the sensors. The edge effect appears when only part of spotlight illuminates a surface edge. For optical triangulation sensors which use analog PSD detectors, it is known that their measurement results depend on arrangement of these sensors in relation to the surface edge. Therefore, in the datasheet of the GP2Y0A02YK0F sensor and other similar ones, it is explained how this kind of sensor should be set for different cases of measurements. In order to check the effect of sensor arrangement on the results obtained for these cases, two GP2Y0A02YK0F sensors were set in the sensor head in two characteristic orientations (see Fig. 2a). The experiments were performed for three cases. These cases are as follows

- a narrow plain surface placed parallel to movement of the sensor head,
- an oblique plain surface inclined to the linear scanner at an angle of 45°,
- a cylinder object.

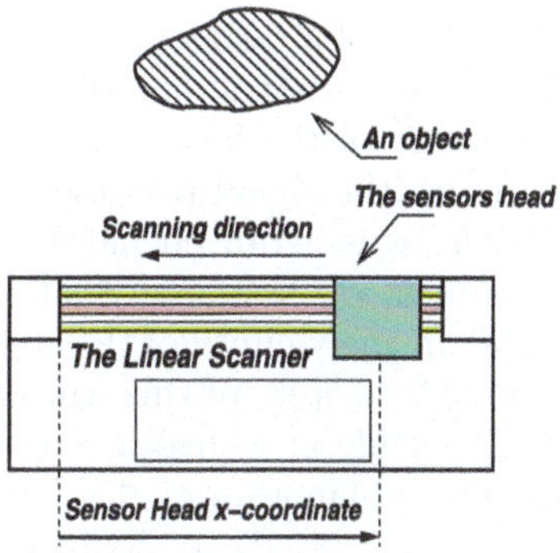

Fig. 5. The schema for performance of measurements in all experiments

For each case, the measurements were done for three distances. The schema for performing all experiments was the same. A selected object was placed in front of the linear scanner (see Fig. 5). Then, the line scanner was started and measurements were being performed during the movement of the sensor head along a distance of 45 cm. Measurements were made every 1 mm of the sensor head movement.

4.1 Small Plain Surface

In the first series of experiments, a narrow rectangular surface was used. It was placed at a distance of 40 cm (see Fig. 6), 90 cm (see Fig. 7) and 140 cm (see Fig. 7). For a distance of 40 cm, the measurement data obtained from the vertical triangulation sensor are as expected and easy to interpret. Along the plain surface, readings are almost the same. Because of the small light spot of this sensor, in the obtained measurements the edges of the surface are simple to distinguish. It cannot be said the same about the TOF sensor. Due to the larger light spot, values of measured distance are gradually increasing near the edges. The results obtained from the horizontal triangulation sensor have the most peculiar form. But explaining it is quite simple. For horizontal sensor arrangement when a part of a visible light spot vanishes, it causes the change of centroid position of light spot image created on the PSD detector. Consequently, in vicinity of right edge (i.e. when the PSD detector is on the left and the emitter on the right), this results in smaller values of measured distance. In the vicinity of left edge, the effect is opposite. Due to the hyperbolic relation expressed by Eq. (1), the change on the right side is larger than the change on the left side. The final effect also depends on the shape of the light spot. Its size is crucial as well. When the light spot is larger, the observed effect is also enlarged. It is well visible in the diagrams presented in Fig. 7 and in Fig. 8. The measurement results obtained from the TOF sensor for distances of 90 and 140 cm are in line with expectations. The larger the light spot, the greater the effect of the edges on the measurement results. For a distance of 140 cm, no part of the smooth surface can be distinguished. It looks rather like the surface of a cylinder. It can be also noticed that for greater distances, variation of measurement values increases. This is the consequence of the fact, that the larger the distance, the weaker the received signal. Thus, the signal-to-noise ratio is lower.

When the triangulation sensor is arranged parallel to the surface edge, the edge effect should not be manifested. Assuming that the light spot is perfectly round then for any visible part in the vicinity of the edge its centroid position on the PSD detector is not changed. In spite of this, the edge effect for the vertical triangulation sensor is clearly visible at a distance of 90 and 140 cm (see Fig. 7 and Fig. 8). The most probable explanation of this situation is the deformation of light spot (see Fig. 3). This corresponds to the fact that the more visible the deformation, the more distinct the edge effect.

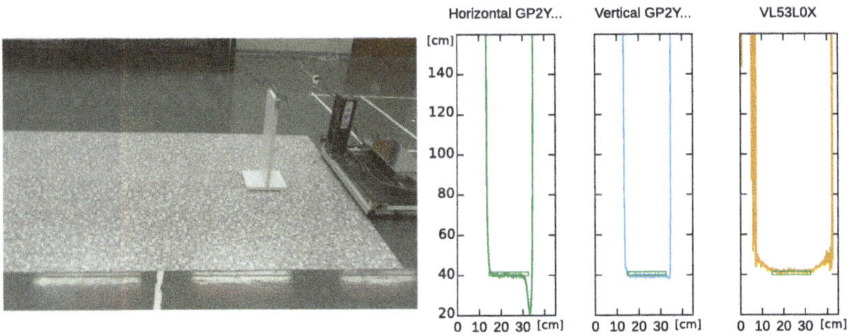

Fig. 6. Measurement results for the small wall placed at a distance of 40 cm

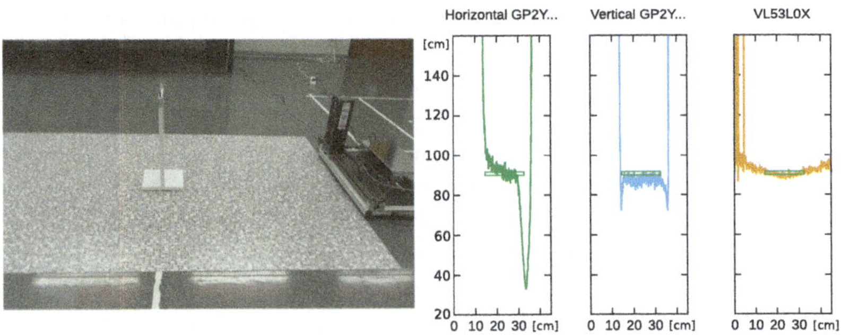

Fig. 7. Measurement results for the small wall placed at a distance of 90 cm

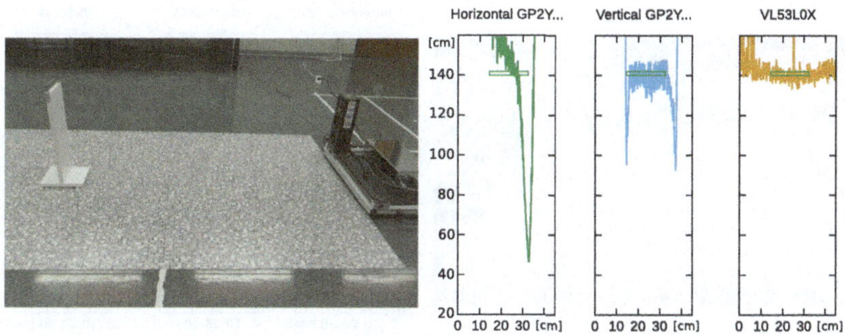

Fig. 8. Measurement results for the small wall placed at a distance of 140 cm

4.2 Oblique Plain Surfaces

The next series of measurements were performed for the plain surface which was inclined to the linear scanner at an angle of 45°. It was observed the edge effect. For the triangulation sensors it was the same as one previously discussed (see Fig. 9, 10 and 11) and it can be explained in the same way. The change that can

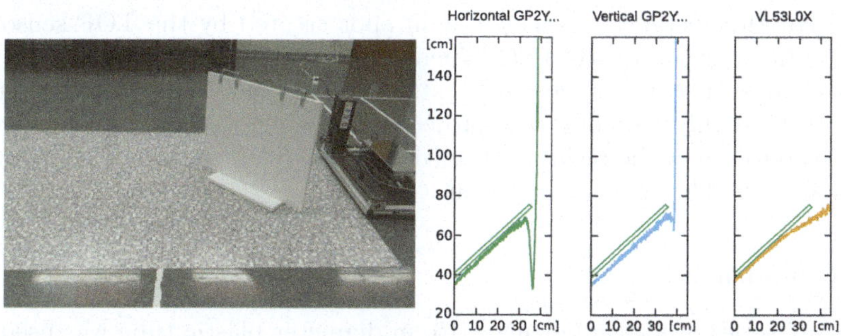

Fig. 9. Measurement results for the oblique surface inclined to the left placed at a distance of 30 cm

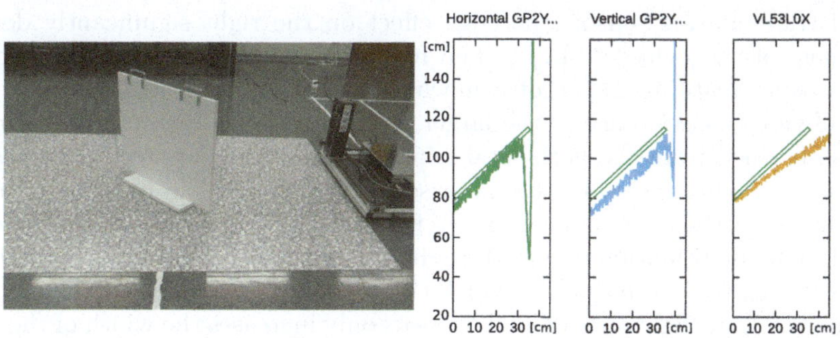

Fig. 10. Measurement results for the oblique surfaces inclined to the left placed at a distance of 70 cm

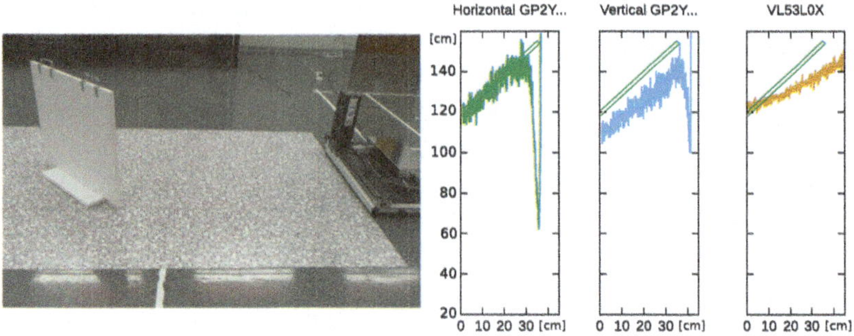

Fig. 11. Measurement results for the oblique surfaces inclined to the left placed at a distance of 110 cm

be noticed is the distance values determined by the vertical triangulation sensor. They are slightly smaller then real ones (see Fig. 9). The difference increases with the distance to the object (see Fig. 10 and 11). This suggests that this change is related to the increase of the light spot with distance. Another important difference was observed for the data obtained from the TOF sensor. Near the edge, the measured distance was a slightly smaller than the real one. This type of results was obtained when the test object was close to the linear scanner (see Fig. 9). For this distance, the entire light spot created by the TOF sensor was on the surface of the test object. When the test object was further, the same kind of changes in distance measurements were observed. However, the light spot was larger and only part of it was projected onto the object surface. Therefore, the interpretation of the results obtained from the TOF sensor and presented in Fig. 10 and Fig. 11 is not unambiguous.

4.3 Cylindrical Surface

In the last series of measurements a 16 cm diameter plastic tube was used. The tube had a uniform light brown color. The measurement results obtained for a distance of 40 cm (see Fig. 12) show that the best mapping of the object's surface was obtained for the vertical triangulation sensor. In the case of a horizontal triangulation sensor, the edge effect on the right significantly deforms the image of the cylinder shape. The measurement results obtained from the TOF sensor create an image of a much wider object than it actually is. This is the obvious consequence of the larger light spot. For a distance of 90 cm, the results obtained from the horizontal triangulation sensor appear to be peculiar (see Fig. 13). This is due to the increase in diameter of the light spot. In this case, the larger the light spot, the greater the impact on the measurement results caused by its deformation due to the cylindrical surface. This also applies to the vertical triangulation sensor, for which the edge effect is enhanced. The increase in diameter of the light spot of TOF sensors only increases the width of the cylinder outline obtained from this sensor. For the distance of 140 cm, the described

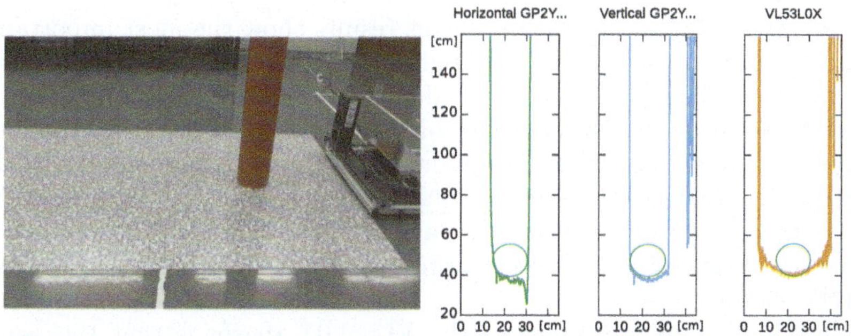

Fig. 12. Measurement results for the cylinder placed at a distance of 40 cm

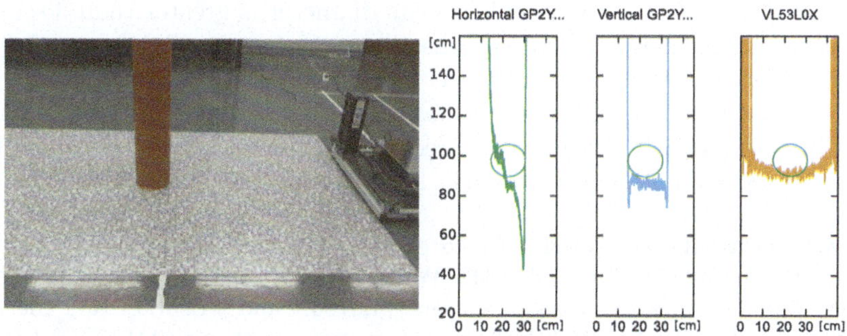

Fig. 13. Measurement results for the cylinder placed at a distance of 90 cm

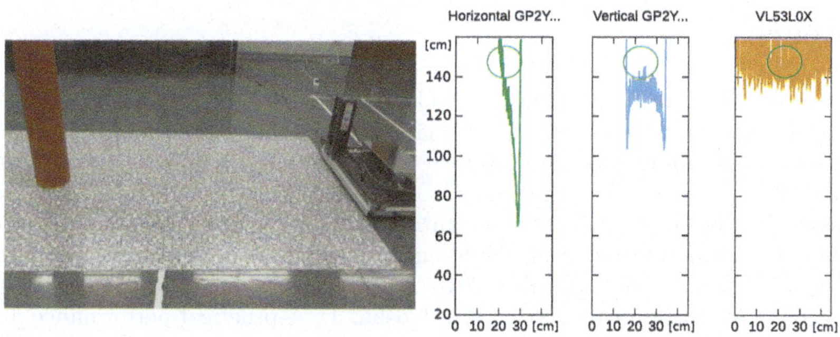

Fig. 14. Measurement results for the cylinder placed at a distance of 140 cm

above effects are enhanced (see Fig. 14). Moreover, as in the previous cases, a greater distance causes a greater effect of noise on the measured values.

5 Conclusion

The paper presents and discusses the results of experiments using the triangulation and TOF sensors. In this case, these sensors are GP2Y0A02YK0F and

VL53L0X respectively. The measurement results show the most important features of these sensors and allow their performance to be compared. Both sensors are active ones. Their features strongly depend on the width of the infrared beam emitted. In addition, measurement results obtained from the triangulation sensor strongly depend on its setting in relation to the curvature and discontinuity of the object's surface. A very important advantage of the GP2Y0A02YK0F sensor is that its beam is relatively narrow and coherent. This enables much more accurate detection of object surface discontinuities than when using the VL53L0X sensor.

A very important advantage of the VL53L0X sensor is that this sensor is compact in size. Its emitted light beam is not so coherent as the light beam emitted by the GP2Y0A02YK0F sensor. The light spot created by VL53L0X e.g. at a distance of 90 cm is very large (its diameter is greater than 30 cm). On the one hand, this can be seen as a disadvantage because it is not possible to measure a distance to a specific object. But on the other hand, this can be the important advantage. This feature makes it possible to detect obstacles in a wide sector. In this sense, it can be compared with ultrasonic sensors. Unlike them, however, there are not their drawbacks such as multiple or specular reflections.

Acknowledgements. We would like to express our great thankfulness to Igus for providing the module DryLin SLW-1040. We also want to thank very much ST Microelectronics for delivering samples of sensors VL6180X and VL53L0X and the microcontroller modules NUCLEO-F401RE. Finally, we want to thank ZAP-KOOPERACJA Sp. z o.o. for making the device's housing.

References

1. Abah, C., Orekhov, A.L., Johnston, G.L.H., Yin, P., Choset, H., Simaan, N.: A multi-modal sensor array for safe human-robot interaction and mapping. In: 2019 International Conference on Robotics and Automation (ICRA), pp. 3768–3774, May 2019
2. Ahsan, F., Hasan, K.M.U.: Seeker: autonomous maze-navigating and ball-potting robot. In: 2015 International Conference on Open Source Systems Technologies (ICOSST), pp. 52–57, December 2015
3. Hasan, O.A., Rashid, A.T., Ali, R.S., Kosha, J.: A practical performance analysis of low-cost sensors for indoor localization of multi-node systems. In: 2017 Internet Technologies and Applications (ITA), pp. 284–289, September 2017
4. Henderson, M., Kelly, S., Horne, R., Gillham, M., Pepper, M., Capron, J.: Powered wheelchair platform for assistive technology development. In: 2014 Fifth International Conference on Emerging Security Technologies, pp. 52–56, September 2014
5. Jans, R.M., Green, A.S., Koerner, L.J.: Characterization of a miniaturized IR depth sensor with a programmable region-of-interest that enables hazard mapping applications. IEEE Sens. J. **20**, 5213–5220 (2020)
6. Lentini, G., et al.: Alter-ego: a mobile robot with a functionally anthropomorphic upper body designed for physical interaction. IEEE Robot. Autom. Mag. **26**(4), 94–107 (2019)

7. Li, S., De Wagter, C., de Croon, G.C.H.E.: Unsupervised tuning of filter parameters without ground-truth applied to aerial robots. IEEE Robot. Autom. Lett. **4**(4), 4102–4107 (2019)
8. Lin, Y., Eskandarian, A.: Experimental evaluation of cooperative adaptive cruise control with autonomous mobile robots. In: 2017 IEEE Conference on Control Technology and Applications (CCTA), pp. 281–286, August 2017
9. Malheiros, P., Gonçalves, J.A.C., da Costa, P.J.C.G.: Towards a more accurate infrared distance sensor model (2009). http://gecad.isep.ipp.pt/iscies09/Papers/19November/iscies09_sharp_model.pdf
10. Sharp: GP2Y0A02YK0F – distance measuring sensor unit measuring distance: 20 to 150 cm. https://global.sharp/products/device/lineup/data/pdf/datasheet/gp2y0a02yk_e.pdf. Accessed January 2020
11. STMicroelectronics: VL53L0X – world smallest time-of-flight ranging and gesture detection sensor. https://www.st.com/resource/en/datasheet/vl53l0x.pdf. Accessed January 2020
12. STMicroelectronics: Time of flight: Principles, challenges, and performance. https://www.st.com/content/dam/technology-tour-2017/session-1_track-4_time-of-flight-technology.pdf. Accessed March 2020
13. Mehanovic, D., Rancourt, D., Desbiens, A.L.: Fast and efficient aerial climbing of vertical surfaces using fixed-wing UAVs. IEEE Robot. Autom. Lett. **4**(1), 97–104 (2019)
14. Minati, L., Frasca, M., Yoshimura, N., Koike, Y.: Versatile locomotion control of a hexapod robot using a hierarchical network of nonlinear oscillator circuits. IEEE Access **6**, 8042–8065 (2018)
15. Mustapha, B., Zayegh, A., Begg, R.K.: Microcontroller based wireless obstacle detection system for the elderly. In: 2014 4th International Conference on Artificial Intelligence with Applications in Engineering and Technology, pp. 325–329, December 2014
16. Papa, U., Del Core, G., Giordano, G., Ponte, S.: Obstacle detection and ranging sensor integration for a small unmanned aircraft system. In: 2017 IEEE International Workshop on Metrology for AeroSpace (MetroAeroSpace), pp. 571–577, June 2017
17. Portasiak, P.K.: Laboratory stand for testing distance sensors. Engineering thesis, Wrocław University of Science and Technology, February 2018. (in Polish)
18. Sundram, J., Van Nguyen, D., Soh, G.S., Bouffanais, R., Wood, K.: Development of a miniature robot for multi-robot occupancy grid mapping. In: 2018 3rd International Conference on Advanced Robotics and Mechatronics (ICARM), pp. 414–419, July 2018
19. Thamrin, N.M., et al.: Tree diameter measurement using single infrared sensor for non-stationary vehicle context in agriculture field. In: 2013 IEEE 4th Control and System Graduate Research Colloquium, pp. 38–42, August 2013
20. Verbeke, J., Vantilt, J., Vanthienen, D., Vochten, M., Debruyne, S., De Schutter, J.: A constraint-based flight control system architecture for UAVs using the iTASC framework. In: 2016 International Conference on Unmanned Aircraft Systems (ICUAS), pp. 310–319, June 2016
21. Wnuk, M.: Measurement of deviation from the vertical of a balancing robot trunk. Technical report SPR nr 28/2008, Inst. Inform. Autom. Robot. Politechniki Wrocławskiej, November 2008. (in Polish)

Image-Based Monitoring of Glass Melting Process

Piotr Garbacz(✉) and Piotr Czajka

Łukasiewicz Research Network – Institute for Sustainable Technologies, 6/10 K. Pulaskiego Street, 26-600 Radom, Poland
{piotr.garbacz,piotr.czajka}@lukasiewicz.gov.pl

Abstract. The paper presents an image-based method for monitoring glass melting process in a glass furnace. To ensure the high reliability of the monitoring system and compliance with the industrial conditions, a robust vision head module was designed and integrated with the cooling and refraction systems. The vision system structure was verified in the course of long-term tests, which allowed for the identification of the risk factors and impact of operating conditions. Based on the images recorded during the operation, algorithms for determining selected process parameters were implemented. Additionally, self-diagnostics algorithms were proposed to identify potential undesired system behaviours in the future operation. The integration into the glassworks network and knowledge database was also an important element of the described system. The article ends with the presentation of the benefits of an ongoing image-based monitoring of glass melting processes and of the capabilities of the statistical data analysis employing a machine learning technology.

Keywords: Glass industry · Industrial process monitoring · Machine vision

1 Introduction

Nowadays, industrial process monitoring is gaining more attention in the majority of the manufacturing sectors. Due to its economic competitiveness and given applicable environmental regulations, resource management has become a critical task in most industries. Glass manufacturing generally can be described as a medium or even low-margin branch. Therefore, new technological solutions are sought in order to adapt to the changing market trends.

In automated glassworks, the daily demand for raw glass material can be as large as several hundred tons [1]. The melting process monitoring is crucial for the finished product quality and thus for the company's financial result. Apart from the typically monitored process parameters, like temperature, pressure or energy consumption, there are also other aspects of the process important for the finished product quality that can be monitored only visually. Although on the market there are vision systems specially designed to monitor the inside of the glass furnace, the visual evaluation of melting process is, in most cases, carried out by human operators [2, 3].

The glass preparation process consists of a few stages. Once the raw materials have been supplied at one end of the tank, the melting process begins as a result of firing.

R. Szewczyk et al. (Eds.): AUTOMATION 2022, AISC 1427, pp. 296–305, 2022.
https://doi.org/10.1007/978-3-031-03502-9_30

Depending on the location of the burners, the furnaces are divided into side-port (Fig. 1a) and end-port furnaces (Fig. 1b). The glass melt creates convection currents which allow the batch to mix. The length of the conveying path along which the glass mass is transported before reaching the other side of the bath should fall within a precisely defined range [4, 5].

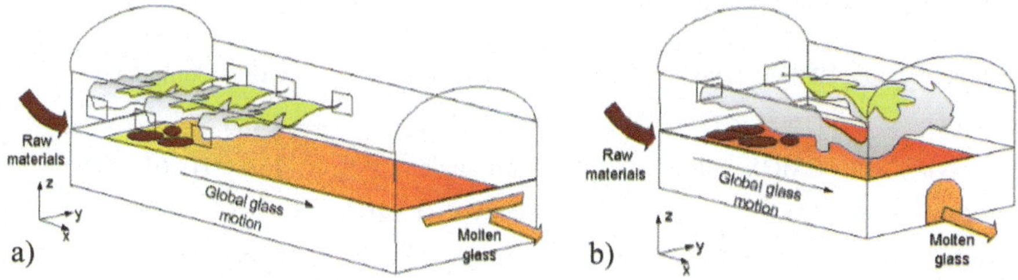

Fig. 1. Diagrams of glass furnaces with different flame paths [4]: a) side-port furnace, b) end-port furnace

Other important process elements which can be visually inspected are the flame and the bubbling system. As regards the flame, the monitoring may concern its brightness, location, direction, shape or geometrical parameters (e.g. length or width) [6–8]. As for the bubbling system, on the other hand, the inspection may help in the identification of process disturbances such as unevenness or bubbler failure [8, 9].

2 Vision System Architecture

The structure of the designed vision system structure [8] is shown in Fig. 2. In the front wall there is an opening for inserting the vision head. However, due to extreme

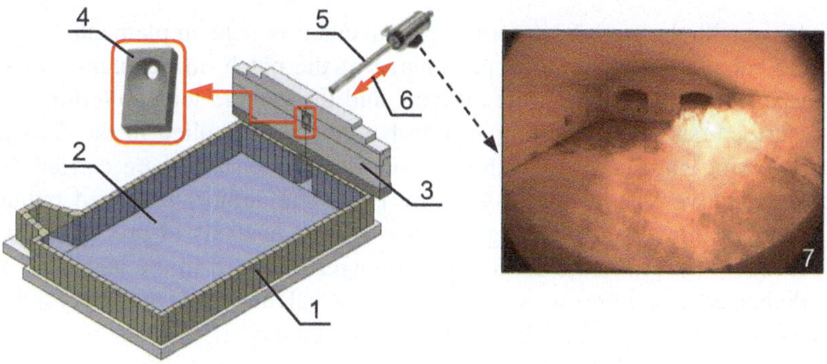

Fig. 2. Concept of a vision monitoring system inside a glass furnace [8]: 1 – simplified model of the glass melt tank, 2 – glass melt bath, 3 – furnace front wall, 4 – fitting with an opening for the vision head, 5 – example of a vision head with a furnace endoscope, 6 – retraction system for the endoscope, 7 – sample image of the furnace interior captured by an RGB camera

temperatures inside the furnace (1,500 °C), only an endoscope is inserted. It consists of an internal image transmission system and a water jacket cooling system.

The vision head consists of a digital visible light camera and a wide angle pinhole lens. The main specifications of the vision system of the camera and the lens are presented in Tables 1 and 2 below.

Table 1. Main parameters of the camera

Manufacturer	Allied vision
Resolution	(1600 × 1200 px)
Sensor type	CMOS, colour
Sensor size	1/1.8″
Interface	GigE PoE
Mounting type	C-mount
Temperature	5 °C–45 °C

Table 2. Main parameters of the lens

Manufacturer	CBC (Computar)
Focal length	4 mm
View angle	76.3° (H), 61.6° (V)
Max sensor size	1/2″
Aperture	F2.5 – F32C
Mounting type	C-mount
Temperature	−20 °C–50 °C

Figures 3a and b show the vision head with the mounting system, with which it is fastened to the structure of the furnace, and the head retraction system with a pneumatic actuator. The operator's stand for a glass melt surface monitoring system was developed and manufactured (Fig. 3c). The dedicated industrial casing protects the following elements placed inside: an industrial computer, an LCD monitor, a programmable PLC, and an emergency UPS power supply.

Long-term tests were performed at the glassworks to record sample images and verify mechanical and electrical systems under operating conditions.

3 Image Processing and Analysis

In accordance with the guidelines for the system end users, the implemented algorithms include the calculation of the total percentage of the batch mix content, the analysis of the shape of the flames from the burners, and the camera lens pollution detection. In the described case, the bubbling system has not been applied in the furnace. The image processing and analysis in the designed vision system were carried out with the use of Adaptive Vision Studio software [10]. The authors reviewed publications relating to algorithms used for image processing and analysis of processes occurring in glass furnaces [2, 3, 9, 10]. Based on the images recorded in the period of over six months, dedicated algorithms were proposed for the calculation of the important process parameters and for the system self-diagnostics procedures.

3.1 Image Rectification

The first stage of image processing consists in the rectification of the captured images of the glass surface to obtain a top-view (Fig. 4b) and a back-wall Region of Interest (ROI) (Fig. 4c).

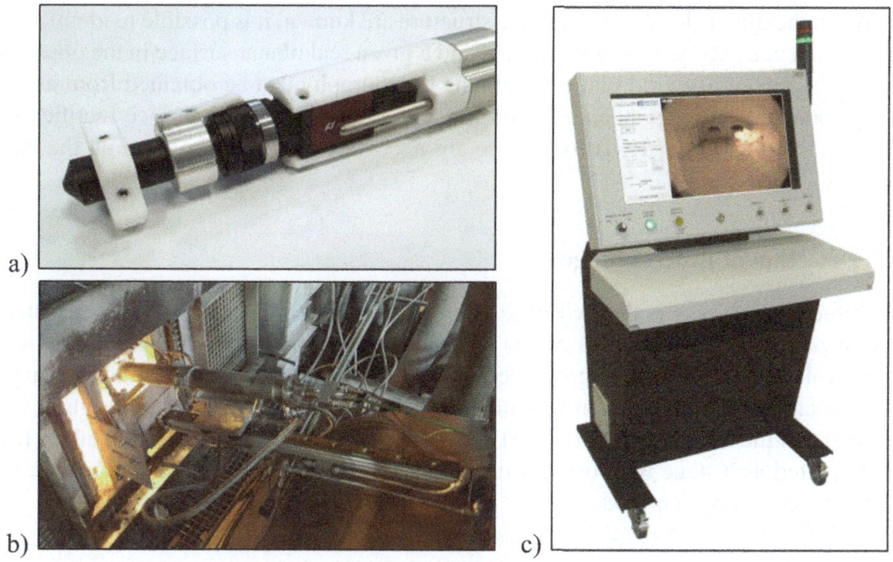

Fig. 3. Main elements of the vision system [8]: a) detailed view of the developed image transmission system which is placed inside the cooler with a water jacket, b) mounting system for fixing the vision head to the furnace structure and the endoscope retraction system, c) developed operator's stand to be placed in the furnace control room

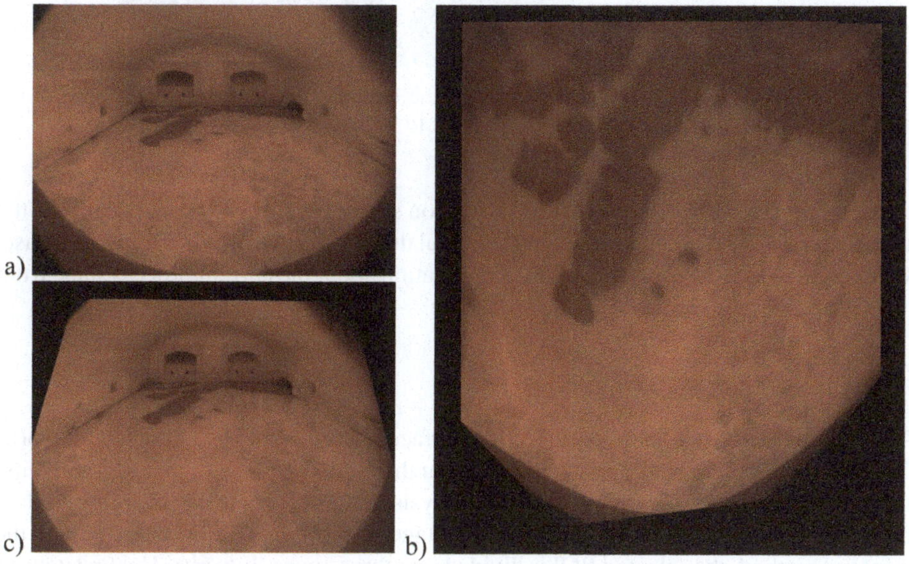

Fig. 4. Sample image of the furnace interior: a) captured by the camera, b) rectified to glass surface, c) rectified to back-wall surface

When the dimensions of the furnace structure are known, it is possible to identify the correspondence between the image plane and a given real planar surface in the observed space. The image-to-world plane coordinate homography can be obtained from as few as four points of correspondences. Additionally, the correct glass surface rectification and the identification of ROIs to analyse are necessary to detect the edge on the back wall of the furnace.

3.2 Detection of Process Stage

Batch melting process is divided into several stages. In order to achieve the assumed process parameters, the burners on the left and right are switched on alternately with a short break for the reversion stage. To detect the stage of the process, the colour raw image from the camera is converted to a monochromatic image, and the average brightness is calculated in the selected ROIs (Fig. 5). The result below the set threshold in either ROI is interpreted as a stage with burners turned off. If the burners on each side are turned off, the reversion is identified.

Fig. 5. Detection of process stages by assessing the brightness; the average pixel value in the right ROI (blue) is approximately 65 and in the left ROI (orange) – approximately 225.

By using the described algorithm, the vision system can work independently of the furnace control system. This is especially useful during the testing and pre-launch phase. An alternative solution is to receive the information about the current process stage from the burner control system.

3.3 Segmentation of Batch Material

The distribution of the batch on the glass surface is studied in several research works [2, 3, 10]. To distinguish the raw material from the melted glass, a proper segmentation method needs to be applied (Fig. 6a). The analysis of the process correctness is based on the calculation of the batch mix cover percentage over the melted glass surface. In the proposed method, the selected ROI is divided into three zones (Fig. 6b). The first zone in which the feeder system operates can be almost entirely covered. In the second (yellow) zone the percentage of the surface coverage should be low. No raw material should be

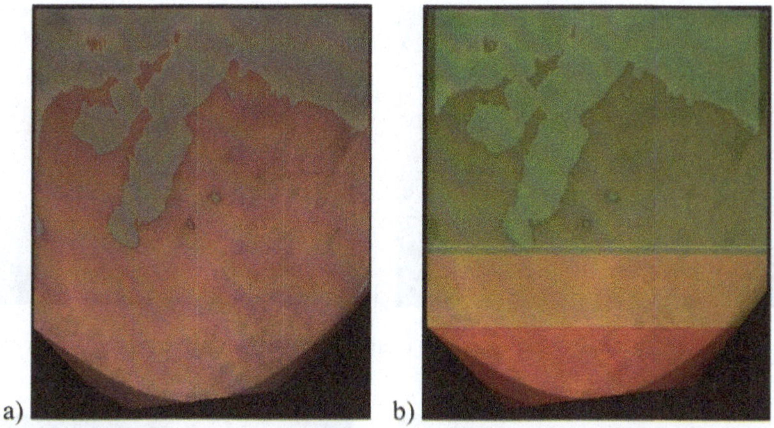

Fig. 6. Glass surface inspection: a) detected raw material, b) ROI divided into zones

detected in last (red) zone. Operators can change the sizes of those zone depending on the production type.

The main difficulty in the detection of the raw material are uneven lightning conditions. The surface closer to the camera is much brighter than that near the feeder system. Setting a global threshold level is not an effective solution. In the proposed algorithm, adaptive thresholding methods were used.

3.4 Flame Analysis

The monitoring of the flame shape by human operators may not be an easy task due to the rapidly changing view (Fig. 7a). With the use of the rectified top view image (Fig. 7b) it is possible to analyse the shape of the flame in every captured image and the trends of the moving average values of the calculated parameters. Due to the perspective, it is not possible to map the whole flame correctly. Nevertheless, the front boundary can be measured. In the proposed algorithm, three variables are analysed: A – the outermost point, B – the point closest to the centre, and C – the outermost point on the side wall edge.

3.5 Self-diagnostics Algorithms

During the experimental test carried out at the glassworks, the camera recorded several undesired events. After about one year of operation, a cooling system broke down (Fig. 8a). Although the vision head control system is equipped with a temperature sensor and the temperature of the camera sensor is monitored, the above-described situation could lead to overall system failure. Therefore, a dedicated leak detection algorithm based on the background estimation in the image sequence using the Mixture of Gaussians model was implemented [12].

It was observed that in the long term operation dirt is easily deposited on the vision head lens (Fig. 8b), most probably as a result of an improper air circulation. Such a

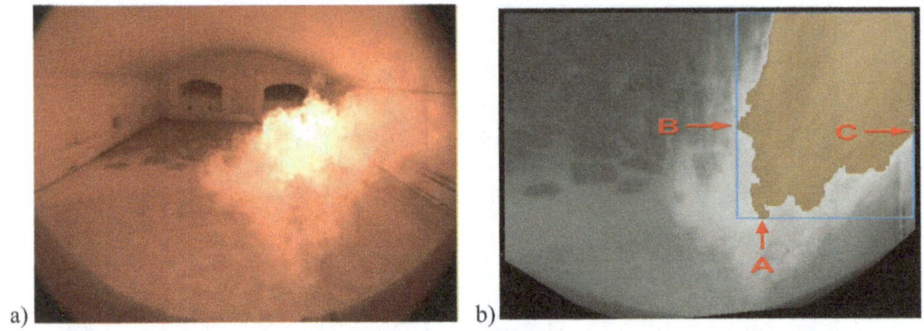

Fig. 7. Flame analysis: a) camera image, b) rectified image with flame segmentation observed

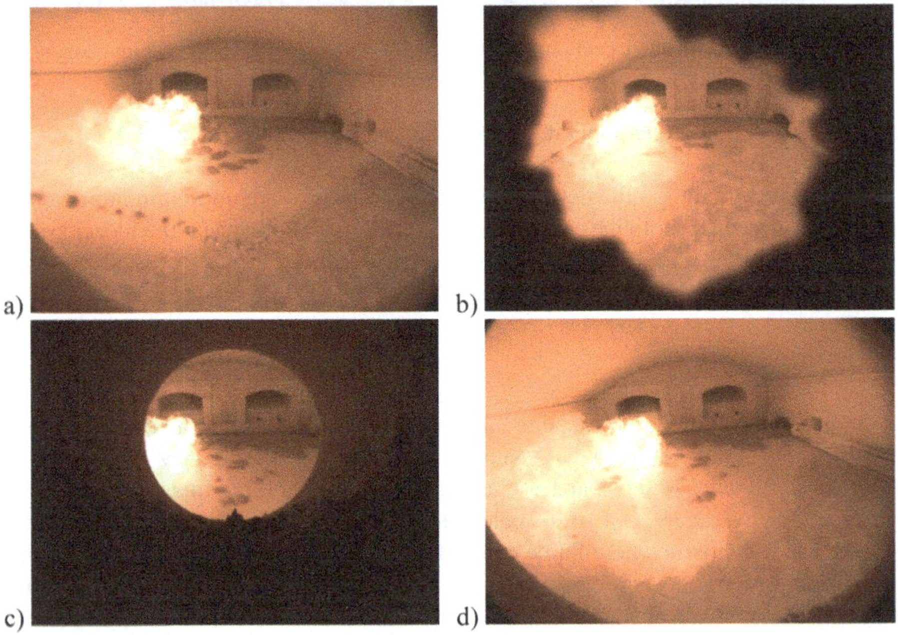

Fig. 8. Incorrect vision system state: a) leak in cooling system, b) dirty frontal lens, c) incorrect position of vision head, d) rotated vision module

situation can be detected with the use of the image brightness statistics analysis. Similar images can be observed when the camera module system is not in the correct position once the refraction system has been activated (Fig. 8c). However, in this case, the control system sent an appropriate message to the analysis software to enable these two issues to be detected. Another problem that may occur concerns the rotation of the vision head (Fig. 8d) at the time of its insertion into the furnace. This can be detected by measuring the angle of the glass surface level on the front wall of the furnace. Optionally, such a situation can be interpreted as a system error or the camera image can be rotated by the software by the calculated angle.

4 Data Acquisition System

The development of Industry 4.0 and big data technologies means that business executives are looking for new solutions to increase production efficiency. Engineers can use big data analytics for predictive maintenance, self-service systems or even to automate production. The Internet of Things is one of the most important features of the fourth industrial revolution. In this concept any sensor that produces valuable data should be capable to communicate with the factory network [11]. Ensuring interoperability between the communication protocols existing in the domains of industrial applications is a key for big data analysis systems. With the development of the OPC UA (Open Platform Communications Unified Architecture), the standard integration of devices running on different platforms is simplified.

The vision system for monitoring glass melting processes is a part of the developed system (Fig. 9) dedicated to improve the manufacturing effectiveness and product quality in the glass industry [8]. Each module of the system carries out its assigned task that consists in the monitoring of the different stages of the production process and product quality inspection, including the preparation of the tools for product forming, oversight of the melting process, and the quality inspection of glass products.

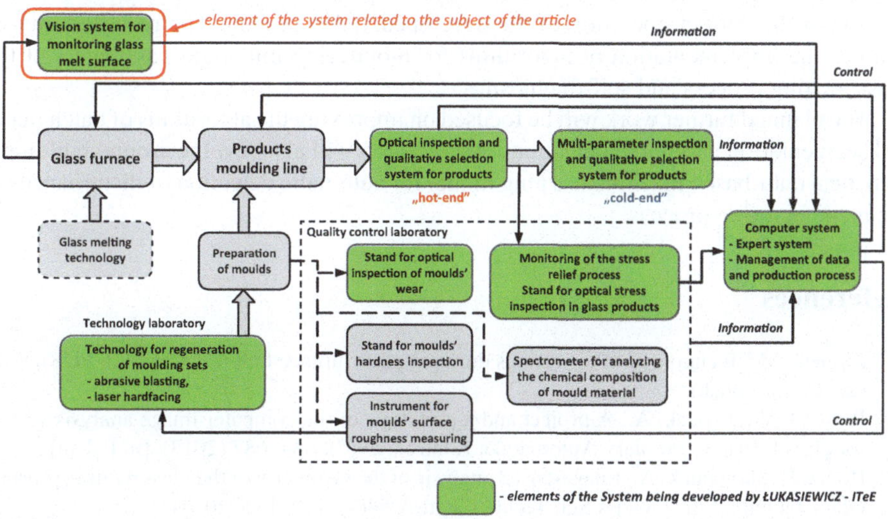

Fig. 9. Integrated, multifunctional system for improving effectiveness of manufacturing process and product quality – general structure (developed by Łukasiewicz Research Network – Institute for Sustainable Technologies) [8]

In the design network system, the client-server UPC UA model is used to feed information to the main database which is based on GE Proficy Historian software [12]. The aggregated historical data can be visualised and used by the glassworks process technologist to find production disturbances and analyse the root cause of the problems identified. Additionally, the integration with GE Proficy CSense [13] machine learning software is available for supporting historical data analysis to identify problems, discover root causes and predict the future performance [14].

5 Conclusions

The visual evaluation of the glass melting process occurring inside the furnace at the time of manufacture is a standard procedure at glassworks. However, in most cases, no automatic systems are still used and the analysis of the process correctness is based on human operators' subjective and periodic assessment. The use of image-based monitoring methods enables the acquisition of objective and continuous data which can be used as input source for the glassworks knowledge database. Additionally, machine vision also provides instant feedback if selected parameters exceed the tolerance value.

The proposed system structure is based on low-cost visible light camera which can significantly improve return of investment and still provide useful information to technologists. Alternative infrared camera-based solutions can provide additional data but the cost-to-benefits ratio does not always justify their application. In line with the Industry 4.0 concept, the OPC UA standard is used to transfer inspection results to the glassworks main database.

Tests carried out under real conditions on the production line allowed for the verification of the correct operation of the developed vision system. The recorded images enabled the implementation of algorithms for monitoring important parameters of the glass melting process and self-diagnostics.

The planned further work will be focused on improving the algorithms of batch material segmentation and flame detection. The research will also revolve around building a statistical data-based machine learning model for automatic detection of disturbances in the manufacturing process.

References

1. Ziomek, M.: Technology of operations during a glass furnace failure. W akcji **3**(2018), 8–13 (2018). (in Polish)
2. Rotter, P., Skowiniak, A.: A project and a prototype of the computer image analysis system for glass furnaces. Pomiary Automatyka Kontrola **59**(7), 684–687 (2013). [in Polish]
3. Rotter, P., Skowiniak, A.: Image-based analysis of the symmetry of the glass melting process. Glass Technol. Eur. J. Glass Sci. Technol. Part A **54**(3), 119–131 (2013)
4. Auchet, O., Riedinger, P., Malasse, O., Iung, C.: First-principles simplified modeling of glass furnaces combustion chambers. Control. Eng. Pract. **16**(12), 1443–1456 (2008)
5. Piech, J.: Ceramic and gas furnaces. AGH Uczelniane Wydawnictwa Naukowo-Dydaktyczne, Cracow (2001). (in Polish)
6. Lu, G., Yan, Y., Huang, Y., Reed, A.: An intelligent vision system for monitoring and control of combustion flames. Meas. Control **32**(6), 164–168 (1999)
7. Shenzhen Enerson Technology Company: Endoscopic flame monitoring system for heating furnace. https://cctv-ax.com/endoscopic-flame-monitoring-system-for-heating-furnace/
8. Czajka, P., Mizak, W.: A method for monitoring glass melt surface in a glass furnace. J. Mach. Constr. Maintenance **4**(2019), 37–48 (2019)

9. Glass Service: Expert System ES III^{TM} - Batch monitoring. https://www.gsl.cz/services-products/smart-process-control/advanced-control-expert-system-es-iii/batch-monitoring/
10. Adaptive Vision: Adaptive Vision Studio - machine vision software. https://adaptive-vision.com
11. Klemiato, M., Rotter, P., Skowiniak, A.: Analysis of batch asymmetry and batch line position for the decision support in the glass melting process. Prod. Eng. Res. Devel. **15**(5), 725–734 (2021). https://doi.org/10.1007/s11740-021-01053-3
12. Marcomini L., Cunha, A.: A Comparison between Background Modelling Methods for Vehicle Segmentation in Highway Traffic Videos. arXiv preprint arXiv:1810.02835 (2018)
13. Cognex Corporation: Industry 4.0 and Machine Vision - The Promise of the Smart Factory and the Industrial Internet of Things (2017). https://www.cognex.com
14. GE Digital: Proficy Historian - data historian software. https://www.ge.com/digital/applications/proficy-historian
15. GE Digital: Proficy CSense - industrial analytics software. https://www.ge.com/digital/applications/proficy-csense
16. Cavalieri, S.: A proposal to improve interoperability in the Industry 4.0 based on the open platform communications unified architecture standard. Computers **10**(6), 70 (2021). https://doi.org/10.3390/computers10060070

The Effect of Camera Exposure on the Results of Spatial-Frequency Processing and the Quality of the Obtained Amplitude Images

Piotr Miś(✉) and Przemysław Szulim

Faculty of Automotive and Construction Machinery Engineering, Warsaw University of Technology, Narbutta 84, 02-524 Warsaw, Poland
{piotr.mis.dokt,przemyslaw.szulim}@pw.edu.pl

Abstract. The paper presents the results of experiments which purpose was disclosure the influence of image recording parameters, such as exposure time, on the quality of the resulting amplitude image. An amplitude image is a condensed data representation which is the result of spatial-frequency processing. It is an author's method that uses spectral analysis and focuses on the study of changes in pixel intensity levels in time. The amplitude image, proposed by the authors, is something new in relation to known solutions using pulsating visual markers. The main motivation to start work in this area was the need to design a localization system that will generate position information from several independent sources. The method can be used in the issue of localization and navigation. The ability to search visual markers located at a great distance from the receiver is its special feature. The authors present the results of research into the impact of the recording camera exposure time on the quality of the resulting images in order to select the best value. The beginning of the manuscript presents a description how the exposure time of the camera can potentially affect the quality of the obtained images. The description of the problem was supported by a literature analysis of papers related to the issue of localization based on the visual markers and the camera. The next section presents the description of spatial-frequency processing and the main purposes of generating amplitude images. The research part describes the assumptions, equipment and research procedure. Then, the results of the research into the influence of the camera exposure time on the quality of the resulting amplitude images are presented.

Keywords: Exposure time · Image processing · Spectrum analysis · FFT · Visual markers

1 Introduction

The use of image recording as a signal and information carrier is becoming more and more common by the development of computers. The high computing power of today's processors and graphics cards enables the use of image analysis in real-time systems.

The paper focuses on the issue of localization based on the extraction of information from visual images. This issue is widely used and described in many publications. A

R. Szewczyk et al. (Eds.): AUTOMATION 2022, AISC 1427, pp. 306–316, 2022.
https://doi.org/10.1007/978-3-031-03502-9_31

different approach to processing information contained in the image is new in compared in to existing solutions based on visual markers and images from camera. The proposed algorithm should allow solve the localization problem using visual markers located at a large distance from the camera (compared to systems composed of similar component components) which is a significant novelty. The system will not use radio waves. This feature increases its resistance to deliberate interference (contrary to case of GNSS system).

The presented image processing algorithm can be used to build a vision system that determines the position of objects in the reference system. The vision system performs tasks by searching visual markers on recorded images. If the coordinates of the visual markers in the reference system are specified, the vision system is able to estimate the position of the recording camera and the object equipped with it. It is also possible to determine the position of the visual markers if the location of the image recording system is known.

A visual marker is a characteristic object with a specific feature or parameter which is used by the vision system in order to recognize it in the recorded images. Vision systems based on markers are widely used in the localization task. Some information about using visual systems were presented in [1, 3, 4]. Examples of characteristic features of the visual markers are shape or color. This kind of markers are used in conventional solutions. This paper is an extension of the research on the proprietary method of localization. Authors use visual markers emitting pulsating light with a specific range of wavelength, flickering frequency and phase.

The use of visual markers is not a new solution. The reviews [1, 3, 4] present localization methods that use LEDs emitting radiation with a specific frequency. However, the markers are used with photodiodes as a receiver in most of the methods. The papers [9–11] present examples of localization methods using photodiodes and visual markers emitting visible or infrared light.

A special feature of the presented algorithm is the use of the camera (as a receiver) to locate the pulsating visual markers. The markers pulse with specific frequency. This approach requires the use of other processing methods in relation to the measured signal by the photodiode. The algorithm processes the signals of change of the pixel intensity level which are generated from the recorded images. The method of determining the position of the markers is different in relation to solutions that use photodiode.

The presented method concentrates on the analysis of the change in the intensity level of pixels in the recorded images. The signal of change of intensity level is generated for each pixel. Then, the generated signals are subjected to spectral analysis.

The algorithm performs a one-dimensional Fourier transform for each pixel. The result are amplitude spectra. Theirs number is consistent with the number of pixels of the recorded single image. It is different processing than the two-dimensional Fourier transform. The algorithm can generate an image that represents the amplitude values for the pulsation frequency of the visual marker. Knowing the marker pulsation frequency it can read amplitudes from the spectra for each individual pixel. This is a new representation of the data that the authors call working an amplitude image. It should not be mistaken with the amplitude component of the 2D Fourier transform. The resulting amplitude image is a condensed result of the spectral analyzes of the change in intensity levels of individual pixels.

The combination of the pulsating visual markers and the camera is not an innovative solution. The paper [2] presented the possibilities of combining pulsating visual markers and a camera. However, the discussed algorithm is distinguished by the idea of compensated data recording in the form of an amplitude image in relation to the solution from [2]. Section 2 presents more accurate description of generating of the amplitude image.

The vision system should include a more number of factors that can affect the quality of image recording in the case of recording by a camera. It is worth focusing on improving the quality of the amplitude images in order to reduce the probability of the system not recognizing the marker. The camera exposure time is a parameter that defines the time for which the camera shutter is open and the light reaches the matrix. The papers [6–8] present various methods of selecting the exposure time parameter depending on the specific problem to solve. The authors of article [6–8] try to find an algorithm for the automatic selection of the exposure time depending on the application of the system and the results they want to achieve.

The aim of the manuscript is to investigate the influence of the exposure time on the results of the spatial-frequency processing and on the quality of the obtained amplitude images. The spatial-frequency analysis of images is aimed at locating a marker in the image – a pulsating LED with a specific frequency. The spatial-frequency method is described in more detail in Sect. 2. Section 3 describes the research procedure, assumptions and the measuring equipment. Section 4 presents example of the obtained amplitude image and the result of the measurement. The last two chapters contain conclusions from the research and subsequent development paths of the processing algorithm.

2 Description of the Processing Algorithm

The main task of the algorithm is to locate the pulsating visual markers on the recorded images. The visual marker is a light source that pulses with a specific frequency. The first step, The algorithm generates signals of change of intensity level in time for each pixel from the set of successively registered N images. If the image has size of $W \times H$ pixels, the algorithm generates $W * H$ signals with a length of N elements. Then, each of the signals is subjected to the spectral analysis. The number of amplitude spectra is the same as the number of previously generated signals. The next step, the algorithm checks the amplitude value on each spectrum for the marker pulsation frequency and saves it on a new image. The coordinates of the amplitude value in the new image correspond to the coordinates of the pixel whose spectrum is actually checked. High values of the read amplitudes mean that the change in intensity on a specific pixel coincides with the frequency of the marker pulsation. This means that the visual marker has been registered on checking the pixels. Consequently, the algorithm generates an amplitude image in which the high-value pixels show the position of the visual marker on the registered images. Figure 1 shows a diagram of the algorithm's operation, which was previously described. More detailed information about described method can be found in [5].

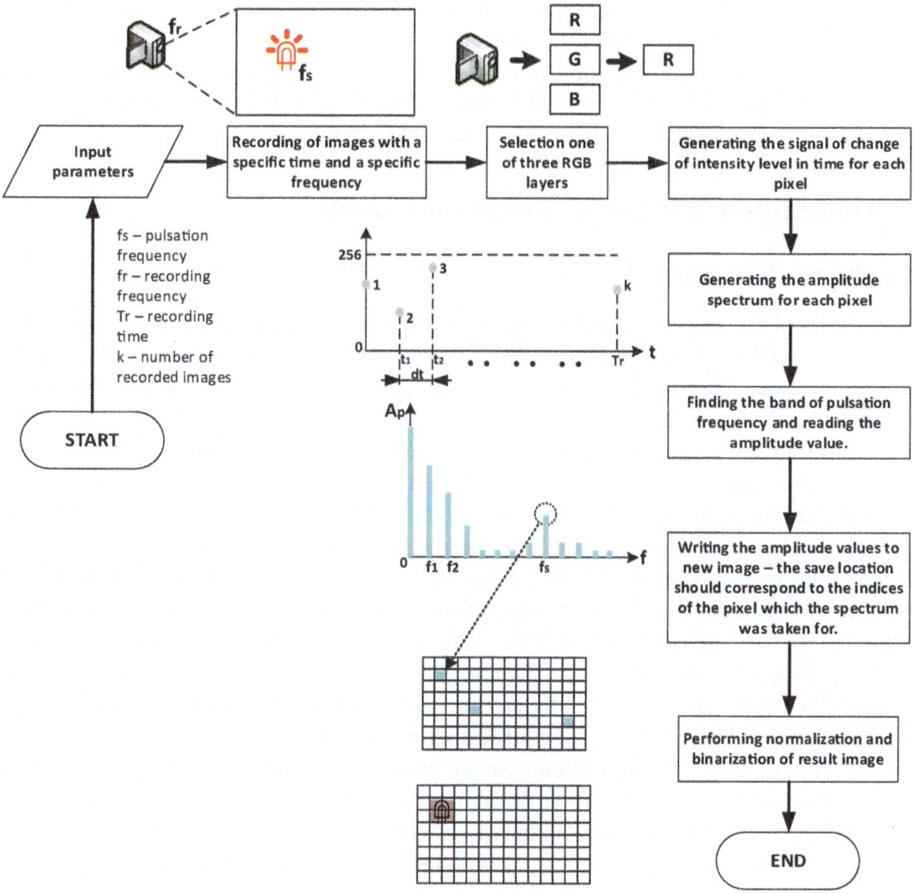

Fig. 1. Block diagram showing the individual steps of the operation of the spatial-frequency processing algorithm [5]

The normalization of the amplitude image can be performed in order to better present the effects of the algorithm's operation. It consists of dividing the pixel values of the amplitude image by the largest recorded value. Consequently, the pixel values assume a value in the range <0, 1>. The pixels that the change corresponding to the marker pulsation frequency are amplified and the pixels which are responsible for the image background are suppressed. The normalized amplitude image indicates the area in the image where the marker has been recorded.

The algorithm requires recording a few images to perform a correct operation. Their number depends on the camera recording frequency, image size and the computing power of the processing computer. The bigger number of images, the greater the accuracy of the generated amplitude spectra. On the other hand the computation time also increases. Moreover, the position of the visual markers is averaged over a larger number of frames so accuracy of positioning is smaller. It is calculated for a larger time sample. The above options should be taken into account while determining the number of processed frames in a single step of the algorithm's operation.

The operation time of the algorithm is not the main topic of this. The authors assume a constant number (100) of processed frames in a single step of the algorithm's operation in order to ensure high resolution of the generated amplitude spectra.

3 Description and Purpose of the Research

3.1 Research Problem

The results of the discussed algorithm depend on the value of the intensity around the pulsating visual marker. Too high exposure time can introduce the effect of saturation of the channels recording a pixel color. The signal of change of the pixel intensity can achieve too large mean component. The amplitude corresponding to the marker pulsation frequency can be suppressed in this case. Similar consequences are observed when the exposure time is too short. Therefore, it would be optimal if the exposure time was selected in such a way that the time signal of the intensity changes caused by the pulsating light of the marker would have the maximum amplitude. The spatial-frequency processing can also be unreliable for an outdoor localization systems, especially in high background sunlight. Too high exposure of the camera can cause that too much sunlight is captured. The algorithm might have problem to locate the visual marker correctly because of suppression of the light from the pulsating visual marker (the negative impact of strong lighting can be reduced by controlling the light power of the pulsating visual marker, but it will not be the subject of further considerations).

The results of the above problems can be reduced by selecting the appropriate exposure time of the camera. Therefore, the main aim of the research is check the impact of the exposure time of the camera on the quality of the resulting amplitude images.

Two parameters were used to assess the quality of the amplitude images:

- the ratio of the amplitude value for the marker pulsation frequency to the mean amplitude all image,
- the normalized mean value of background intensity level of the amplitude image.

The first parameter indicates the ratio of the intensity of the pulsating visual marker to the mean intensity level for the recorded images. If this parameter is high, the pulsating visual marker should be better visible in the recorded images. The parameter is determined from the amplitude spectrum for those pixels which the pulsating marker has been registered on.

The second parameter assumes linking the quality of the resulting images with the noise value of the background. This is the mean value of the background intensity level (without markers) of the normalized amplitude image. It allows to estimate the noise level and indirectly check the quality of the generated image. The greater value of index means more noises and lower quality of the resulting images. A detailed description of the calculation of the presented parameter is found in [5].

The above two indicators were used to examine and evaluate the exposure time of images.

3.2 Equipment Used for Tests

The following components were used for the tests:

- three 1W LEDs in red, green and blue were used as markers; additionally, a white plastic ping-pong ball was placed on the LEDs in order to bring the shape of the registered marker closer to the ball (The LED current was controlled so that the results could be related to the catalog power of the emitted radiation),
- Basler acA1600-60gc cameras with CMOS matrix were used as the recording device; its parameters are available in [12],
- development kit with an STM microcontroller to generate a sinusoidal current signal with a specific frequency, which powered the LEDs,
- GigaSwitch device for connecting the camera with a computer and for sending recorded images,
- a computer equipped with the Pylon Viewer camera control software and the MATLAB environment for data processing.

The use of a ping-pong ball is not necessary to correct operation of the algorithm. However, Using ping-pong ball makes the light from the LEDs is more visible in the recorded images. The visual marker has a shape of sphere in this case.

3.3 Input Parameters

The values of the input parameters for the operation of the spatial-frequency processing algorithm were as follows:

- resolution of recorded images: 800 × 600 pixels,
- recording frequency: 60 Hz (60 frames per second),
- number of processing frames per sequence: 100,
- LED colors: red, green, blue,
- diode pulsation frequency: 20 Hz,
- tested camera exposure times [μs]: 1000, 1500, 2000, 2500, 3000, 3500, 4000, 4500 and 5000.

Some of the assumed values of the parameters are the result of the research described in [5]. A high quality of the amplitude images was obtained for the marker pulsation frequency of 20 Hz. Additionally, it was necessary to remember about the Nyquist theorem concerning the relation between the recording frequency and the frequency of the measured signal. Therefore, the recording frequency of the camera is more than twice the marker pulsation frequency of the visual markers. The mounting of the camera and the pulsating visual marker is shown in Fig. 2. The current signal powering the LEDs was sinusoidal in order to eliminate higher order harmonics and the risk of aliasing. The sinusoidal current signal was also ensuring that the change of the intensity of the pulsating markers in similar to a sinusoidal. Additionally, the LED current was controlled in order to ensure possibly similar and controlled light emission conditions for all LED colors.

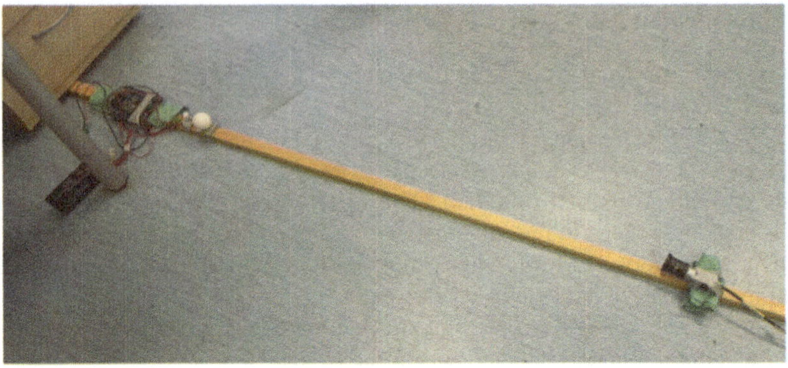

Fig. 2. Photo showing the position of the camera in relation to the light marker during the discussed research

4 Research Results

This chapter presents the results of research into the influence of exposure time on normalized amplitude images. The results were presented by using column charts with the division into diode colors and types of processing layers (in according to the RGB recording). The camera displayed a registration error and stopped working during the registration for the blue LED and the 2.5 ms exposure time value. Such an event occurred only for such those parameters. The reason for this has not been found. That is the way, this case was omitted in the results presented.

Figure 3 shows the image seen by the camera and a well-quality amplitude image generated by the algorithm.

Fig. 3. The normalized amplitude image of the searched marker

4.1 Comparison of Amplitude Values

Figure 4 shows bar graphs comparing the values of the first quality parameter for the processed red layer. It is the ratio of the amplitude value of the marker pulsation frequency to the mean value of amplitude all image. The parameter is calculated for pixels where the change of the intensity level coincides with the marker pulsation frequency.

The highest values of the index were achieved for the green diode, the red layer and the low value of the camera exposure time. They were noted for the exposure times of 1000, 1500 and 2000 μs in Fig. 4. Their values are greater than 1. It means that the amplitude of the 20 Hz band was higher than the amplitude of the zero band. The index exceeded the value of 1 only for the discussed case – the green diode and the red layer.

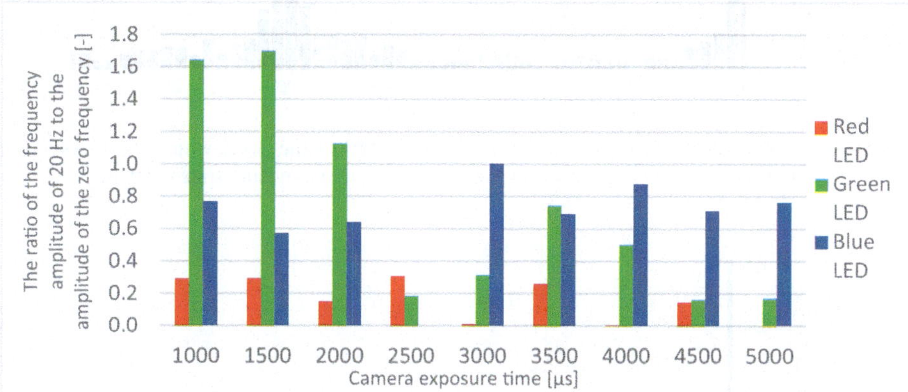

Fig. 4. The ratio of the frequency amplitude of 20 Hz (diode pulsation frequency) to the amplitude of the zero frequency (mean intensity level of the image) for red layer of RGB

Spatial-frequency processing requires that the input data is not a single image but a sequence of images recorded over time. It is also important that the images are saved in RGB format. The algorithm reads each of the RGB recording layers (red, green and blue) and processes the data separately for each of them.

The weakest results were achieved when the color of the visual marker was the same as the color of the processed layer. Figure 4 shows the results for the red layer. The lowest ratios were obtained for the red diode in this case. The quality of the amplitude images is higher for the compatibility of the diode-layer colors than for the case of incompatibility. This fact was also stated in [5]. The above conclusions are duplicated for the remaining colors of the processing layers. Using low exposure times causes higher quality of amplitude images.

Figures 5 and 6 show two examples of amplitude spectra generated from signals of change of the intensity level of a pixel with to better understanding of the presented bar graphs. Both spectra was generated for pixels whose change of intensity level was similar to the marker pulsation frequency. One example shows the spectrum at a low exposure time and the other at a higher exposure time.

4.2 Comparison of Background Level

Figure 7 shows the values of the normalized mean value of the background intensity level depending on the LED color, processing layer and exposure time.

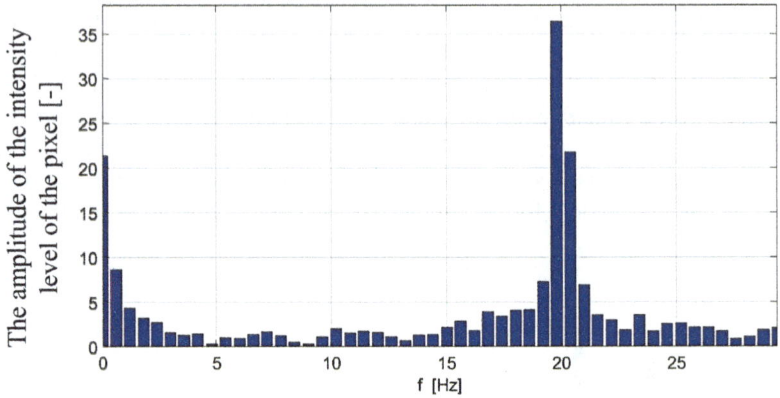

Fig. 5. Amplitude spectrum for the intensity level change of the pixel which the marker was detected. Marker color – green, processing layer – red, camera exposure time – 1.5 ms

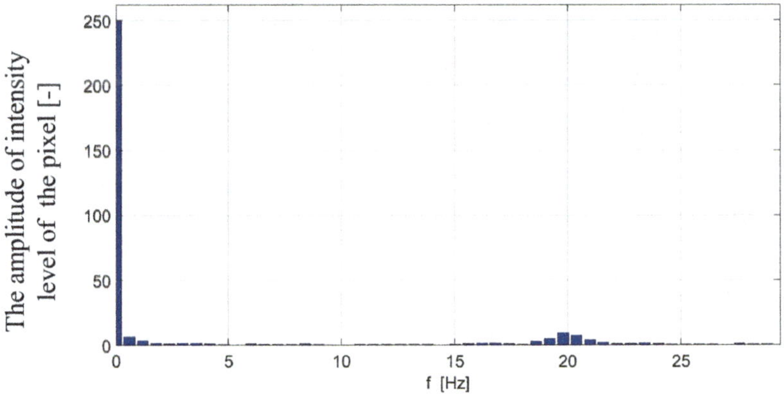

Fig. 6. Amplitude spectrum for the intensity level change of the pixel which the marker was not detected. Marker color – green, processing layer – red, camera exposure time – 4.5 ms

The presented graphs do not indicate that the change in the exposure time significantly influenced the level of background intensity. The value of the normalized mean background intensity level is less than 0.03 in the case of color incompatibility. The values are much higher for compatible colors.

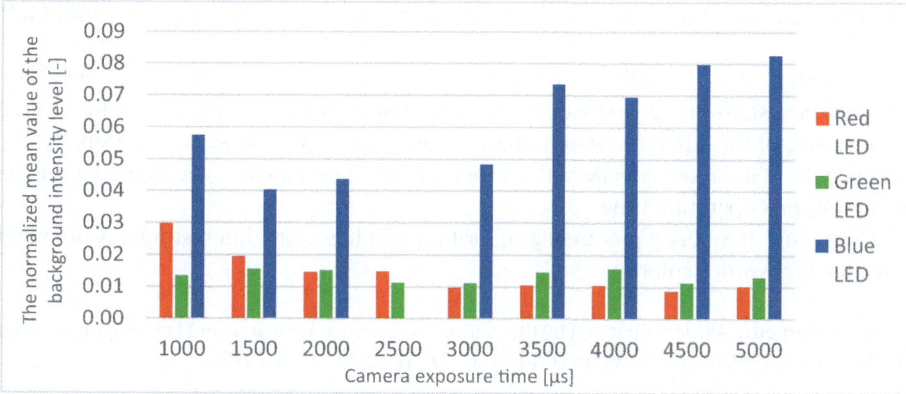

Fig. 7. The normalized mean value of the background intensity level of the amplitude image for green layer of RGB form depending on the camera exposure time

5 Conclusion

The presented research showed that the exposure time of the recording camera has an influence on the quality of the amplitude images. The exposure time had the greatest impact on the relation between amplitude value for the marker pulsation frequency and the amplitude of the mean value of background. The amplitude spectra with better quality (from the point of view of the algorithm) are generated for a low exposure time. It comes from the analysis of the ratio of the amplitude of frequency close to the marker pulsation frequency to the amplitude of the mean intensity level of the images. The important fact is that Authors obtained index values above 1 (Fig. 4, Sect. 4.1). It means that the amplitude of the band with a frequency close to the marker pulsation frequency was higher than the amplitude of the mean intensity level of the image. Such an index value was achieved for the green diode, the red layer and the low exposure time (1000, 1500 and 2000 µs).

The amplitude of frequency closed to the marker pulsation frequency was suppressed the smallest in the case the spectra generated from combination green LED and red processing layer.

The exposure time has no significant effect on the normalized mean value of the background intensity level of images. The characteristics show the relationship between the color of the LED and the color of the processing layer (Fig. 7 is an example). The parameter is higher in case of the same color of the pulsating visual marker and the processing layer.

6 Summary

The algorithm of spatial-frequency processing allows to obtain well-quality normalized amplitude images, which can be the base for determining the position in coordinate system. The research into the exposure time has shown that the algorithm should work with low exposure times because of increase the quality of the resulting amplitude images.

The authors want to further develop the processing algorithm and use it in the vision localization system in the next steps. The algorithm will be tested in an outdoor to see the influence of the sunlight on amplitude images in further research. Additionally, the authors want to use more advanced tools for time-frequency analysis in order to optimize the algorithm's operating time.

The spatial-frequency processing algorithm still has many interesting development areas that are worth exploring.

Acknowledgment. Project financed by the National Center for Research and Development under the "Security and Defense" program (Project no. DOB-BIO9/04/02/2018).

References

1. Hassan, N.U., Naeem, A., Pasha, M.A., Jadoon, T., Yuen, C.: Indoor positioning using visible LED lights: a survey. ACM Comput. Surv. **48**(2), 20:1-20:32 (2015)
2. Kuo, Y.-S., Pannuto, P., Dutta, P.: Demo—Luxapose: indoor positioning with mobile phones and visible light. In: Proceedings of the 20th Annual International Conference on Mobile Computing and Networking, pp. 447–458. ACM, New York (2014)
3. Luo, J., Fan, L., Li, H.: Indoor positioning systems based on visible light communication: state of the art. IEEE Commun. Surveys Tuts. **19**(4), 2871–2893 (2017)
4. Maheepala, M., Kouzani, A.Z., Joordens, M.A.: Light-based indoor positioning systems: a review. IEEE Sens. J. **20**(8), 3971–3995 (2020)
5. Miś, P., Szulim, P.: Analysis of the possibility of using markers emitting pulsating light in the task of localization. Appl. Comput. Sci. **17**(1), 26–39 (2021)
6. Rao, L., Da, F.: High dynamic range 3D shape determination based on automatic exposure selection. J. Vis. Commun. Image Represent. **50**, 217–226 (2018)
7. Rehder, J., Nikolic, J., Schneider, T., Siegwart, R.: A direct formulation for camera calibration. In: Proceedings of the IEEE International Conference on Robotics and Automation, pp. 6479–6486 (2017)
8. Yang, X., Zhong, S.: The implementation of fast light-adjusting based on PID system. In: Proceedings of the Chinese Automation Congress, CAC 2015, pp. 1828–1833 (2015)
9. Yoshino, M., Haruyama, S., Nakagawa, M.: High-accuracy positioning system using visible LED lights and image sensor. In: Proceedings of the IEEE Radio and Wireless Symposium, pp. 439–442, Orlando (2008)
10. Zhou, Z.: Indoor positioning algorithm using light-emitting diode visible light communications. Opt. Eng. **51**(8), Article no. 085009 (2012)
11. Zhuang, Y., et al.: A survey of positioning systems using visible LED lights. IEEE Commun. Surveys Tuts. **20**(3), 1963–1988 (2018)
12. Parameters of Basler camera. https://www.baslerweb.com/en/products/cameras/area-scan-cameras/ace/aca1600-60gc/. Accessed 29 Oct 2021

Development of a Bioreactor Design for Cultivation of Cell Cultures

Igor Korobiichuk[1(✉)], Serhii Semeniuk[2], Vladyslav Shybetskyi[2], Sergii Kostyk[2], and Vadym Povodzinsky[2]

[1] Łukasiewicz Research Network – Industrial Research Institute for Automation and Measurements PIAP, Warsaw, Poland
igor.korobiichuk@piap.lukasiewicz.gov.pl

[2] Igor Sikorsky Kyiv Polytechnic Institute, National Technical University of Ukraine, 37 Avenue Peremogy, Kyiv 03056, Ukraine
{kostyksergey,vnpovodzinski}@ukr.net

Abstract. A new bioreactor design for the cultivation of anchorage-dependent cell cultures in a monolayer was developed. Computational simulation of hydrodynamic in the developed apparatus and the reference bioreactor Xpansion (Pall) were carried out. The modeling was carried out in the CFX module (Ansys) using the k-e turbulence model. For two constructions results were obtained for the velocity of fluid, the rate of shear forces, and the shear stress on the walls of the anchorage of the cells. The obtained results show better hydrodynamic characteristics of the developed apparatus design.

Keywords: Bioreactor · Fluid flow · ANSYS · Modeling · Hydrodynamics · Cells · Vaccines

1 Introduction

Pharmaceutical biotechnology is a new and advanced industry that has developed rapidly over the past decades. The topmost developed pharmaceutical companies consider this industry as one of the most important strategic directions for their development.

Biotechnological products are an increasing share of all pharmaceutical sales. According to the analytic data presented in "Evaluate Pharma", significant growth is predicted in the pharmaceutical market from 30% in 2020 to 37% by 2026, and in the top 100 best-selling medicines, up to 50% of sales will be provided by biotechnological medicines. Also, according to the analytics on sales volume in 2024, the top 5 the most promising medicines will be products of biological origin [1].

On November 25, 2020, the European Commission presented a pharmaceutical strategy for Europe. The main goals are supporting competitiveness, innovation and sustainability of the EU's pharmaceutical industry and the development of high-quality, safe, effective and greener medicines. This includes the development and research of the production of vaccines and other medicines in the biotechnology sector as one of the most important industries [2].

R. Szewczyk et al. (Eds.): AUTOMATION 2022, AISC 1427, pp. 317–328, 2022.
https://doi.org/10.1007/978-3-031-03502-9_32

The most advanced therapeutic areas (in terms of medicines sales) are: oncology, diabetes mellitus, antiviral therapies and vaccines [1]. A significant part of medicines for these areas is obtained by the technology of biological synthesis using biological agents.

The way from development of medicines to their industrial production is very laborious, expensive and time-consuming [3, 4]. Typically, active pharmaceutical ingredients (APIs) of biological origin are developed in several stages, including: understanding of the basic concepts that underline human health and disease; fundamental molecular mechanisms regulating the function of related biomolecules; synthesis and purification of molecules; determination of the shelf life, stability, toxicity and immunogenicity of the product; drug delivery systems; patenting; and clinical trials [5, 6].

The production of APIs of biological origin involves usage of biological agents (microorganisms, cells of plants, animals, humans, etc.) to obtain target products (molecules, proteins, viruses, etc.). The main and critical, in terms of quality and quantity of APIs obtained, is the stage of biological synthesis, which is carried out using special equipment - bioreactors (fermenters) [7].

Using of cells culture as biological agents makes it possible to obtain vaccines, hormones, monoclonal antibodies, to research new drug molecules and their effect on certain organ tissues.

Cell culture applications can be divided into:

1. Products derived from cells (therapeutic proteins, antibodies, vaccines, etc.).
2. Cells as a final product (suspension cells for cell therapy, e.g. keratinocytes, bone marrow cells; functional tissue for transplantation, e.g. chondrocytes; extracorporeal organelles, e.g. artificial liver and kidneys).
3. Testing cells (effect and dosage of medicines, cosmetic products) [8].

In comparison with microorganisms, the size of animal and human cells are large, with a thin cytoplasmic wall, so they are more sensitive to the shear stress of the liquid and are damaged by contact with air bubbles.

Most cells do not grow in suspension and are anchorage-dependent, that is, they need a certain surface for growth.

Therefore, the development and research of new types of bioreactors that will allow efficient cultivation of cell cultures from the stage of laboratory research to industrial synthesis appear especially relevant.

In general, bioreactors for cell cultures can be classified by the type of energy input into the bioreactor – static systems (no energy input for mixing) and dynamic systems (energy for mixing the medium is introduced mechanically, liquid phase or gas phase) Fig. 1 [8].

It should be noted that cultivation using static systems is used at the stages of development and research in laboratory conditions (for example, for obtaining primary cultures and subculturing, obtaining permanent cell lines). Laboratory bioreactors volume is in the range from a few milliliters to several liters and are suitable for in vitro cultivation of anchorage-dependent cells and cells in suspension.

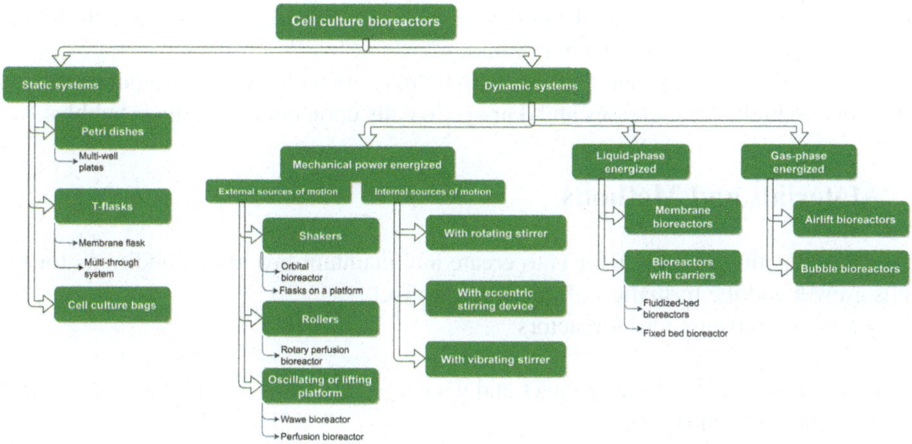

Fig. 1. Classifications of bioreactor for cell cultivation

Bioreactors with mechanical energy input can be classified into:

- external sources of motion – flasks or test tubes on a movable platform; roller bioreactors - horizontally located cylindrical glass or plastic containers with a culture medium rotating around its axis; wave bioreactors - a polymer bag with a culture liquid located on a platform performing oscillatory movements.
- internal sources of motion – as a rule, capacitive devices are located horizontally and have a shaft motor and a mixing device located in the culture medium. The mixing device (concentric, eccentric) can perform rotating, reciprocating or oscillatory movements, thus ensuring the mixing of the medium inside the apparatus.

The disadvantages of bioreactors with mechanical energy input can be attributed to significant shear stresses upon contact of volumes of liquid with a mixing device or the limitation of the industrial use of devices (in the case of devices with an external energy source).

Liquid phase energy bioreactors: these are membranes on which surface cells are fixed. Supply of nutrients and the removal of metabolites is provided by pumping the nutrient medium from the opposite side of the membrane. Also, the cells can be fixed on the surface of porous microcarriers located in a suspension medium, or carriers that are motionlessly fixed in the volume of the apparatus, and mixing occurs with the help of a pump that pumps the nutrient medium in a closed loop.

These bioreactor designs are the most promising from the point of view of ensuring optimal cultivation conditions. But improvements are required given the need to develop and manufacture new types of membranes with a certain selectivity or microcarriers. Also, it is necessary to develop of a method for separating cells from the attachment surface.

Gas phase energy bioreactors are containers with a culture medium having an aerator in the lower part of the apparatus and an additional circulation loop inside. Mixing is achieved by pushing volumes of liquid from the lower part of the apparatus to higher using a stream of air bubbles from the aerator.

Not all cells can withstand cultivation in bioreactors with the power input by the gas phase due to high shear stresses and damage to cells upon contact with air bubbles.

2 Materials and Methods

The main function of bioreactor is to create and maintain favorable conditions for the cells growth and the formation of the target product.

Main requirements for bioreactors:

1. Formation of intercellular contact and detachment of cells during the cultivation of anchorage-dependent cells.
2. Homogeneous circulation with low shear stress rate.
3. Sufficient turbulence for efficient heat transfer.
4. Sufficient degree of aeration.
5. Avoidance of substrate segregation.
6. Measurement and control of critical process parameters
7. The possibility of scaling.
8. Long-term stability and sterility.
9. Operational simplicity and maintenance requirements.

Therefore, the goal was to develop and research an improved design of a bioreactor, which would provide optimal cultivation conditions. A dynamic system with energy input by the liquid phase is taken as a basis for developing a new design. Xpansion (Pall) bioreactor was chosen as a reference design of the bioreactor to check the results obtained for compliance with the cultivation of cell cultures [9].

The design of an apparatus for anchorage-dependent growth of animal cells cultivation (2D) was developed Fig. 2. The apparatus consists of a cylindrical body 1 made of a hydrophobic polymer on a frame, a cover 2 containing a nutrient medium supply pipe 3, an inoculation supply pipe 4, an air vent 5, a dissolved oxygen sensor sleeve 6, and a pH meter sensor sleeve. The bottom of the apparatus 8 has a pipe 9 for the circulation of the nutrient medium through the peristaltic pump 10.

The aeration device is a cylindrical body 1 with a tube made of a hydrophobic polymer 11, in the form of a coil, located in the outer body 12. The aeration part have a branch pipe for supplying a nutrient medium 13 for aeration, a return pipe for a nutrient medium into the apparatus 14, a pipe for supplying aeration air 15 and a pipe for exhaust air 16. The plates, on the surface of which biological agents 17 are fixed and grow, located on the shaft 18 with a certain pitch (H) – this arrangement allows to provide a directed flow of the nutrient medium, flowing about the entire surface of each plate.

This bioreactor belongs to the Single Use Bioreactor category – that is, it is supplied already assembled, sterilized and ready for use. After the completion of the process, the device is disposed of.

The bioreactor works as follows. The device is ready for use, it is placed on the stand. Sterile nutrient medium is fed through the pipe 3 to the specified filling level.

Then, the inoculation is introduced through the pipe 4. When filling the apparatus, the air vent 5 must be open to avoid increasing pressure inside the bioreactor. Biological agents adhesion and growing on the plates 17. Plates 17 are located on the shaft 18, so that channels are formed through which the circulating nutrient medium in a downward flow. The direction and circulation of the medium streamflow are shown in Fig. 3. The aeration air supply through the branch pipe 15 and the outlet through the branch pipe 16. After filling and adhesion of the cells, the peristaltic pump 10 is turned on to circulate the nutrient medium from the branch pipe 9 located on the bottom 8, and return to the inside of the apparatus through the branch pipe 14. Circulating along the channels formed by the polymeric hydrophobic oxygen-selective tube 11 and the body of the apparatus 1, the nutrient medium is saturated with oxygen and returns to the upper part of the apparatus. On the cover 2 there are sensors for dissolved oxygen 6 and acidity pH 7.

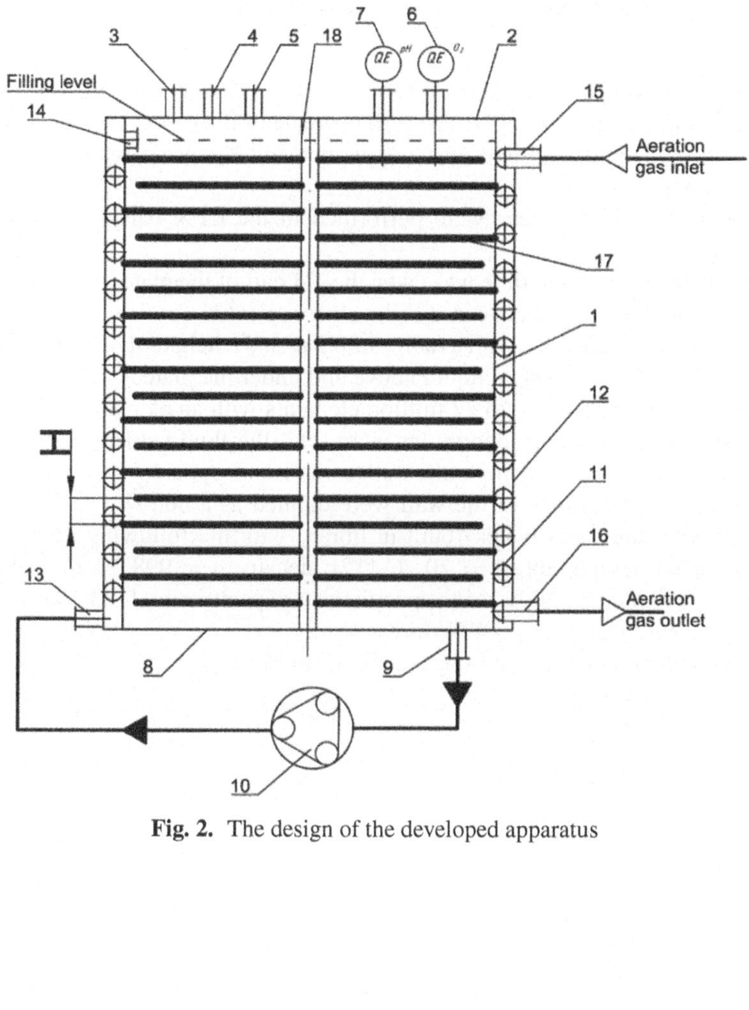

Fig. 2. The design of the developed apparatus

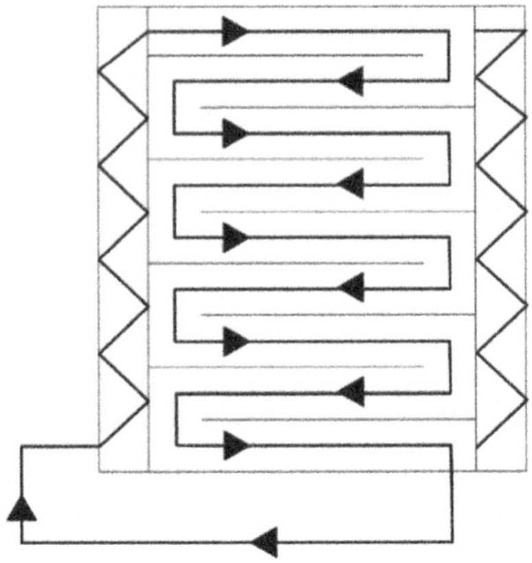

Fig. 3. The flow of the nutrient medium in the apparatus

The use of the computational fluid dynamics tool allows evaluating the hydrodynamic characteristics in the apparatus [7] and on the surface of the disks [10].

For the developed design and of the apparatus and Xpansion bioreactor (Pall) 3D models were built.

Fluid dynamics simulations were performed in the CFX Ansys based on Finite element analysis.

To reduce the calculation time was taken the 1/4 part of the plate. For computational modeling, 2 liquid volumes between the plate were selected.

The sector has dimensions with a radius of 0.2 m and a height of 4.2 mm – the height of the plate is 1 mm, the layer of liquid above and under the plate is 1.6 mm.

The volume was meshed into 22 million elements with an element size of 0.3 mm. For each sector, the boundary conditions were set – the fluid velocity at the inlet was 0.05 m/s and the relative pressure at the outlet was 1 atmosphere.

The boundary conditions of the wall were defined as a non-slip surface. Material used as a model liquid is «Water» built-in library with the following thermophysical parameters at the temperature t = 20 °C [12]: density $\rho = 998.2$ kg/m^3; kinematic viscosity coefficient $\nu = 0.001$ m^2/s; specific heat capacity c = 4183 J/kg K; thermal conductivity coefficient $\lambda = 0.599$ W/m K.

Scheme with boundary conditions are shown in Fig. 4.

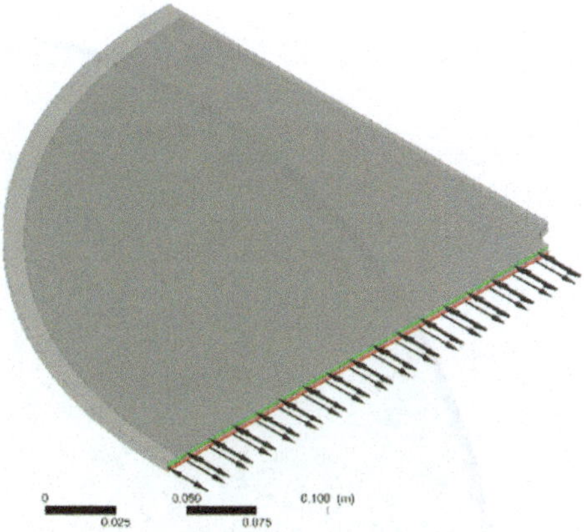

(a)

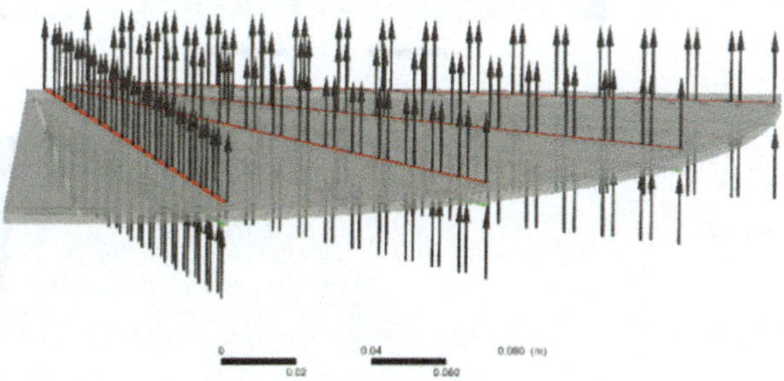

(b)

Fig. 4. Scheme of the boundary conditions (gray volume – liquid, green – inlet flow; red – outlet flow): a – developed design; b – Xpansion (Pall) bioreactor

3 Results and Discussion

The hydrodynamic environment for the developed bioreactor and the Xpansion (Pall) bioreactor have been modeled.

On Fig. 5 the contours of the velocities distribution on the plate surface in the zone of the flow supply are shown.

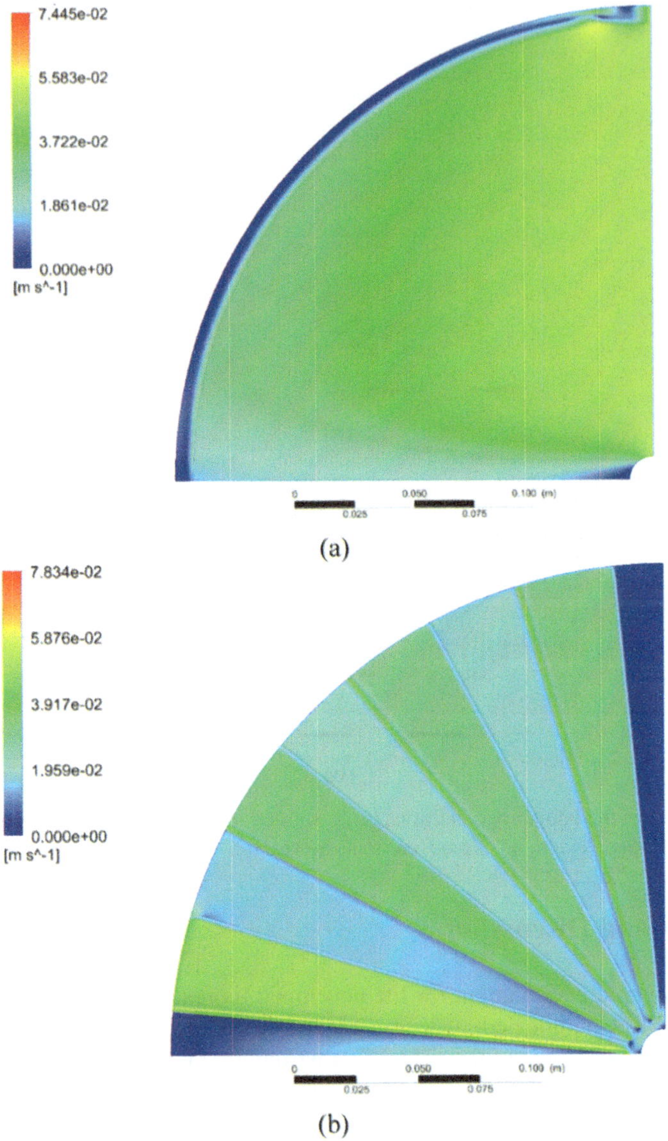

Fig. 5. Fluid velocity on the plate: a – developed bioreactor; b – Xpansion (Pall) bioreactor

For the researched bioreactor (Fig. 5a), the velocity distribution is more steady over the entire surface of the plate and keep between values from $5.58 \cdot 10^{-2}$ m/s to $1.8 \cdot 10^{-2}$ m/s; in the case of the Xpansion bioreactor (Fig. 5b), the highest velocities are observed in radial sections – $5.87 \cdot 10^{-2}$ m/s and adjacent areas of the plate – $3.91 \cdot 10^{-2}$ m/s.

On Fig. 6 presented contours of the shear strain rate of the fluid flow.

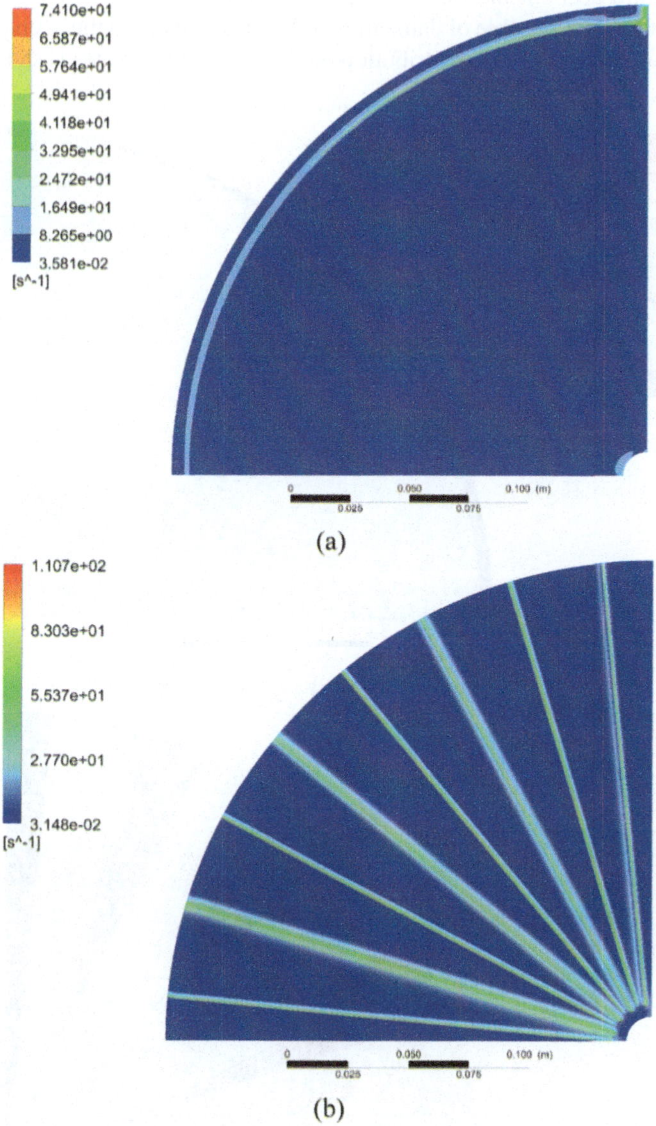

Fig. 6. Shear strain rate of the fluid flow on the plate: a – developed bioreactor; b – Xpansion (Pall) bioreactor

The shear strain rate of the flow on the plate for the developed bioreactor (Fig. 6a) has a minimum value $3.58 \cdot 10^{-2}$ s^{-1} and increases only in the zone of flow transition between the plates to 16.49 s^{-1}. For the Xpansion (Pall) bioreactor, the shear rate is maximum in the input flow zone 55.37 s^{-1} and distributed over the plate from 27.7 s^{-1} to the $3.14 \cdot 10^{-2}$ s^{-1} in the middle of the plate.

The key indicator of the efficiency and acceptance criteria of the developed bioreactor design in this study was the shear stress on the plate surface (the cell attachment zone), since taking into account the mechanical effect during the circulation of the nutrient medium, the flows are capable of damaging or flushing cells from the anchorage surface.

The contour of shear stress distribution on the walls of the plates is shown on Fig. 7.

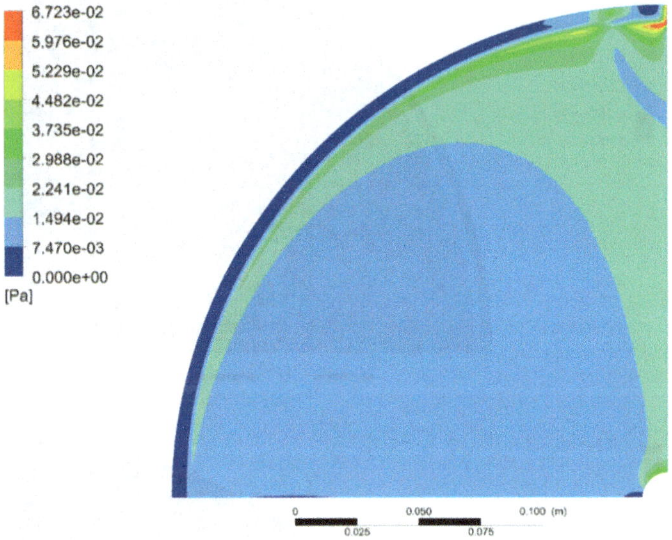

(a)

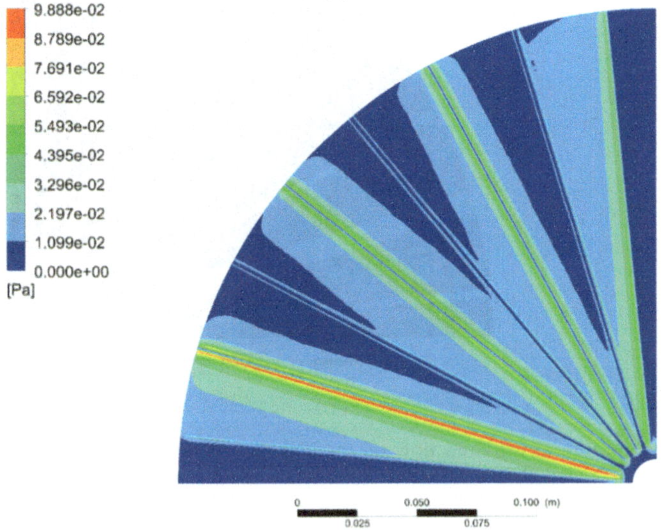

(b)

Fig. 7. Wall shear on the plate: a – developed bioreactor; b – Xpansion (Pall) bioreactor

The wall shear in the researched design of the bioreactor has a steady distribution and ranges from 0.02 to 0.007 Pa, the average shear on the plate becomes 0.014 Pa. The wall stress Xpansion (Pall) has a slightly higher value of 0.07–0.01 Pa, the average shear on the plate becomes 0.037 Pa. It should be noted that in both cases the wall shear has acceptable values and are below 10 Pa [13].

Conclusions

In this research, a new design of the apparatus was developed and first stage verification of the conformity of the developed design for cell cultures cultivation was carried out.

The fluid velocity and shear strain rate for the developed design has a more steady distribution on the plate surface and less value than that of a Xpansion (Pall).

Since the average value of the shear stress rate on the plate for the developed reactor is 10% less, it is possible to increase the consumption of the nutrient medium for intensifying the cultivation process. Increasing the flow rate at the optimum shear stress will allow for homogeneous cultivation conditions.

The developed design has a lower value of shear stresses in the area of cell anchoraging and growth, which makes it possible to provide optimal hydrodynamic conditions for biological agents during cultivation. The stress on the plate walls for the designed stress design is 46% lower compared to Xpansion (Pall) bioreactor.

The use of CFD methods allows confirm reliability developed design of bioreactors or improvement of existing equipment for the cell cultures cultivation at the stage of prototyping and scaling up.

References

1. Evaluate Pharma® Evaluate Pharma World Preview 2021, Outlook to 2026, July 2021 (2021)
2. Pharmaceutical Strategy for Europe, 25 November 2020. https://ec.europa.eu/health/human-use/strategy_en
3. Korobiichuk, I., Ladanyuk, A., Zaiets, N., Vlasenko, L.: Modern development technologies and investigation of food production technological complex automated systems. In: ACM International Conference Proceeding Series. 2nd International Conference on Mechatronics Systems and Control Engineering, ICMSCE 2018, Amsterdam, Netherlands, 21–23 February 2018, pp. 52–57 (2018). https://doi.org/10.1145/3185066.3185075
4. Korobiichuk, I., et al.: Synthesis of optimal robust regulator for food processing facilities. In: Szewczyk, R., Zieliński, C., Kaliczyńska, M. (eds.) ICA 2017. AISC, vol. 550, pp. 58–66. Springer, Cham (2017). https://doi.org/10.1007/978-3-319-54042-9_5
5. Mallela, K.: Pharmaceutical biotechnology – concepts and applications: Walsh Gary Wiley, Chichester, West Sussex, UK. Hum. Genomics **4**, 218 (2010). https://doi.org/10.1186/1479-7364-4-3-218
6. Oliinyk, A., Skrupsky, S., Subbotin, S., Korobiichuk, I.: Parallel method of production rules extraction based on computational intelligence. Autom. Control Comput. Sci. **51**(4), 215–223 (2017). https://doi.org/10.3103/S0146411617040058
7. Semeniuk, S., Povodzinskiy, V.: Risk assessment in development of technical requirements for designing fermentation equipment in accordance with requirements of good manufacturing practice. Technol. Audit Prod. Reserves **3**(2(59)), 46–50 (2021). https://doi.org/10.15587/2706-5448.2021.233536, SSRN: https://ssrn.com/abstract=3880597

8. Eibl, R., Eibl, D., Pörtner, R., Catapano, G., Czermak, P.: Cell and Tissue Reaction Engineering, p. 363. Springer, Heidelberg (2009). https://doi.org/10.1007/978-3-540-68182-3
9. Xpansion® Multiplate Bioreactor System. https://shop.pall.com/us/en/products/systems/zidhw7uq21i
10. Korobiichuk, I., Shybetska, N., Shybetskyi, V., Kostyk, S.: Modeling of systems of automated auxiliary processes in pharmaceutical industry. In: Szewczyk, R., Zieliński, C., Kaliczyńska, M. (eds.) AUTOMATION 2021. AISC, vol. 1390, pp. 128–135. Springer, Cham (2021). https://doi.org/10.1007/978-3-030-74893-7_13
11. Kostyk, S., Shybetskyy, V., Fesenko, S., Povodzinskiy, V.: Revealing special features of hydrodynamics in a rotor-disk film vaporizing plant. East.-Eur. J. Enterp. Technol. **1**(6(97)), 28–33 (2019). https://doi.org/10.15587/1729-4061.2019.156649
12. Vargaftik, N.B.: Handbook on the Thermophysical Properties of Water and Gases. Nauka, Moscow (1972)
13. Shybetskiy, V., Semeniuk, S., Kostyk, S.: Design of construction and hydrodynamic modeling in a roller bioreactor with surface cultivation of cell cultures. ScienceRise **7**, 53–59 (2017). https://doi.org/10.15587/2313-8416.2017.107176

Ways to Reduce the Creation of Vortex During Homogenization of Liquid Products

Igor Korobiichuk[1(✉)], Vladyslav Shybetskyi[2], Sergii Kostyk[2], Myroslava Kalinina[2], and Andriy Tsytsiura[2]

[1] Łukasiewicz Research Network – Industrial Research Institute for Automation and Measurements PIAP, Warsaw, Poland
igor.korobiichuk@piap.lukasiewicz.gov.pl
[2] Igor Sikorsky Kyiv Polytechnic Institute, National Technical University of Ukraine, 37 Avenue Peremogy, Kyiv 03056, Ukraine
{kostyksergey,bottt7}@ukr.net

Abstract. This article is devoted to computer simulation of the hydrodynamics of a bioreactor with a standard open-type turbine mixer, with a two-disc open-type turbine mixer with straight blades and with differently oriented blades at an angle of 45° and 135° using ANSYS software. Comparative analyses of the hydrodynamics of liquid flows that occur in the volume of the bioreactor due to the mixing of the medium by different designs of mixing devices were made. Experimental researches for the investigated designs of mixers for confirmation of adequacy of the offered models are carried out. A new design of an open-type two-disk turbine mixer with differently oriented blades is proposed, which allows reducing the probability of a vortex during mixing. It is established that the proposed turbine practically does not form a vortex, and at a speed of 400 rpm its depth is 0.06 m, and for 600 rpm – 0.012 m.

Keywords: Vortex · Bioreactor · ANSYS · Hydrodynamics · Mixer · Homogenization

1 Introduction

The development of production of typical products of biotechnology – microbial biomass, organic acids, production of amino acids and active pharmaceutical ingredients (API) for the pharmaceutical industry is focused on the exploitation of biological agents (BA) of different phenotypes [1–3].

An integral component of biotechnological production is the stage of cultivation of BA in bioreactors of various design and scale. As a rule, the choice of bioreactors design is determined by the specifics of BA and production productivity [4].

Volumetric apparatus with mixing devices are widely used as reactors for the synthesis of various macromolecular compounds for various processes with the liquid phase. They are used as chemical reactors to obtain emulsions, suspensions, dissolution of the solid phase, mixing liquids. Reactors with mixing devices are most often used in industry [5, 6].

R. Szewczyk et al. (Eds.): AUTOMATION 2022, AISC 1427, pp. 329–343, 2022.
https://doi.org/10.1007/978-3-031-03502-9_33

Modern production of recombinant proteins, vaccines, etc. focuses on the use of animal and human cells, and in this case it is necessary to take into account such phenotypic features of BA as the formation of single or multilayer cell structure on the carrier surface – adhesive structure and the existence of suspension structures multicellular clusters. The main factor that should be constantly accepted as a critical condition in the terms of reference for the design of bioreactors is the level of shear stress (shear forces). Industrial cultivation of hybridoma cells, CHO cells and other cell cultures is significantly complicated by the impossibility of the presence in the culture medium of intense streams of gas-liquid dispersion and bubbles of the aeration phase [7–9].

During liquid media mixing, it is important to homogenize the product through all volume. Typically, high-speed mixing devices are used for mixing liquids, namely turbine mixers of open or closed type. When used, such negative phenomena are observed as significant shear rates of the liquid flow, which leads to the consumption of significant amounts of energy for mixing, and the formation of a vortex, which adversely affects the homogenization of the product throughout the volume [10].

Therefore, it is necessary to improve the existing design of mixing devices to increase the efficiency of use of bioreactors in general, improve product quality and reduce the time of mixing in this apparatus [11].

According to the speed of rotation, the stirrers are conventionally divided into two groups: low-speed (anchor, frame and others, in which the speed of rotation of the blade ends is approximately 1 m/s) and high-speed (propeller, turbine and others, in which the speed of rotation of the blade ends is approximately 10 m/s).

The stirrer is the main structural element, which is designed to mix the fluid. Belt, scraper and auger mixers are used to mix highly viscous media in laminar mode. Anchors and frame mixers are used to mix relatively low viscosity liquids. High-speed turbine and propeller stirrers have the ability to create axial circulating fluid flows. In devices without internal devices, these mixers provide a pumping effect that is twice the pumping effect of conventional mixers [12–14].

However, the known designs of high-speed mixers do not ensure the absence of the formation of a vortex with increasing speed in the apparatus without reflective partitions, which significantly increases time of the product mixing and does not ensure homogeneity of the product in its entire volume. That is why the study was conducted using ANSYS software [15]. The purpose of this study is to reduce the likelihood of a vortex during the process and ensure high homogenization of the product throughout its volume by selecting the optimal design of the high-speed mixing device.

2 Materials and Methods

2.1 Theory of Formation of the Central Vortex

Determining the geometry of the vortex takes into account the following considerations (Fig. 1): in a vessel of radius R, filled with liquid to a height of H_0, placed a rectangular stirrer with radius r_m and width b, rotating with an angular velocity ω. If the viscosity of the liquid is neglected, the angular velocity of the flow flowing from the mixer can be considered equal to the angular velocity of the mixer. In fact, due to the internal friction of the fluid, the angular flow rate flowing from the mixer will be lower. It is assumed that the volume of liquid rotating at the speed of the stirrer has the shape of a cylinder. The radius of this cylinder r_c, however, is less than the radius of the stirrer r_m.

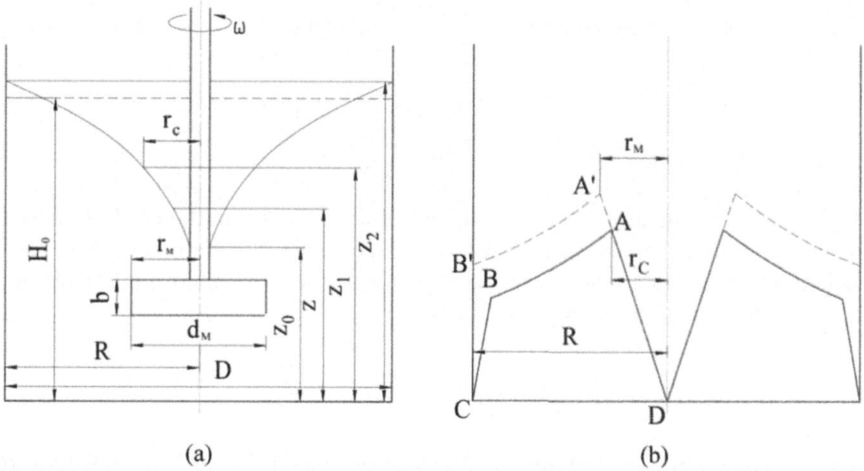

Fig. 1. Parameters of vortex equation: a – geometric parameters of the tank with a mixer and vortex, b – change in the circular velocity of the liquid across the width of the tank

The shape of the free surface of the liquid in the area of the forced vortex describes the following dependence:

$$z = z_0 + \frac{\omega^2 \cdot r^2}{2g} \tag{1}$$

in which z_0 is the height of the liquid layer at the lowest point of the free surface of the liquid.

The equation describing the hyperbolic curve of the distribution of the circumferential velocities AB is shown in Fig. 1a.

$$z = z_1 + \frac{\omega^2 \cdot r_C^2}{2g}\left(1 - \frac{r_C^4}{r^2}\right) = z_0 + \frac{\omega^2 \cdot r_C^2}{2g}\left(2 - \frac{r_C^2}{r^2}\right). \tag{2}$$

The liquid level in this area will be higher compared to the liquid level in the forced vortex area.

On the walls of the vessel there is a laminar layer of liquid, for which the equation will be valid if the value of r is replaced by the radius of the vessel R:

$$w = \frac{\omega \cdot r_C^2}{R}. \tag{3}$$

Between this layer and the wall there is a sharp decrease in circular velocity, shown by a straight line BC (Fig. 1b).

Thus, as a result of the rotational movement of the liquid, the liquid level in the direction of the walls of the vessel will increase by reducing the level in the center of the vortex, which deepens in proportion to the square of the circular velocity of the rotating fluid.

The depth of the liquid rotating near the walls of the vessel is determined by the equation:

$$z_2 = z_0 + \frac{\omega^2 \cdot r_C^2}{2g}\left(2 - \frac{r_C^2}{r^2}\right). \tag{4}$$

If in the above equations enter the height of the mixed fluid layer H_0 (when it is stationary), and this value is easily measurable, we obtain to determine the increase in the height of the liquid column near the vessel wall (i.e. maximum liquid rise) in the formation of the funnel the following equation:

$$z_2 = H_0 + \frac{\omega^2 \cdot r_C^4}{g \cdot R^2}\left(0.25 + ln\frac{R}{r_C}\right). \tag{5}$$

The unknown value of r_c in this equation can be found by the measured values from the equation:

$$\frac{(H_0 - z_0) \cdot g}{R^2 \cdot \omega^2} = \left(\frac{r_C}{R}\right)^2 - \left(\frac{r_C}{R}\right)^4\left(0.75 + ln\frac{R}{r_C}\right), \tag{6}$$

This is solved graphically.

The radius of the static region of rotation of the Rankine vortex:

$$R_0 = R \cdot e^{\left[0.5 + \frac{0.25 \cdot n_1^2}{n_1^2 + 0.278\left(\frac{R}{r_M}\right)^{1.46}} - \frac{R^2}{2R_y^2}\right]}. \tag{7}$$

Average integral total velocity of liquid:

$$w = \frac{\omega \cdot R_0^2}{(R - R_0)n_1}\sqrt{0.278 \cdot \left(\frac{R}{r_M}\right)^{1.46} + n_1^2} \cdot ln\frac{R}{R_0}, \tag{8}$$

where r_M is the radius of the mixer; R is the radius of the vessel; n_1 is the number of sources, which at the usual location of the stirrer in the vessel should be taken equal to 4.

2.2 Materials and Methods for Computer Simulation

3D models of mixing devices were built for modeling: standard open-type turbine mixer with straight blades, open-type double-disc turbine mixer with straight blades, open-type double-disc turbine mixer with differently oriented blades at an angle of 45° (upper level) and 135°. These 3D-models of mixing devices are shown in Fig. 2.

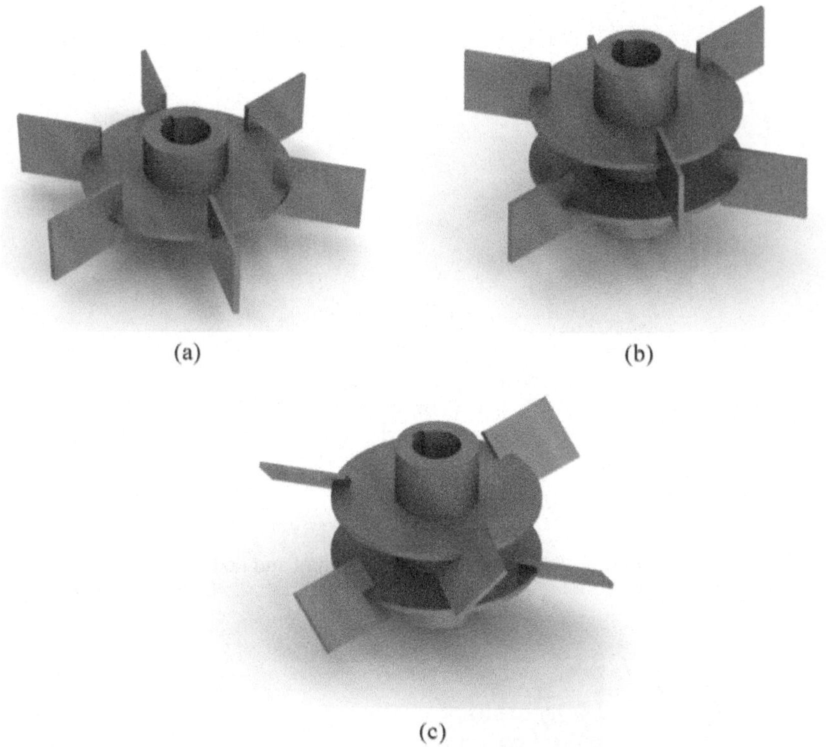

Fig. 2. The design of the mixing device: a – standard turbine open mixer with straight blades; b – turbine open two-disk mixer with straight blades; c – turbine open two-disk mixer with differently oriented blades at an angle of 45° and 135°

Impellers have the following dimensions: diameter $d_m = 35$ mm, disc diameter d = 21 mm, disc thickness mm, blade height b = 7 mm, blade length l = 9 mm, blade thickness 1 mm, distance between discs s = 6 mm (Fig. 3). The impeller is placed in a vessel with diameter D = 105 mm and height H = 250 mm, distance from the bottom $h_m = 35$ mm.

The simulation is performed in the CFX ANSYS analysis system. To implement the finite element method, the total volume was divided into a set of elements in the module Mesh (Fig. 4). The computational mesh was created with a Tetrahedrons element type. Number of nodes 38 876, number of elements 202 624.

Only a third of the actual volume of the device is used to reduce the calculation area and reduce the simulation time. To realize the full volume, the boundary surfaces are connected by the *«Rotational Periodicity»* interface (Fig. 5). The modeling area consists of two volumes: fixed I and moving II. The moving volume is the zone of liquid near the mixer, which rotates at the same frequency as the impeller – 10 rev/s. This statement makes it possible not to consider the mixer itself in the simulation, setting in the surface of its contact with the liquid boundary condition *«Wall»*. Fixed and movable volumes are connected via the *«Frozen Rotor»* interface. The influence of the shaft is not

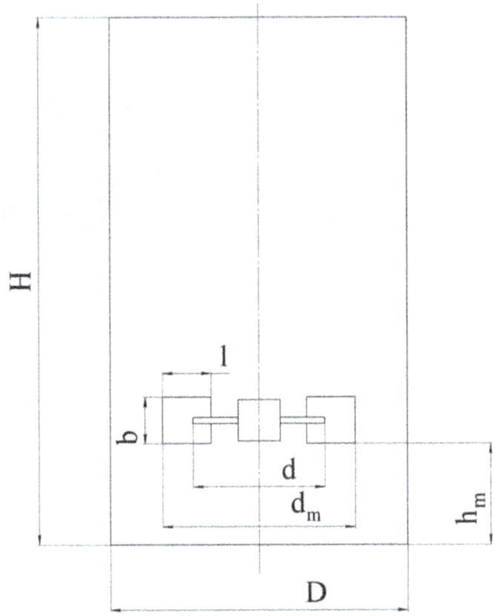

Fig. 3. Dimensions of the impellers and tank

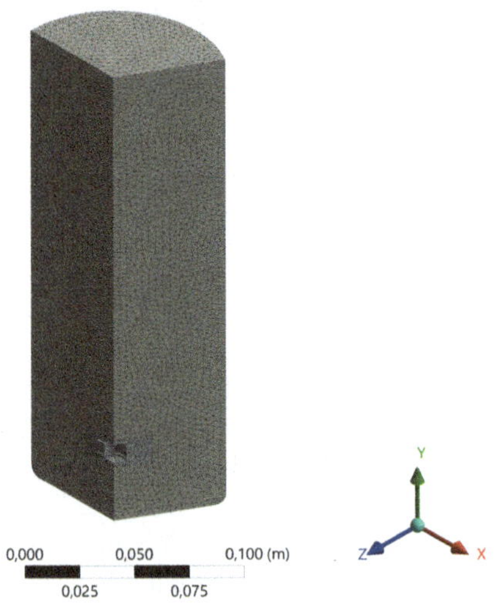

Fig. 4. The computational mesh

taken into account in the solution In Domain settings the "Buoyancy model" is activated with reference density 1000 kg/m^3 and direction of gravitation – y. The "K-Epsilon" turbulence model is used. Transient simulation is performed. For calculation is chosen adaptive timestep: min. timestep – 0.005 s; max. timestep 0.01 s. Number of loops for each timestep is from 10 to 2. Simulation time is 60 s, which is enough to achieve convergence criteria (residual target is 10^{-4}).

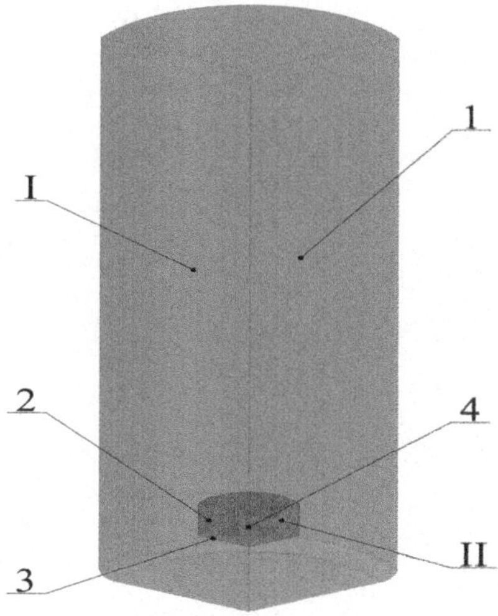

Fig. 5. Area of modeling: I – fixed volume of modeling; II – moving volume of modeling; 1 – «*Rotational Periodicity*» interface, for a fixed volume; 2 – «*Rotational Periodicity*» interface, for the moving volume; 3 – «*Frozen Rotor*» interface; 4 – boundary condition «*Wall*»

For this modeling the Reynolds number is:

$$\mathrm{Re} = \frac{n^2 \cdot d_m}{\upsilon} = \frac{10^2 \cdot 0.035}{10^{-6}} = 3.5 \cdot 10^6$$

where n = 10 rev/s is the frequency of the impeller; $d_m = 0.025$ m is the diameter of the impeller, $\nu = 10^{-6}$ m/s^2 is viscosity of the water.

The Froude number is:

$$Fr = \frac{n^2 \cdot d_m}{g} = \frac{10^2 \cdot 0.035}{9.81} = 0.357$$

2.3 Materials and Methods for Experimental Research

To verify and assess the adequacy of computer modeling, it is proposed to conduct a full-scale study of the hydrodynamic environment, which is implemented in an experimental installation (Fig. 6), equipped with a removable mixing device. To do this, the 3D printer was printed mixers, which are presented in Fig. 1. The installation for the experiment consists of a tank 1 with a capacity of 2.2 L. The working tank is fixed on the frame 4 by means of fastening 5. In the tank 1 on the shaft 3 is fixed a mixing device 2, driven by a DC electric drive 6 which provides frequency range from 0 to 800 rpm and power – 1 kW, equipped with a 7 – speed controller (frequency range 0–5000 rpm).

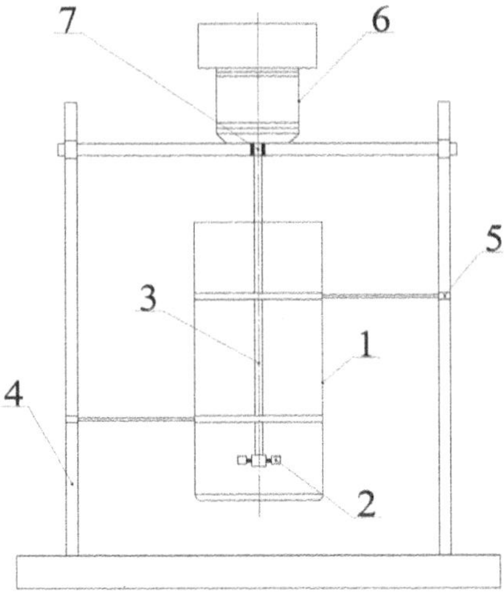

Fig. 6. Experimental installation: 1 – tank; 2 – mixing device, 3 – shaft; 4 – frame; 5 – attachment of the tank to the frame; 6 – electric drive; 7 – digital laser tachometer

3 Results and Discussion

3.1 Computer Simulation Results

Velocity vectors of water are constructed (Fig. 7). They are formed due to the rotational motion of the mixing device with a frequency of 600 rpm. Plot diagrams of YZ and XZ for three simulations are presented. We could see annular, radial and tangential fluid flows. The largest values are observed in the radial direction, which forms the vortex. The changes in the direction and magnitude of the vectors of fluid motion in the volume are significant. The maximum values are in the area near the mixer. The average speed over the entire volume is 0.68 m/s for a standard open turbine agitator (Fig. 7a) and 0.62 m/s for a two-disc agitator with straight blades (Fig. 7b). In the case of a two-disc agitator with divergent blades at an angle, the fluid flows resemble the shape of an infinity symbol, the form of the mixing device is designed so that the ascending fluid flows stimulated by the rotation of the lower blades on the lower disk are immediately compensated by downward flows. The average values of velocity over the entire volume are 0.58 m/s for a standard open turbine mixer, and the distribution of velocity vectors is intensive over the entire volume (Fig. 7c).

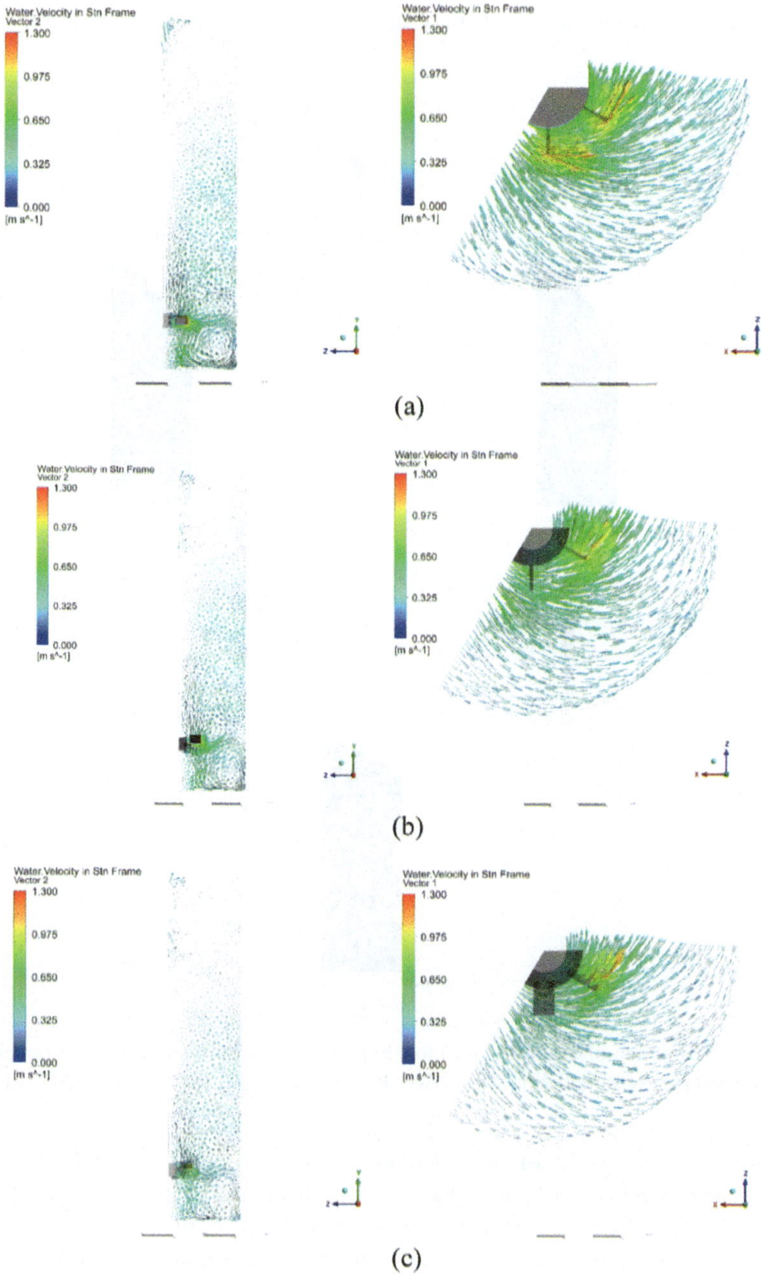

Fig. 7. Velocity vectors: a – for modeling with a standard mixer; b – for modeling with a two-disc stirrer; c – for modeling with a two-disc stirrer with differently oriented blades

Contours are constructed to display the water content in the volume for all three simulations (Fig. 8).

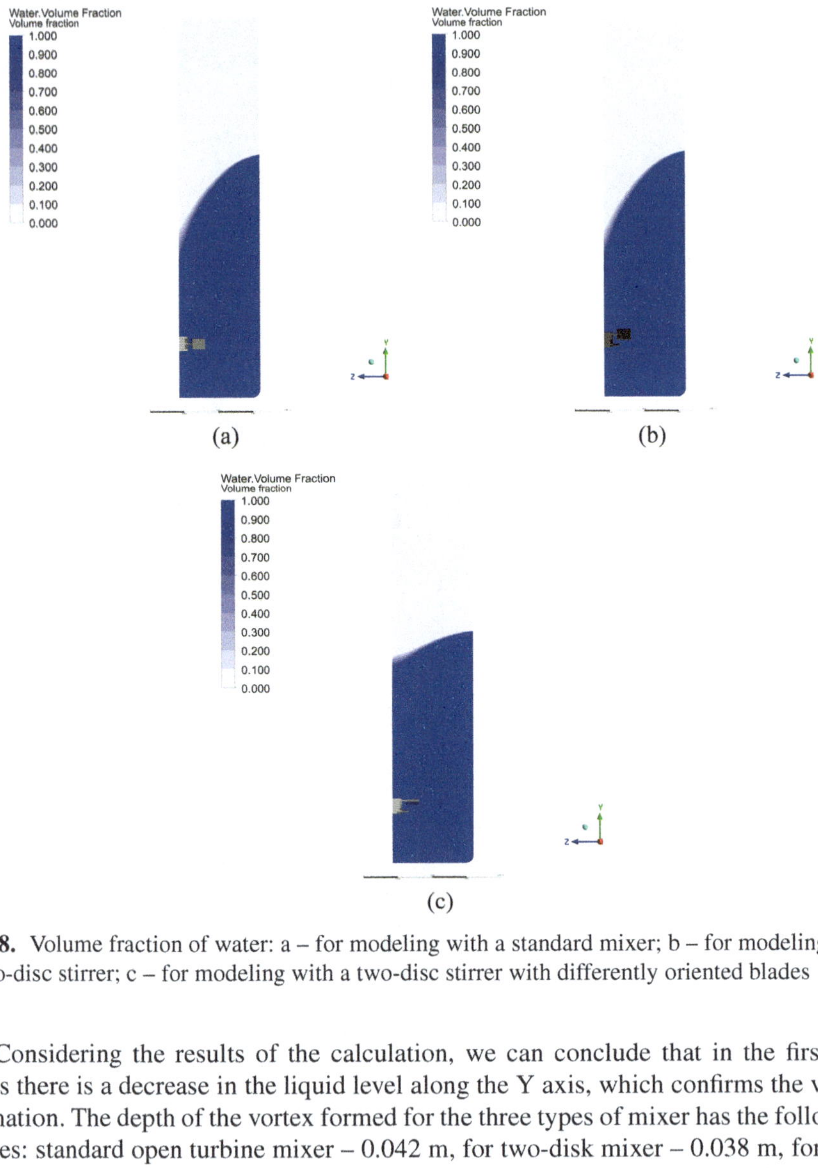

Fig. 8. Volume fraction of water: a – for modeling with a standard mixer; b – for modeling with a two-disc stirrer; c – for modeling with a two-disc stirrer with differently oriented blades

Considering the results of the calculation, we can conclude that in the first two cases there is a decrease in the liquid level along the Y axis, which confirms the vortex formation. The depth of the vortex formed for the three types of mixer has the following values: standard open turbine mixer – 0.042 m, for two-disk mixer – 0.038 m, for two-disk mixer with differently oriented blades – 0.011 m, i.e. does not actually appear. Therefore, the results of this simulation show that this design of the mixing device reduces the likelihood of a vortex in the apparatus without reflective partitions.

Velocity contour is constructed (Fig. 9) for all three simulations in the cross sections YZ and XZ. The constructed contours reflect the range of existing linear flow velocities. We see that the distribution of velocities and regions are almost the same in all three

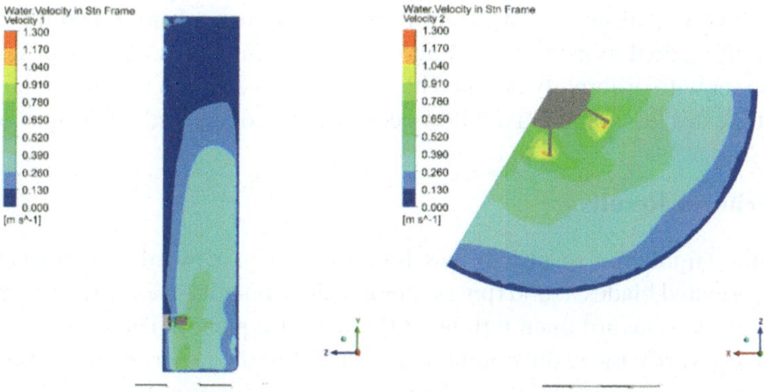

(a)

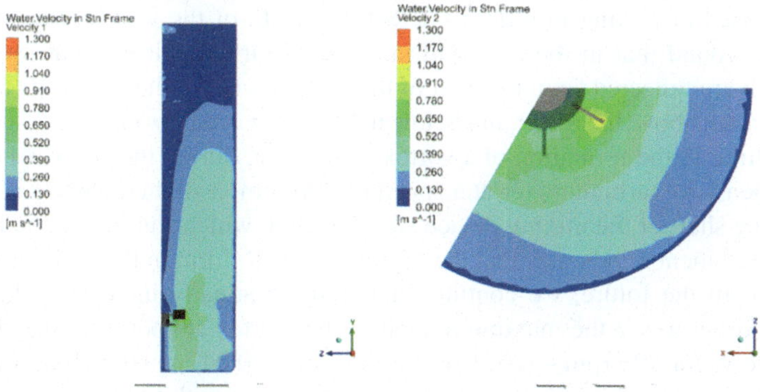

(b)

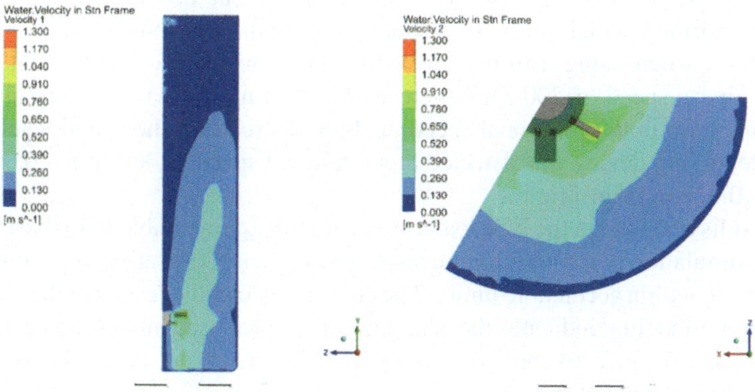

(c)

Fig. 9. Velocity contours: a – for modeling with a standard mixer; b – for modeling with a two-disc stirrer; c – for modeling with a two-disc stirrer with differently oriented blades

cases, which is a confirmation of the uniformity of homogenization and the efficiency of all types of studied mixers. According to the simulation results, we can say that the two-disc mixer with differently oriented blades provides efficient mixing of the medium, although it creates less intense axial flows compared to other studied designs of mixers.

3.2 Experiment Results

Based on the simulation results, it was found that it is a two-disc construction with differently oriented blades at 600 rpm practically does not give the formation of a vortex, in contrast to the standard open turbine at the same frequency. Therefore, the task was to qualitatively verify the results of modeling in real in the experiment, for this purpose on a 3D printer was printed mixers, which are presented in Fig. 1.

The study was performed by increasing the speed of the mixing device to reach the operating speed of mixing in order to detect the depth of the vortex. During the field study it was found that in the tank with a standard turbine stirrer creates radial fluid flows, which are reflected from the wall of the tank moving to the top of the tank and to the bottom, then above the mixer and below it creates a vacuum zone where suction fluid flows, resulting in the formation of a vortex. Therefore, during the study, when using a standard open-type turbine mixer at a speed of 200 rpm, a slight decrease in the liquid level near the shaft of the mixing device was detected, which can be seen in Fig. 8.

This phenomenon indicates that the vortex begins to form in this volume at a speed of 200 rpm. In the future, we continue the study, gradually increasing the speed to 600 rpm in order to see the maximum depth of the vortex at operating speed, which is approximately: for 200 rpm – 0.021 m, for 600 rpm – 0.047 m (Fig. 10a, b). Next, the mixing device was replaced by a turbine two-disc mixer with blades at an angle and a similar study. Not only radial fluid flows are observed, but also axial displacements of the fluid are formed, which promotes better mixing and homogenization.

Therefore, when using a turbine two-disk mixer with differently oriented blades when reaching a speed of 200 rpm, there was almost no decrease in fluid level. With a further increase in the rotational speed, a slight decrease in the liquid level near the shaft of the mixing device was detected at a rotational speed of 400 rpm – 0.006 m, for 600 rpm – 0.012 m (Fig. 11a, b).

The results obtained in the field experiment are fully comparable with those obtained during the simulation in ANSYS for the respective modes of operation of the mixers. The differences are within acceptable limits. The convergence of the results of the experiment and computer modeling indicates the adequacy of the proposed models and confirms the ability to assess the effectiveness of mixing and eliminate undesirable factors, such as the vortex at the design stage.

(a) (b)

Fig. 10. Standard open type turbine mixer: a – speed 200 rpm; b – speed 600 rpm

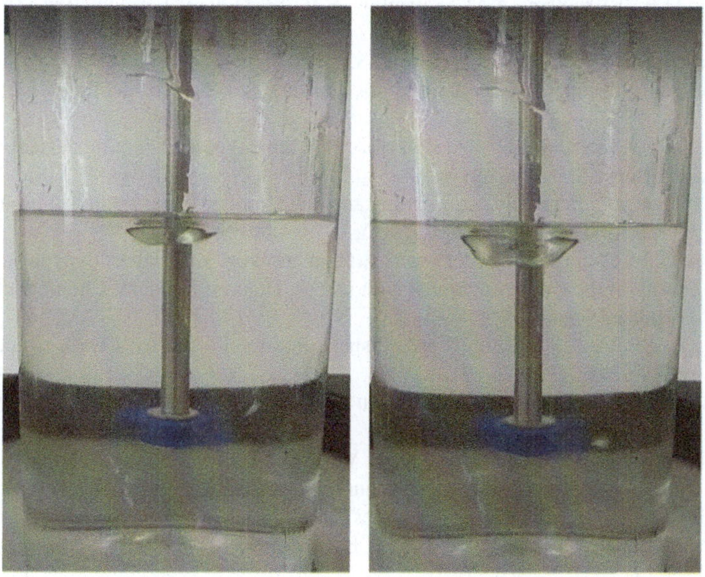

(a) (b)

Fig. 11. Two-disc turbine with differently oriented blades: a – speed 400 rpm; b – speed 600 rpm

Conclusions
The research provides the opportunity to make the following conclusions:

1. The standard design of the mixing device of the open-type turbine mixer with straight blades when used without reflective partitions leads to the formation of a vortex, which does not allow the stability of hydrodynamic parameters of the bioreactor and thus the required mass intensity and sufficient homogeneity of the stirred medium.
2. A new design of an open-type two-disc turbine mixer with differently oriented blades is proposed, which allows to reduce the probability of a vortex during mixing. It is established that the proposed turbine open two-disk mixer practically does not form a vortex, and at a speed of 400 rpm its depth is 0.006 m, and for 600 rpm – 0.012 m.
3. As a result of computer simulation and field research, the design of a standard open-type turbine mixer and a two-disc turbine mixer with differently oriented blades were compared. It was found that the vortex with a standard turbine stirrer occurs at a speed of 200 rpm, and with a two-disc turbine – at 400 rpm.

References

1. Kafarov, V.V., Vynarov, A.Y., Gordeev, L.S.: Modeling of Biomechanic Reactors. Lesnaya Industriya, Moscow (1979)
2. Kedian, W., Xia, D.: Prediction and simulation on the shear stress and mass transfer in perfused bioreactors EL: effect of feed mode and bioreactor geometry. In: Chinese Control and Decision Conference, CCDC 2008, pp. 4338–4343 (2008)
3. Viesturs, U.E., Kuznetsov, A.M., Savenkov, V.V.: Systems of Fermentation. Zynatne, Riga (1988)
4. Zakomorny, D.M., Kutovy, M.G., Kostyk, S.I., Povodzinsky, V.M., Shybetskyy, V.Y.: Hydrodynamics of a fermenter with a multi-shaft stirrer. Sci. J. "ScienceRise" **5/2**(22), 65–70 (2016)
5. Storozhenko, V.Ya., Smirnov, V.A.: Reactors of volume type with mixing devices (calculation and design). Sumy State University, Sumy (2010)
6. Zhang, W., et al.: Parameter estimation and cold model experiments of gastrointestinal simulation reactor. Huagong Xuebao **70**(5), 1879–1886 (2019)
7. Kazenyn, D.A., Chepura, Y.V., Petrov, Y.A., Zhavoronkov, V.A.: Hydrodynamics, mass transfer and energy in the cavity apparatus with stirrer. Theoret. Fundam. Chem. Technol. **42**(2), 128–134 (2008)
8. Frijlink, J.J., Bakker, A., Smith, J.M.: Suspension of solid particles with gassed impellers. Chem. Eng. Sci. **45**(7), 1703–1718 (1990)
9. Sergeev, V.A., Sobko, Yu.A.: Cell cultures in veterinary medicine and biotechnology. Urozhay, Kiev (1990)
10. Korobiichuk, I., Ladanyuk, A., Zaiets, N., Vlasenko, L.: Modern development technologies and investigation of food production technological complex automated systems. In: ACM International Conference Proceeding Series. 2nd International Conference on Mechatronics Systems and Control Engineering, ICMSCE, pp. 52–57 (2018). https://doi.org/10.1145/3185066.3185075
11. Korobiichuk, I., et al.: Identification of technological objects on the basis of intellectual data analysis. In: Szewczyk, R., Krejsa, J., Nowicki, M., Ostaszewska-Liżewska, A. (eds.) MECHATRONICS 2019. AISC, vol. 1044, pp. 487–495. Springer, Cham (2020). https://doi.org/10.1007/978-3-030-29993-4_60

12. Dytnerskyi, Yu.I.: Processes and devices of chemical technology: textbook for universities, 2nd edn. In: 2 Books: Part 1. Theoretical Foundations of the Processes of Chemical Technology. Hydromechanical and Thermal Processes and Devices. Chemistry, Moscow (1995)
13. Kostyk, S., Shybetskyy, V., Fesenko, S., Povodzinskiy, V.: Revealing special features of hydrodynamics in a rotor-disk film vaporizing plant. East.-Eur. J. Enterp. Technol. **1**(6(97)), 28–33 (2019)
14. Korobiichuk, I., Shybetska, N., Shybetskyi, V., Kostyk, S.: Modeling of systems of automated auxiliary processes in pharmaceutical industry. In: Szewczyk, R., Zieliński, C., Kaliczyńska, M. (eds.) AUTOMATION 2021. AISC, vol. 1390, pp. 128–135. Springer, Cham (2021). https://doi.org/10.1007/978-3-030-74893-7_13
15. Korobiichuk, I., Karachun, V., Mel'nick, V., Kachniarz, M.: Modeling of the hypersonic conditions influence on the gyroscopic inertial navigation sensor suspension. Metrol. Meas. Syst. **24**(2), 357–368 (2017). https://doi.org/10.1515/mms-2017-0024

Estimate of the Impact of Pollutant Emissions from Imported Vehicles from the European Union Countries to Ukraine

Igor Korobiichuk[1(✉)], Andriy Ilchenko[2], Volodymyr Shumliakivskyi[2], and Hanna Kireitseva[2]

[1] Warsaw University of Technology, Institute of Automatic Control and Robotics, Warsaw, Poland
igor.korobiichuk@pw.edu.pl
[2] The State University «Zhytomyr Polytechnic», Zhytomyr, Ukraine
avi_77@ukr.net

Abstract. Potential emissions volumes of toxic substances of cars are calculated and the material analysis of their structure is carried out. Estimated of statistical analysis of the parameters of cars that have already been used in EU countries and were imported to Ukraine. These cars were tested by means and equipment of the testing laboratory. The number of studied cars is 1235 units. It is proved that these cars have a significant impact on the environmental situation, their emissions from exhaust gases are 440.6 tons/year, 61% of which is carbon monoxide. Based on the analysis of the parameters of imported cars, as well as the type of fuel used by them, their impact on the state of the air and possible consequences for public health was assessed.

Keywords: Testing laboratory · Car · Emissions · Statistical analysis · Exhaust gases · Pollutant emissions

1 Introduction

The main cause of air pollution are cars which account for 88% of total emissions [1]. The number of cars per thousand inhabitants in different regions of the country varies considerably. The main sales are concentrated in nine regions. They have more cars per thousand inhabitants than the national average. Today, according to the Ukrainian information and analytical group AUTO-Consulting, the number of cars per consumer in Ukraine is much lower than in Europe [2]. On average, we have 187 cars per 1,000 inhabitants across the country, but this number varies greatly from region to region. According to this indicator, Kyiv and Kyiv region are in the lead and they have 343 and 255 cars per 1000 inhabitants respectively.

It is worth noting Chernihiv and Lviv regions, as they have the lowest number of registered cars per thousand inhabitants respectively 131 and 103 cars. The discrepancy between these data [2] can be explained by the fact that a large number of cars are imported to Ukraine from the EU and the USA. Today, Zhytomyr region is not in the top

R. Szewczyk et al. (Eds.): AUTOMATION 2022, AISC 1427, pp. 344–355, 2022.
https://doi.org/10.1007/978-3-031-03502-9_34

ten regions of Ukraine in terms of the number of cars per 1,000 inhabitants. Compared to EU countries, this number is about 565 cars per 1,000 inhabitants in France, 519 – in Germany, 426 – in the UK, 353 – in Estonia. Ukraine ranks about 50th in the world according to this indicator [3].

Today, the share of gasoline consumption by cars compared to 2000 has decreased 1.5 times. The share of cars using liquefied gas and diesel fuel is currently growing in Ukraine. Car park renewal in Ukraine is slow. Further operation of obsolete cars has a negative impact on the environmental situation and requires timely control, diagnosis, carrying out maintenance and repair work [4–7]. According to statistical observations the air received 1358.4 thousand tons of pollutants from cars in 2018 which is 3% less than in 2017. The largest share of emissions from the total amount of pollutant emissions is observed in the city of Kyiv (11%), Dnipropetrovsk (8%), Kyiv (7%), Odesa (6.3%), Poltava (5.8%) and Kharkiv (5.5%) regions. Zhytomyr region ranked 13th in terms of the amount of pollutants emitted by cars into the atmosphere (Fig. 1) [1].

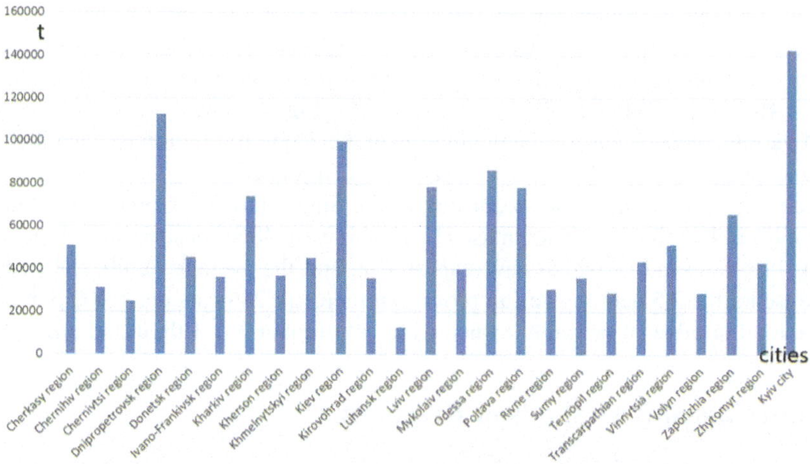

Fig. 1. Emissions of pollutants by cars in cities of Ukraine, t

It is established that air pollution according to the degree of chemical danger for living organisms reaches one of the first places [8, 9]. This is primarily due to the fact that air pollutants are the most widespread and fall into different environments. For example, precipitation gives up to 10% of pollution of water reservoirs, significantly pollutes soils, etc. Carcinogenic compounds, in particular such highly toxic substances as benz(o)pyrene, are of particular danger to the environment. Carcinogenic compounds affect the genome of living organisms and cause cancer [10].

Thus, the large volume of emissions from cars is explained primarily by the increase in the number of private cars, operation of technically obsolete cars, including those imported from EU countries to Ukraine, as well as the use of low quality fuel [11, 12], road conditions, etc. An integrated approach to monitoring car emissions and the state of the country's automotive complex will improve the environmental situation [13].

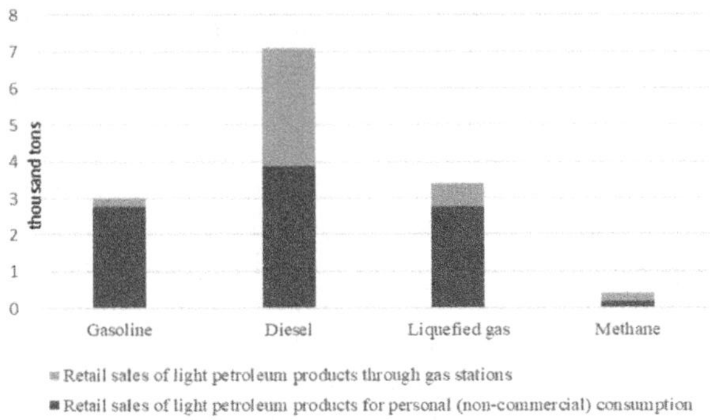

Fig. 2. Average monthly volume of retail and wholesale sales of light oil products and gas to the population through gas stations in Zhytomyr region [7]

According to statistical observations [14], the total retail sales of light petroleum products and gas through gas stations directly to the public for personal (non-commercial) consumption averages 9.7 thousand tons per month and 70% of total sales of light petroleum products and gas in the Zhytomyr region. Sales of gasoline are 29%, diesel – 40%, liquefied gas – 29%, methane – 2% (Fig. 2). Thus, the annual consumption of various fuels by cars of the population in the Zhytomyr region is: gasoline – 34.0 thousand tons, diesel fuel – 47.0 thousand tons, liquefied gas – 34.0 thousand tons, methane – 2, 4 thousand tons. According to the State Statistics Committee, 23.8% of households in Ukraine have cars in use [1]. As there are 486 thousand households in the region, 116 thousand of which have cars in use, and the average number of people in one household is 2.5 people, we can prove that there are 289 cars per 1,000 population, and the total number of privately owned cars will number 351.1 thousand units [14].

2 Materials and Methods of Research

The statistical analysis of the parameters of cars that have already been used in EU countries and were imported to Ukraine in the period from April 2018 to February 2019 was conducted. The number of studied cars is 1235 units.

These cars were tested by means and equipment of the testing laboratory «Researcher» LLC «Veltest» (Zhytomyr) according to the following parameters:

1. Identification features of vehicles (manufacturer, model, category, year of manufacture, mileage from the beginning of operation);
2. General characteristics (number of axles/wheels, geometrical parameters of wheels, weight in the equipped condition, technically admissible maximum weight, technically admissible maximum static weight on the traction coupling device);
3. Power unit and its systems (engine manufacturer, its code and type, number and location of cylinders, displacement, maximum engine power and fuel type, gearbox type, maximum vehicle speed);
4. Axles and suspension (axle track, tire brand and air pressure in them);
5. Brake system (number of circuits, presence of anti-lock braking system);

6. Body type, the presence of a clutch device.

Testing was performed according to the following regulations: UNSS 3649:2010, UNSS 4277:2004, UNSS 4276:2004, UN regulations No. 13 or No. 13-H, UN regulations No. 14, UN regulations No. 16, UN regulations No. 43, UN regulations No. 46, UN regulations No. 48, UN regulations No. 49, UN regulations No. 58, UN regulations No. 73, UN regulations No. 83, UN regulations No. 89 [15].

3 Results and Discussion

The results of the survey and measurements were evaluated by the following parameters:

1. Brake systems (according to UN Rules No. 13, 13H and UNSS 3649): operation of working, parking, reserve brake systems; return of the control bodies of the working and reserve brake systems to the initial position at the end of the action; operation of the backup brake system; tightness of the drive, absence of leakage of liquid, its level; condition of mechanical drive elements; availability and operation of alarm and system status monitoring; efficiency of anti-lock system and anti-slip system of driving wheels; efficiency of the working brake system by the road test method (uniform deceleration and specific braking force); efficiency of the parking brake system (duration of steady state on a slope of 16% and specific braking force);
2. Steering and tires (according to UNSS 3649): no unauthorized movement of the steering wheel and (or) steered wheels; lack of movement and gaps in the joints and their condition; no leakage of working fluid in the hydraulic system of the amplifier and its level; no damage and deformation of parts and improperly repaired parts; condition of alarm systems and control systems; no increased effort, jerks and jams when turning the steering wheel; compliance of tire parameters installed on one axle of the vehicle; no local damage; compliance with the class of repair of tires for buses (UNSS 3649) and their marking; the presence of all fasteners; backlash in steering; the minimum height of the tread pattern;
3. External lighting devices: type of dipped beam distribution ("European beam"); the presence of a corrector of the angle of the dipped beam; the location of the dipped-beam cut-off line, the fog lamps and the brightness of their light rays; frequency of flickering of lamps of indicators of turn, their work in one phase; quantity, color, mode of operation (switching order), marking of headlights, fog lights, daytime running lights, indicators, front, rear and side lights, brake signals, taillights and license plate lighting, reflectors; headlight power;
4. Condition of car windows: identification data (sign of official approval); light transmission of glass.
5. Engine and its systems: no fuel leakage; condition of fuel tanks and fuel shut-off devices; efficiency of exhaust elements, their completeness and condition; absence of burns, mechanical breakdowns and leaks in the elements of the exhaust system; the level of external noise during tests on a stationary vehicle (according to UNSS 3649); the content of carbon dioxide and hydrocarbons in the exhaust gases of cars with a gasoline engine (according to UNSS 4277); opacity of exhaust gases of cars with diesels (according to UNSS 4276); marking with the sign of official approval (in the presence);

6. Compliance of vehicles with other safety parameters (according to UNSS 3649): windshield wipers and windshield washers (presence, operation, mode of operation); rear-view mirrors (presence, marking, adjustment); sound signal (presence, marking with the appropriate sign); sun protection devices (presence, construction); no cracks on the windshield in the area of the wipers; body or cab door locks; mechanisms for adjusting and fixing the driver's and passengers' seats; windshield heating and blowing devices; anti-theft device; speedometer and (or) tachograph (presence, operation, illumination); devices that protect against the ejection of foreign particles and dirt from under the wheels of vehicles, bumpers; fixing and condition of electrical networks and electrical fuses; the condition of the working surfaces of the parts in the clutch and safety devices; seat belts (presence, marking with the appropriate sign); headrests (presence, design). The results of the analysis of 1235 research protocols show that 88% are cars with diesel engines, and 9.9% – cars with gasoline engines (Fig. 3).

The main number of cars (87.7%) has a mileage of 100–500 thousand km from the start of operation (Fig. 4). It is a well-known fact that the mileage of cars from the beginning of their operation and years of manufacture have a significant impact on the reliability and environmental safety [16]. To maintain their technical condition it is necessary to conduct systematic technical control and diagnosis [17]. As age and mileage increase, the amount of maintenance and repair work increases, so it needs more and more relevant maintenance companies in the region. Therefore, it is important to note that mainly imported cars were used, which were exploited from 4 to 9 years (Fig. 5). The highest share of surveyed cars (17%) belongs to cars manufactured.

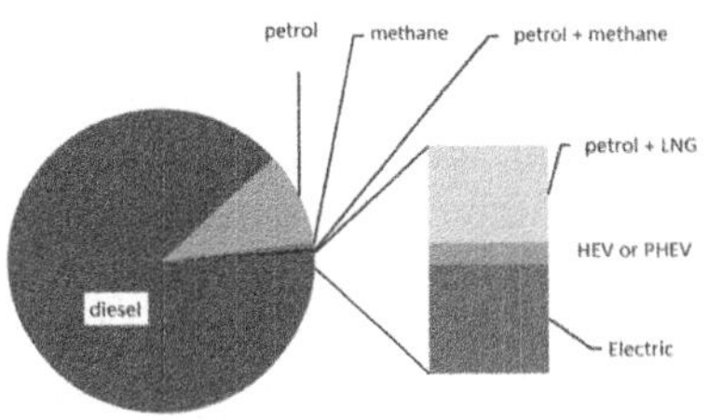

Fig. 3. Distribution of cars imported into the Zhytomyr region by type of engine (fuel)

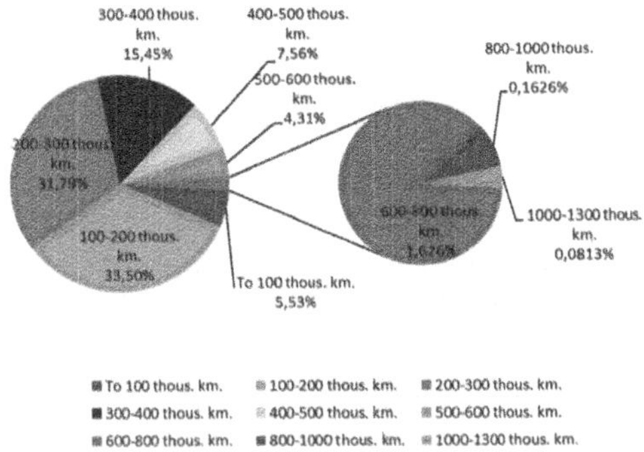

Fig. 4. Distribution of cars by mileage from the beginning of operation, thousand km

The volume of the internal combustion engine will affect the fuel consumption and, consequently, the amount of exhaust gases [18]. The analysis showed that it was mainly imported cars with engine capacity from 1700 up to 2000 cm^3 (their share was 38%). 18% of cars with an engine capacity were also imported from 1200 up to1500 cm^3, 20% cars with engine capacity from 1500 up to 1700 cm^3, 18% cars with engine capacity from 2000 up to 2500 cm^3 (Fig. 6).

According to the type of body imported mainly cars of the category M1 (77,41%) and N1 (22,27%) (Fig. 7). Of course, such cars are imported for individual use, but control over their technical condition is the responsibility of the owner, which will also affect their environmental performance during further operation. It is not difficult to establish that the total number of cars during the study period increased by 0.5–1% due to the number of imported. This number mainly applies to cars aged 4–9 years, which is approximately 64.9% of the total number of imported cars.

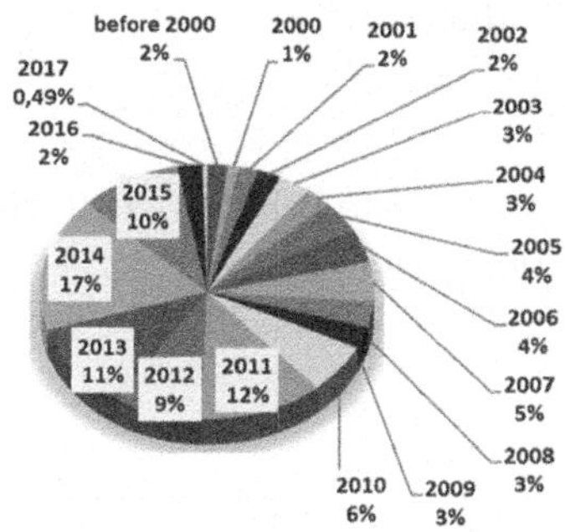

Fig. 5. Distribution of imported cars by years of production

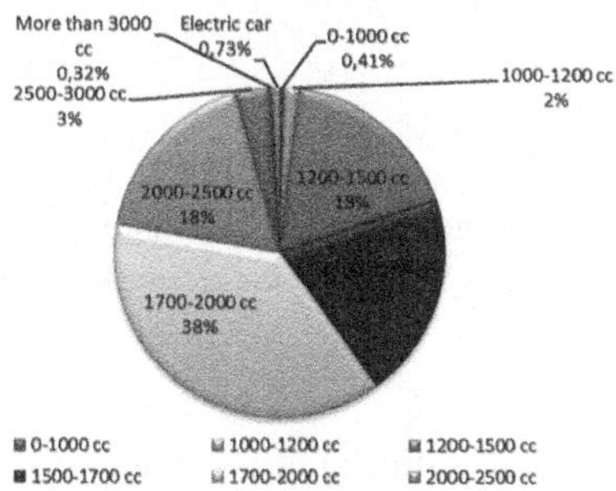

Fig. 6. Distribution of imported cars by engine capacity

Based on the analysis of the parameters of imported cars, as well as the type of fuel used by them, their impact on the state of the air and possible consequences for public health was assessed.

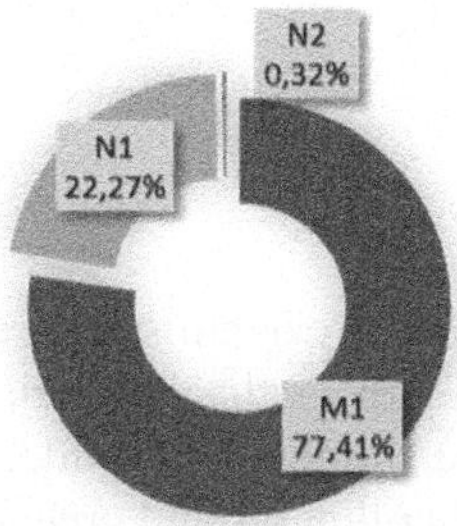

Fig. 7. Distribution of imported cars by categories

Thus, the assessment of emissions of air pollutants from cars was carried out using the recommendations of international methods, in particular the methods of the UN Economic Commission of Europe for the inventory of emissions into the air – CORINAIR [19]. Based on the recommendations of this methodology, Ukraine has developed a Methodology for calculating emissions of pollutants and greenhouse gases into the air from vehicles, approved by the order of the State Statistics Committee No. 452 by 13.11.2008, repealed by 19.02.2015 [20]. The calculation of pollutant emissions for different types of car fuels, which were registered from April 2018 to February 2019, is performed according to the following method [20]: if you take j = 1, 2, 3, 4, … 10 – number of pollutant and greenhouse gas, i = 1, 2, 3, 4 – fuel type number, so:

1. Average annual consumption of another type of fuel (gasoline, diesel, methane or liquefied gas) by one car F_{a1}, thousand tons:

$$F_{a1} = \frac{F_a}{N}, \quad (1)$$

where F_a – the amount of fuel sold to the population of another type in the region, thousand tons; N – total number of privately-owned cars in the region, thousand units;

2. Annual consumption of another type of fuel F_{ar}, thousand t/y:

$$F_{ar} = N_r \times F, \quad (2)$$

where N_r are number of cars registered from April 2018 to February 2019, units; F are average annual fuel consumption per car, thousand tons.

The results of the calculations are given in Table 1.

Table 1. Calculation of consumption of different types of fuel by cars that were certified from April 2018 to February 2019

Fuel consumption	Fuel	Diesel fuel	Liquefied gas	Methane
Average annual consumption of defined type of fuel by one car F_{a1}, thousand tons	0,1	0,13	0,1	0,007
Annual consumption of other fuels F_{ar}, thousand t/y	12	140,4	0,9	0,06

3. The main pollutants contained in car emissions are: carbon (II) oxide, nitrogen (IV) oxide, sulfur (IV) oxide, non-methane volatile organic compounds, methane, soot, ammonia, benz(o)pyrene, carbon (IV) oxide, nitrogen (II) oxide. Volumes of emissions of each of the listed pollutants and greenhouse gas j from the use of other fuels by cars that were certified from April 2018 to February E_{ji}, t/y:

$$E_{ji} = \frac{F_a \times N_{ja}}{1000} \times N_{jatc}, \quad (3)$$

where N_{ja} – average specific emissions of pollutants and greenhouse gas j by cars from consumption of another fuels, kg/t (Table 2); N_{jatc} – coefficients of influence of technical condition of the car on emissions of polluting substance and greenhouse gas j by cars from use of another type of fuel (Table 2).

4. Emissions of pollutants and greenhouse gases j from the use of all fuels E_j, t/y:

$$E_j = \sum_{i=1}^{4} E_{ja} \quad (4)$$

5. Total emissions of all pollutants and greenhouse gases (except carbon (IV) oxide) from cars certified from April 2018 to February 2019, E_{tot}, t/y:

$$E_{tot} = \sum_{j=1}^{9} \sum_{i=1}^{4} E_{ja} \tag{5}$$

The results of calculations of total emissions of all pollutants and greenhouse gas from cars that were certified from April 2018 to February 2019 are given in Table 3. Analyzed the emissions of pollutants by cars for different types of fuels, it was found that a significant impact on the environment is exerted by cars running on diesel fuel. The share of their emissions is 74% of the total emissions into the atmosphere. Methane-powered cars have the least impact on environmental safety. The share of emissions from this type of fuel is only 0.02%.

Table 2. Initial data for calculating pollutant and greenhouse gas emissions j from the use of another type of fuel

Name of the substance	Specific emissions of pollutants and greenhouse gases N_{ja}, kg/t of fuel				Coefficient of influence of technical condition of vehicles on specific emissions N_{jatc}			
	Gasoline	Diesel fuel	Liquefied gas	Methane	Gasoline	Diesel fuel	Liquefied gas	Methane
Carbon (II) oxide	201.80	36.20	201.8	87.70	1.5	1.5	1.5	1.7
Nitrogen (IV) oxide	21.00	31.40	21.00	27.40	0.9	0.95	0.9	0.9
Sulfur (IV) oxide	1.00	4.30	1.0	1.00	1.0	1.0	1.0	1.0
Non-methane volatile organic compounds	53.00	3.10	25.7	22.70	1.0	1.0	1.0	1.0
Methane	0.90	0.10	0.96	1.0	1.5	1.4	1.5	1.8
Benz(o)pyrene	–	0.03	–	–	–	1.0	–	–
Ammonia	0.004	–	–	–	1.0	–	–	–
Soot	–	3.80	–	–	–	1.8	–	–
Nitrogen (II) oxide	0.188	0.165	–	–	1.0	1.0	–	–
Carbon (IV) oxide	3183	3138	–	–	1.0	1.0	–	–

6. Greenhouse gas emissions of carbon monoxide are calculated only for oil-based fuels (gasoline and diesel), as the combustion of liquefied gas and methane emits the

least amount. Emissions of carbon (IV) oxide from the use of gasoline and diesel fuel E_{gg}, t/y:

$$E_{gg} = \sum_{i=1}^{2} E_{agg} \quad (6)$$

where E_{agg} – emissions of carbon (IV) oxide from the use of another fuels, t/y.

The species composition of pollutant and greenhouse gas emissions was assessed according to the fuel used by the cars under study (Fig. 8). It was found that among the toxic components emitted by cars 61% is carbon (II) oxide, 6% – non-methane volatile organic compounds, 24% – nitrogen (IV) oxide, 3% – sulfur (IV) oxide, 5% – soot, all other pollutants account for 1% of emissions. A separate category is carbon (IV) oxide emissions. The share of carbon dioxide emissions from cars running on gasoline is 4%, and of those running on diesel – 96%. This indicates that diesel fuel is the most environmentally hazardous. Thus, when burning fuel based on oil from cars that were

Table 3. The results of calculations of total emissions of all pollutants and greenhouse gas (E_{tot}) from the studied cars

Type of the substance	Emissions of pollutants and greenhouse gas *j* from the use of another fuels E_{ja}, t/y				Total emissions of all pollutants and greenhouse gases, E_{tot}, t/y
	Gasoline	Diesel fuel	Liquefied gas	Methane	
1	2	3	4	5	6
Carbon (II) oxide	3.62	7.62	0.27	0.01	11.52
Nitrogen (IV) oxide	0.23	4.2	0.02	0.0015	4.45
Sulfur (IV) oxide	0.01	0.6	0.0009	0.00006	0.61
Non-methane volatile organic compounds	0.64	0.44	0.023	0.0014	1.1
Methane	0.02	0.02	0.0013	0.0001	0.04
Benz(o)pyrene	–	0.004	–	–	0.004
Ammonia	0.00005	–	–	–	0.00005
Soot	–	0.96	–	–	0.96
Nitrogen (II) oxide	0,002	0.23	–	–	0.232
Emissions of pollutants and greenhouse gases *j* from the use of all fuels E_j, t/y	4.52	14.07	0.32	0.013	18.9

certified from April 2018 to February 2019, the atmospheric air receives carbon (IV) oxide in the amount of (E_{gg}) 478.8 t/y, of which: 38.2 t/y are emitted when burning gasoline, 440.6 t/y – when diesel fuel burning.

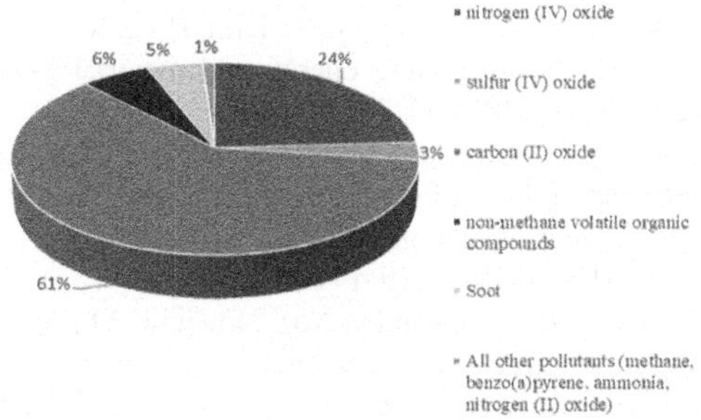

Fig. 8. The ratio of pollutants in car emissions that were certified

Thus, after assessing emissions of pollutants from cars imported from EU countries to Ukraine from April 2018 to February 2019, it can be argued that they will have a significant adverse impact on both the environment and health of country population.

Conclusions

The analysis of the obtained data shows that Ukraine, in particular the Zhytomyr region, recently imported cars with diesel engines (88.8%) with a mileage of 100–500 thousand km (87.7%) since the beginning of operation. The service life of most cars is in the range of 4–9 years, which is 64.9% of the total number of cars imported from EU countries. The application of the preferential regime of customs clearance of used cars imported into the territory of Ukraine for further operation from EU countries has increased the purchase of cars of categories M1, N1 with an engine capacity 1200–2500 cm^3 for individual use.

It is established that cars imported to Ukraine from EU countries and registered in the Zhytomyr region in the period from April 2018 to February 2019 emit 18.9 t/y of pollutants with exhaust gases, 61% of which is carbon monoxide, 6% –non-methane volatile organic compounds, 24% – nitrogen (IV) oxide, 3% – sulfur (IV) oxide, 5% – soot, 1% – all other pollutants, as well as greenhouse gas carbon (IV) oxide in quantities of 478.8 t/y. It is established that diesel fuel is used by cars that were certified from April 2018 to February 2019 in the largest amount (88%). Its emissions contain pollutants such as carbon (II) oxide, nitrogen (IV) oxide, sulfur (IV) oxide and the carcinogenic substance benzo (o) pyrene. Thus, the import of used European cars causes an increase in the volume of exhaust gases, causes air pollution in the region and Ukraine, and thus increases the risk to public health and the environment as a whole.

References

1. Official site of the State Statistics Service of Ukraine. http://www.ukrstat.org.ua. Accessed 02 Jan 2022

2. Osieczko, K., Zimon, D., Płaczek, E., Prokopiuk, I.: Factors that influence the expansion of electric delivery vehicles and trucks in EU countries. J. Environ. Manag. **296**, Article no. 113177 (2021). https://doi.org/10.1016/j.jenvman.2021.113177
3. Ranking of countries by number of cars per 1000 inhabitants. http://autonews-ua.info/more.html?id=5545
4. Aguilar Esteva, L.C., Kasliwal, A., Kinzler, M.S., Kim, H.C., Keoleian, G.A.: Circular economy framework for automobiles: closing energy and material loops. J. Ind. Ecol. **25**(4), 877–889 (2021). https://doi.org/10.1111/jiec.13088
5. Korobiichuk, I., Bezvesilna, O., Ilchenko, A., Trostenyuk, Y.: Thermoanemometric flowmeter of biofuels for motor transport. In: Jabłoński, R., Szewczyk, R. (eds.) Recent Global Research and Education: Technological Challenges. AISC, vol. 519, pp. 443–448. Springer, Cham (2017). https://doi.org/10.1007/978-3-319-46490-9_59
6. Korobiichuk, I., Kachniarz, M., Shavursky, Yu., Nowicki, M., Szewczyk, R.: Algorithms on improvement of accuracy of biofuel temperature measurement in thermo-anemometric flowmeter. In: 2017 4th International Conference on Control, Decision and Information Technologies (CoDIT), IEEE Conference Publications, pp. 0964–0968 (2017). https://doi.org/10.1109/CoDIT.2017.8102723
7. Korobiichuk, I., Kachniarz, M., Bezvesilna, O., Ilchenko, A., Szewczyk, R.: Calorimetric flow meter of motor fuel with inlet temperature regulation. In: 2017 4th International Conference on Control, Decision and Information Technologies (CoDIT), IEEE Conference Publications, pp. 0975–0979 (2017). https://doi.org/10.1109/CoDIT.2017.8102725
8. Attard, M.: Mobility justice in urban transport-the case of Malta. Transp. Res. Proc. **45**, 352–359 (2020). https://doi.org/10.1016/j.trpro.2020.03.026
9. Perovich, L.M., Vanchura, R.: The impact of road transport on the pollution of land resources. Geodesy Cartography Aerial Photogr. **73**, 102–109 (2010). (in Ukrainian)
10. Lee, C.-C., Chen, C.S., Wang, Z.-X., Tien, C.-J.: Polycyclic aromatic hydrocarbons in 30 river ecosystems, Taiwan: sources, and ecological and human health risks. Sci. Total Environ. **795**, Article no. 148867 (2021). https://doi.org/10.1016/j.scitotenv.2021.148867
11. Korobiichuk, I., Mel'nick, V., Karachun, V., Shybetskyi, V.: Investigation of optimization of combustion processes in the engine of combat vehicles by use of disk structure. Energies **14**(21), 7039 (2021). https://doi.org/10.3390/en14217039
12. Korobiichuk, I., Mel'nick, V., Karachun, V.: Modeling of voltaic pile surface formation using current-carrying cassini ovals. Eng. Sci. Technol. Int. J. **22**(1), 353–358 (2019). https://doi.org/10.1016/j.jestch.2018.08.004
13. Korobiichuk, I., Mel'nick, V., Karachun, V., Shybetskyi, V., Kostyk, S.: Numerical modeling of the possibility of scale power growth of the internal combustion engine in real time. In: 2021 25th International Conference on Methods and Models in Automation and Robotics (MMAR), pp. 309–314 (2021). https://doi.org/10.1109/MMAR49549.2021.9528472
14. Official site of the Main Department of Statistics in Zhytomyr region. http://www.zt.ukrstat.gov.ua/. Accessed 02 Jan 2022
15. UN Vehicle Regulations – 1958 Agreement. https://www.unece.org/trans/main/wp29/wp29regs1-20.html
16. Gutarevich, Yu.F., Zerkalov, D.V., Govorun, A.G., et al.: Ecology and road transport. Nat. Transp University: Aristeas, Kyiv, 292 p. (2006). (in Ukrainian)
17. Galimshina, G.V., Sobakar, A.A.: State control over road transport in Ukraine: monograph, onets. jurid. Inst. of Lugano. state untu vnutr. affairs them. E.O. Didorenko, Donetsk, 237 p. (2011). (in Ukrainian)
18. Research and development of recommendations for reducing the specific consumption of energy by road: report of the State Enterprise "DerzhavtotransNDIproekt" (intermediate), 0115U006026, 154 p. (2016)

19. EMEP/EEA air pollutant emission inventory guidebook. https://www.eea.europa.eu/themes/air/air-pollution-sources-1/emep-eea-air-pollutant-emission-inventory-guidebook
20. Methodology for calculating emissions of pollutants and greenhouse gases into the air from vehicles, approved by the Order of the State Statistics Committee No. 452 of 13.11.2008

Correction of Temperature Influences in Moisture of Bulk Materials Measurement by Capacitance Method

Mykhaylo Dorozhovets[1,2], Roman Ivakh[2], and Zygmunt L. Warsza[3(✉)]

[1] Rzeszow University of Technology, Rzeszow, Poland
michdor@prz.edu.pl
[2] National University – Lviv Politechnic, Lviv, Ukraine
[3] Łukasiewicz Reasearch Network – Industrial Research Institute of Automation and Measurements PIAP, Warsaw, Poland
zlw1936@gmail.com

Abstract. In this paper the methodical and instrumental aspects of the correction of temperature influences in moisture of bulk material express measurement are considerate. Bilinear coefficients of capacitance dependence on moisture and temperature were determined to provide a correction based on the performed measurements of sensor capacitance for 4 reference moisture specimens of the test material with simultaneous measurement of their temperature. Based on obtained relationship, the temperature influence correction function was determined. The combined standard uncertainty of the correction was estimated as a function of the uncertainty of the 8 influencing factors. It was shown that the correction reduces the systematic influence of temperature by 10 to 40 times.

Keywords: Moisture · Temperature · Systematic · Approximation · Correction · Uncertainty

1 Introduction

Moisture is one of the most important characteristics of grain quality [1–3]. The value of moisture can be used to quantify the proportion of nutrients in the grain, as well as to determine the duration of its storage. According to the Food and Agricultural Organization at the UN, annual grain losses account for more than 10% of total production, and in some less developed countries 30–50% [4, 5]. To the above it should be added that most often these losses occur at high humidity and high grain temperature. Measuring the moisture content of bulk materials and designing express moisture meters is an urgent task for major cereal producing countries and for Ukraine especially. Because the major direction in the agricultural branch of economics of these countries is the production of grain and products of its processing. Precise control of the moisture content in the grain helps to reduce the costs of drying and storage.

Due to excessive humidity, unwanted physical and chemical processes are activated, which lead to negative consequences: swelling and germination of grain; activation of

R. Szewczyk et al. (Eds.): AUTOMATION 2022, AISC 1427, pp. 356–369, 2022.
https://doi.org/10.1007/978-3-031-03502-9_35

enzymes (fermentation processes); reducing flow ability and increasing vulnerability to mechanical damage; rapid development of parasites – microbes, mites, pests [6]. If the grain remains excessively wet for a long time, it leads to the impossibility of its processing and subsequent storage. Even if the humidity does not exceed the allowable norm, grain yield is significantly reduced, and the quality of products made from it deteriorates. That is why it is so important to accurately determine the moisture content of grain, carried out by various methods [7, 8].

The moisture content of bulk materials significantly depends on many factors, such as the environmental temperature and relative humidity, particle size distribution (shape, size, etc.), grain structure, chemical composition, and so on [9, 10]. The chemical composition of organic matter, such as grain, significantly depends on the region and climatic conditions of its cultivation, on mineral fertilizers (organic and inorganic), which are added to the soil to obtain a better harvest. The dielectric constant of bulk materials of inorganic origin also depends on similar factors, in particular humidity and temperature. In general, it is difficult to consider all these factors when developing secondary measuring instruments, but they should be considered in measurements and the systematic errors caused by their influence should be corrected [11, 12].

Among such non-informative parameters, the most significant are the temperature of bulk materials, as well as its biochemical composition [9, 13]. The temperature of the medium can be controlled during the measurement and to correct the temperature error it is enough to know the temperature coefficients. In contrast to temperature, it is extremely difficult to control the biochemical composition of bulk materials during measurement, and therefore it is almost impossible to correct errors caused by ambiguity in biochemical composition by conventional methods. To do this, you must use other methods, which will be discussed later in this section.

The ability to accurately determine the moisture content of grain at the right time, and especially during harvest – is the factor that separates profits from losses. Express moisture meters are used when receiving grains at elevators, to control the preservation during grain processing. The ability to determine the humidity not only of cereals and legumes, but also the products of their processing (flour, etc.), allows significantly expand the range of moisture meters. In contrast to a number of known methods for measuring the moisture content of bulk products, we focused on the dielectric method for measuring humidity. The latter is best suited to determine the moisture content of biomaterials and products [3, 8].

The fastest and most convenient of all known methods of measuring grain moisture is the dielectric method. Measurement of moisture content of materials by their dielectric properties is the most common electrical method. The practical application of the method, which began in the late 1920s, was based on measuring the capacitance of a capacitor filled with the material under study, with dielectric constant as the moisture-related parameter. This is why the method is often called "capacitive" [13]. In fact, the measured values (output values of the transducers) in most cases are the impedance of the transducers and less often the individual components of this resistance. Therefore, the term "dielectric method" is more correct. This method uses medium and shortwave frequency bands ($f = 0.3$–30 MHz); most often the range from hundreds of kilohertz to 10 MHz [8, 14].

Dielectric method of moisture determination involves the assessment of moisture content by measuring the dielectric constant and the tangent of the angle of dielectric loss of wet materials and substances in a wide range of frequencies – from sound to microwave. Figures 1 and 2 show the dependences of ε and tgδ on humidity for different materials [9].

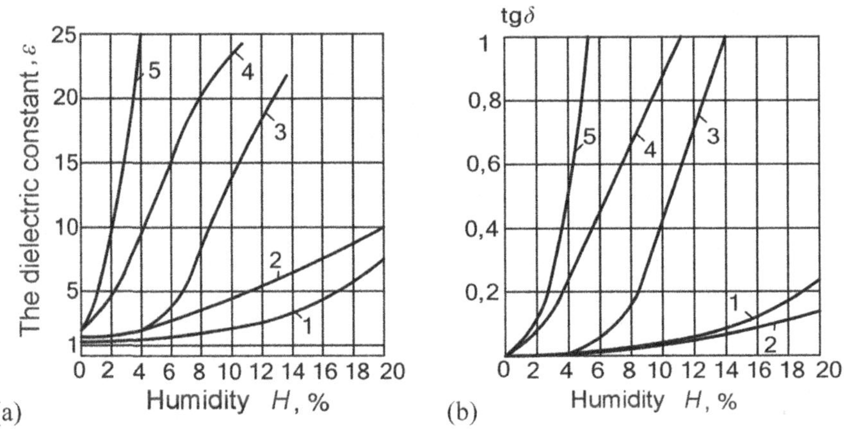

(a) (b)

Fig. 1. Dependence of dielectric constant ε (a), and tgδ (b) on humidity: 1 – grain; 2 – activated alumina; 3 – coal; 4 – sand; 5 – gravel

Measurement of humidity by dielectric method is based on the difference between the values of the dielectric constant of the solid base, air and water. The main characteristics of the test substance are the dependence of dielectric constant and dielectric loss on moisture content, frequency of the alternating field, temperature, density, etc. Compared with the conductometric method, dielectric one is inherent in a number of advantages: lower temperature sensitivity, the ability to compensate for the impact of significant losses (up to tg$\delta = 100$) with the appropriate choice of measurement scheme, reduced moisture distribution by volume of test object, and high accuracy measurements. Taken together, it leads to the fact that, despite the variety of existing methods for measuring the moisture content of bulk materials, today preference is given to the dielectric method.

Humidity characteristics are the basis for calibration of moisture meters and largely determine their metrological properties. If the electrical parameters change according to the proportional-additive law, the considered characteristics are described by the equations [15]:

$$\varepsilon = \varepsilon_c + A \cdot H \text{ and } \mathrm{tg}\delta = \mathrm{tg}\delta_c + B \cdot H, \tag{1}$$

where ε_c and tgδ_c are the parameters of absolutely dry material, A and B are factors that depend on properties of the material and measurement conditions.

Only some materials have linear moisture characteristics. These include some liquid dielectrics – oil, fuel oil and other petroleum products, in which a linear dependence of ε on humidity in the range up to 10–12% has been established. In the vast majority of materials ε and tgδ increase with humidity faster than linearly; the slope of the characteristic tg$\delta(H)$ is usually greater than the characteristic $\varepsilon(H)$. The shape of the moisture characteristics strongly depends on the frequency; therefore, the combined frequency-humidity characteristics are considered more frequently for certain materials [9].

Shape of the frequency characteristics of wet materials is associated with the inhomogeneity of the inner field of these materials due to the difference in electrical properties of their components. The influence of frequency on the electrical properties of an inhomogeneous dielectric is determined by its unequal impact on different types of polarization. In the low-frequency region, most materials demonstrate a significant increase in tgδ due to an raising the relative value of the through conductivity. For materials in which the losses are mainly due to through conductivity, the dependence of tgδ (f) is close to hyperbolic, that is, with increasing frequency tgδ fall down, moving to a certain limit.

As for the dielectric constant, in the wet materials with increasing frequency there is noted a gradual decrease in ε. These general properties of the frequency characteristics may vary depending on the degree of inhomogeneity of the dielectric, in particular, the presence of conductive inclusions, etc. The most important of the factors influencing the nature of the frequency curves is the humidity of the material. In this regard, it is advisable to consider the frequency-humidity characteristics, i.e. the families of curves $\varepsilon(f)$ and tg$\delta(f)$ for different values of H = const.

Dielectric humidity meters are capacitive. They carry out the same conversion algorithm, which can be represented as: $H_d \rightarrow \varepsilon_d \rightarrow C_d \rightarrow B_d$, where H_d; ε_d; C_d; B_d are respectively, the measured humidity, dielectric constant, electrical capacity of the measured material in the primary transducer and the readout of the measuring instrument. Thus, the output value is proportional to the measured capacitance of transducer, filled with the measured material. Then the capacity becomes a simplified form for all structural particles in it: $C_d = \varepsilon_d\, K$; here K is the design parameter of the primary transducer, which depends on the shape and size of both the electrodes and the interelectrode space in which the bulk material is located.

2 Determination of the Coefficients of the Temperature Influence onto Humidity Measurement

2.1 Temperature Influence in Measurement of the Moisture Content of Bulk Materials by Capacitance Method

Humidity of bulk materials significantly depends on many factors, such as temperature, particle size distribution (shape, size), grain structure, biochemical composition, etc. Among such non-informative parameters, the most significant are the temperature, as well as biochemical composition of studied materials. The temperature of the medium can be additionally defined during the measurement and it is enough to know the temperature coefficients for the temperature error adjustment. In contrast to temperature, it is extremely difficult to control the biochemical composition of materials during measurement; therefore, it is almost impossible to correct errors caused by ambiguity of composition by conventional methods.

It is clear that to correct temperature influence into the humidity measurement the temperature of the environment has to be controlled during the measurement. But this correction is possible only when the temperature coefficients, which determined temperature influence onto measured value of humidity, are previously evaluated. So, to determine it, is necessary to conduct experimental studies.

2.2 Experiment

Previously, 4 containers of wheat grains taken from one field were prepared, i.e., it is of the same type, using a pneumatic separator to select grains according to specific size limits. These containers were moistened with different amounts of water. It took about 24 h at controlled temperature to absorb the moisture in the whole grain and to achieve balance. The moisture content in the each containers was determined by thermogravimetric method according to [16] (Ukrainian standard Grain Method of moisture content determination (i.e.: method of drying and weighing) and [17–19].

One 50 g sample from each container of varying moisture content was weighed on the accurate laboratory analytical balances. Then they were placed in a special drying chamber SESH-3, which maintained the specified temperature for this bulk crop 130 °C according to [16]. Maximal permissible error of this method: ±0.5% (millet, buckwheat, rice-grain, sorghum, wheat, rye, barley, oats), ±0.7% (vetch, peas, chickpeas, chickpeas, beans, lentils, corn in grain) [16].

The temperature in the oven was maintained by means of an automatic regulator with an accuracy of ±1 °C. Drying was carried out for 40 min, from the moment of setting the temperature to 130 °C. After it was all transferred to the environment, the material to be analyzed was weighed again and the percentage of moisture content was determined by the difference in weight before and after drying. The drying method is a laboratory method that provides high accuracy. Therefore, it is used as a model, control, i.e. it tests all other methods and instruments for measuring humidity.

As a result of the carried out activities, reference values of wheat grain moisture were determined in four contejnents: 10,87%; 14,63%; 19,30%; 23,42%.

Then, 5 grain samples of each moisture content were prepared according to the method described above, which were alternately filled with EPP and placed in a TV 1000 thermal chamber [20], where the temperature was set from +10 °C to +40 °C in steps of 5 °C with a permissible accuracy of ±1 °C.

Technical characteristics of the TV 1000 thermal chamber [20]:

- temperature range –70 ... +120 °C;
- temperature stability ±1 °C;
- chamber dimensions (mm): width: 3850, height: 3000, depth: 4000.

The output capacity of the sensor was measured with a pre-calibrated LCR E7-12 m with characteristics [21].

- Test frequency: –1 MHz;

Ranges (number of ranges 5):

- Capacitance (C): (0.001 nF–100 nF);

- Conductance (G): 0.01 μS–1 S;
- Resistance (R): 0.1 mΩ–10 kΩ;
- Dielectric losses (tgδ): $1{\cdot}10^{-4}$–1;

Maximal Permissible Errors (MPE):

- Capacitance, conductance, resistance: 0.3%+4 digit;
- Dielectric losses (tgδ): $\pm$ (0.1 tgδ + 5^{-3}).

The capacitances of five grain samples (for each moisture content) were measured. The measurement results given in Table 1 and Fig. 2 represent the mean values of the capacitance for each temperature. The disperse of capacitance values around the averages (less than 1 pF) was largely less than the effect of temperature on capacitance: depending on the moisture content from about 114 pF to about 357 pF for a temperature change of 30 °C. In contrast, the effect of grain size was not investigated because it was not the purpose of the work. Although compared to the effect of temperature, grain size is not significant either. The effect of grain size and grain alignment would only need to be considered for moisture testing at a constant and known temperature and a relatively small sample volume.

Table 1. Values of capacitance (pF) at 7 of temperatures and 4 humidity of investigated material

Temperature, °C	Humidity, %			
	$H_1 = 10.87\%$	$H_2 = 14.63\%$	$H_3 = 19.30\%$	$H_4 = 23.42\%$
10.0 °C	17.52	42.65	73.42	100.94
15.0 °C	36.39	73.87	120.08	161.01
20.0 °C	55.38	104.54	166.29	219.56
25.0 °C	74.19	136.21	211.67	278.83
30.0 °C	93.50	166.53	259.17	339.01
35.0 °C	112.03	197.83	305.48	400.02
40.0 °C	131.28	228.79	350.25	458.36

2.3 Approximation of the Dependence of Capacitance on Humidity and Temperature

From Fig. 2 we can see that obtained by experiments dependences of capacitance (C) for the different temperature (Θ) and humidity (H) can by approximated by linear functions. To establish the parameters of the dependence $C = f(H, \Theta)$ and further analysis, we apply the method of least squares [11, 22, 23], which allows define a bilinear mathematical model (regression) for two factors: humidity and temperature.

A general linear model, given known non-random values of the input 4 values of humidity H_i (i = 1, 2, 3, 4) and 7 values of temperature Θ_j (j = 1, 2, .., 7) that are

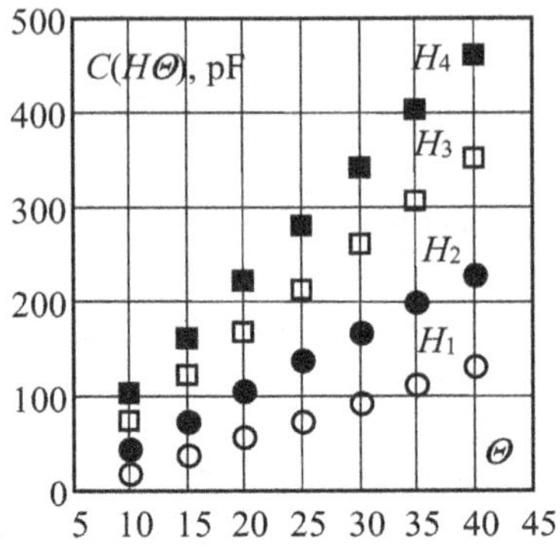

Fig. 2. Measurement results of primary transducer capacity at different humidity depended on temperature changing

matched by 28 values of the output quantity – capacitance $C_{i,j}$ ($i = 1, .., 4; j = 1, 2, .., 7$) can be described by bilinear function:

$$C_{i,j} = \beta_0 + \beta_1 \cdot H_i + \beta_2 \cdot \Theta_j + \beta_3 \cdot H_i \cdot \Theta_j + \varepsilon_{i,j} \quad i = 1,\ 2, .., 4;\ j = 1,\ 2, .., 7, \tag{2}$$

where $\beta = (\beta_0,\ \beta_1,\ \beta_2,\ \beta_3)^T$ is a vector of function coefficients, $\varepsilon_{i,j}$ – are uncorrelated random deviations of a normal distribution $p_N(\varepsilon|0, \sigma)$ with zero expected value ($E(\varepsilon) = 0$) and an unknown standard deviation σ:

$$E(\varepsilon) = 0, \qquad V(\varepsilon) = \sigma^2\mathbf{I}, \tag{3}$$

where **I** is a unit diagonal matrix.

The basis for determining the coefficients of dependence (2) are the results of the experiment (see Table 1). We have totally $n = 28$ experimental values of capacity. The estimates of the regression coefficients – vector $\mathbf{b} = (b_0,\ b_1,\ b_2,\ b_3)^T$ and estimate of standard deviation S:

$$\sum_{i=1}^{4}\sum_{j=1}^{7}\left(b_0 + b_1 \cdot H_i + b_2 \cdot \Theta_j + b_3 \cdot H_i \cdot \Theta_j - C_{i,j}\right)^2 \rightarrow \min, \tag{4}$$

To simplify solution of the values of humidity and temperature should by shifted by arithmetic mean of humidity $\overline{H}$ and temperature and $\overline{\theta}$:

$$h_i = H_i - \overline{H},\ i = 1, 2, 3, 4;\ \theta_j = \Theta_j - \overline{\Theta},\ j = 1, 2, .., 7\ , \tag{5}$$

where:

$$\overline{H} = \frac{1}{4}\sum_{i=1}^{4} H_i = 17.055\%, \qquad \theta = \frac{1}{7}\sum_{i=j}^{7} \theta_j = 25\,^{\circ}\mathrm{C}, \tag{6}$$

Then the dependence of the sensor capacitance on humidity and temperature will be described by a relation with another coefficients $\beta' = \left(\beta_0', \beta_1', \beta_2', \beta_3'\right)^T$:

$$C = \beta_0' + \beta_1' \cdot h + \beta_2' \cdot \theta + \beta_3' \cdot h \cdot \theta, \tag{7}$$

Using denotation (5) and (6) the problem of determination of $\mathbf{b}' = \left(b_0', b_1', b_2', b_3'\right)^T$ in (4) can be presented in the matrix form as:

$$n \cdot \mathbf{M} \cdot \mathbf{b}' = n \cdot C, \tag{8}$$

where

$$\mathbf{M} = \begin{bmatrix} 1 & 0 & 0 & 0 \\ 0 & \overline{h^2} & 0 & 0 \\ 0 & 0 & \overline{\theta^2} & 0 \\ 0 & 0 & 0 & \overline{h^2} \cdot \overline{\theta^2} \end{bmatrix}, \qquad \mathbf{Y} = \begin{bmatrix} \overline{C} \\ \overline{Ch} \\ \overline{C\theta} \\ \overline{Ch\theta} \end{bmatrix} \tag{9}$$

$$\overline{h^2} = \frac{1}{4} \sum_{i=1}^{4} h_i^2 = 22.42; \qquad \overline{\theta^2} = \frac{1}{7} \sum_{j=1}^{7} \theta_j^2 = 50; \tag{10}$$

$$\overline{C} = \frac{1}{28} \sum_{i=1}^{4} \sum_{j=1}^{7} C_{i,j} = 175.53; \qquad \overline{Ch} = \frac{1}{28} \sum_{i=1}^{4} \sum_{j=1}^{7} C_{i,j} \cdot h_i = 366.97; \tag{11}$$

$$\overline{C\theta} = \frac{1}{28} \sum_{i=1}^{4} \sum_{j=1}^{7} C_{i,j} \cdot \theta_j = 779.10; \quad \overline{Ch\theta} = \frac{1}{28} \sum_{i=1}^{4} \sum_{j=1}^{7} C_{i,j} \cdot h_i \cdot \theta_j = 1454.52. \tag{12}$$

From (8) we can see that matrix **M** is orthogonal, therefore solution (8)

$$\mathbf{b}' = \mathbf{M}^{-1} \cdot \mathbf{Y}, \tag{13}$$

can be presented directly in the form:

$$b_0' = \overline{C} = 175.53, \; b_1' = \frac{\overline{Ch}}{\overline{h^2}} = 16.37, \; b_2 = \frac{\overline{C\theta}}{\overline{\theta^2}} = 7.79, \; b_3 = \frac{\overline{Ch\theta}}{\overline{h^2} \cdot \overline{\theta^2}} = 0.649. \tag{14}$$

Therefore, dependence of capacitance from humidity when temperature influenced is given as follow:

$$C(H, \Theta) = b_0' + b_1' \cdot (H - H_m) + b_2' \cdot (\Theta - \Theta_m) + b_3' \cdot (H - H_m) \cdot (\Theta - \Theta_m). \tag{15}$$

If temperature is not measured and its value is taken equal to the mean value $\Theta = \Theta_m$, the result of moisture measurement:

$$H_{meas} = \frac{C_{meas} - b_0'}{b_1'} + H_m, \tag{16}$$

will be affected by a systematic error of value

$$\Delta H(H, \Theta)_{syst} = \left(b_2' + b_3' \cdot (H - H_m)\right) \cdot \frac{(\Theta - \Theta_m)}{b_1'}. \tag{17}$$

The dependences of the absolute and relative systematic temperature errors for the different humidifies are given in Fig. 3.

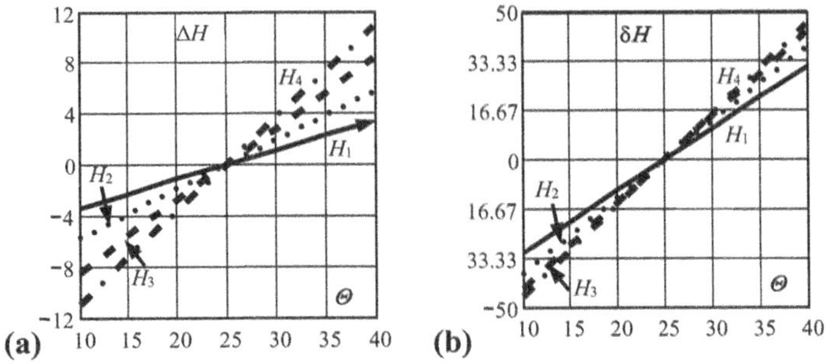

Fig. 3 Absolute (a) and relative (b) systematic temperature effects on humidity measurements

As can be seen from (13) and Fig. 3, when the temperature deviation from the mean value is up to ± 15 °C, the systematic influence on the humidity measurement result depending on its value can reach an unacceptably large value, even in the range of ± 4% (at humidity $H = 10.87\%$) to ± 12% (at humidity $H = 23.42\%$), the relative systematic influence can be from ± 33% even to about ± 50%.

3 Correction of the Systematic Temperature Effect on Humidity

3.1 Correction Function

To correct for the effect of temperature during the moisture measurement, the temperature of the test material contained in the sensor must be measured. Based on the measurement of the sensor capacitance and the measured temperature, the corrected value of the measured moisture can be determined from the determined approximation function $C_{meas} = C(H, \Theta)$ of the measurement chain:

$$H_{cor} = \frac{C_{meas} - b_0' - b_2' \cdot (\Theta - \Theta_m)}{b_1' + b_3' \cdot (\Theta - \Theta_m)} + H_m \cong H, \tag{18}$$

There are 8 quantities in this relationship: the measured capacitance of the sensor C_{meas}, the temperature Θ of the test material, the average values of temperature Θ_m and moisture H_m, and four coefficients: b_0', b_1', b_2', b'. Each of these quantities is not exact and causes uncertainty in the moisture measurement.

3.2 Evaluation of Uncertainty of the Systematic Temperature Effect Correction

According to relation (14), uncertainty of corrected moisture value has 8 mutually uncorrelated components: $u(C)$ – output sensor capacitance; $u(\Theta)$ – temperature of material; $u(\Theta_m)$ – of mean temperature and $u(H_m)$ – of mean humidity used in the determination of function (11); $u(\beta'_0)$, $u(\beta'_1)$, $u(\beta'_2)$, $u(\beta'_3)$ – uncertainties of coefficients β'_0, β'_1, β'_1, β'_3 – respectively. So, in this task we must consider two kinds of uncertainties: which are evaluated by method A (statistical) and method type B (unknown systematic, randomized) [24, 25]. From correction function (14) the sensitivities coefficients for these

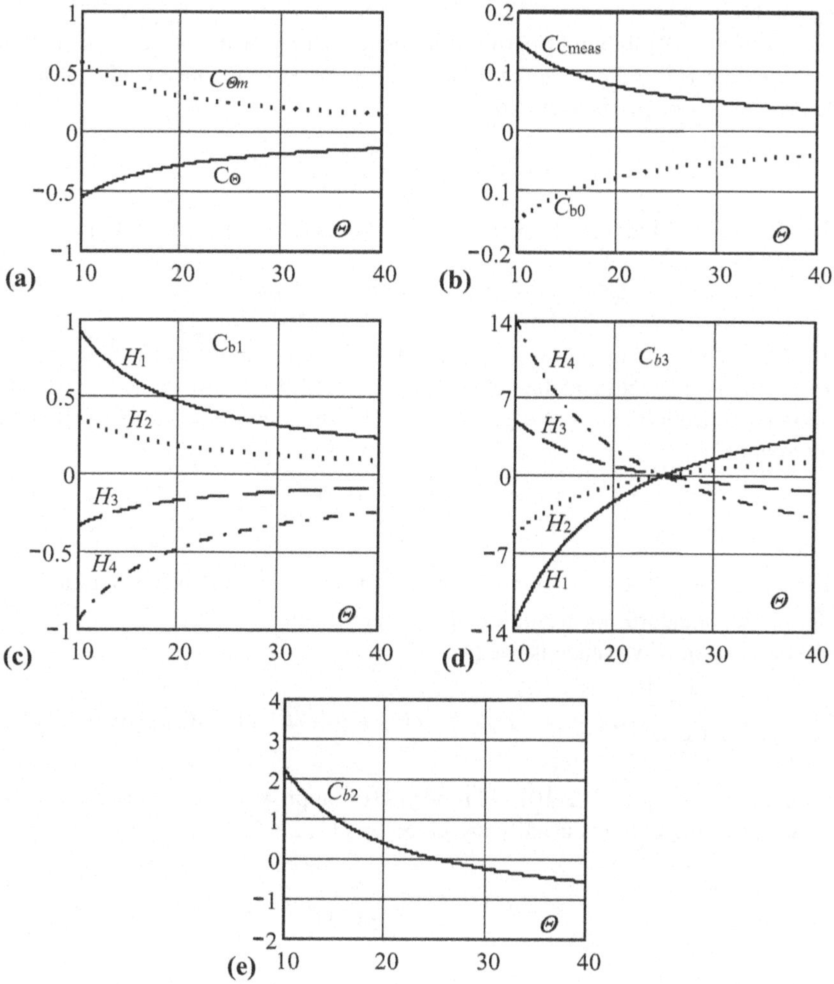

Fig. 4. . Dependence of influence coefficients on temperature and measured humidity

components of combine standard uncertainty are:

$$C_{Cmeas} = \frac{1}{b_1' + b_3' \cdot (\Theta - \Theta_m)}; \qquad C_{Hm} = 1; \qquad C_\Theta = -\frac{b_1' b_2' + b_3'(C_{meas} - b_0')}{(b_1' + b_3' \cdot (\Theta - \Theta_m))^2}; \tag{19}$$

$$C_{\Theta m} = \frac{b_1' b_2' + b_3'(C_{meas} - b_0')}{(b_1' + b_3' \cdot (\Theta - \Theta_m))^2}; C_{b0} = -\frac{1}{b_1' + b_3' \cdot (\Theta - \Theta_m)}; C_{b2} = -\frac{\Theta - \Theta_m}{b_1' + b_3' \cdot (\Theta - \Theta_m)}; \tag{20}$$

$$C_{b1} = -\frac{C_{meas} - b_0' - b_2' \cdot (\Theta - \Theta_m)}{(b_1' + b_3' \cdot (\Theta - \Theta_m))^2}; \; C_{b3} = -\frac{C_{meas} - b_0' - b_2' \cdot (\Theta - \Theta_m)}{(b_1' + b_3' \cdot (\Theta - \Theta_m))^2} (\Theta - \Theta_m), \tag{21}$$

The dependences of influence coefficients on temperature and measured humidity are shown in Fig. 4.

According to [18] maximal permissible error (MPE) of moisture measurement of reference standards is ±0.5%, so the standard uncertainty of the mean value of moisture H_m determined from the measured 4 values is:

$$u_B(H_m) = \frac{0.5 \cdot H_m}{100\% \sqrt{3 \cdot 4}}, \%, \tag{22}$$

For MPE of the chamber temperature measurement ± 1 °C standard uncertainty:

$$u_B(\Theta) = \frac{1}{\sqrt{3}} {}^oC, \;\; u_B(\Theta_m) = \frac{1}{\sqrt{3 \cdot 7}} {}^oC \tag{23}$$

(the average value of temperature Θ_m is from measured 7 temperatures).

According to the technical data of LCR E7-12 m [21] MPE of capacitance measurement is determined as 0.3% + 4 digit, therefore the standard uncertainty of capacitance measurement C:

$$u_B(C) = \frac{\frac{0.5\% \cdot C}{100\%} + 0.04}{\sqrt{3}} \; pF. \tag{24}$$

Because the variance σ^2 of the random deviations of capacitance is not known, then to determinate standard uncertainties of the coefficients $\beta' = (\beta_0', \beta_1', \beta_2', \beta_3')^T$ the unbiased estimator of variance is used:

$$S^2 = \frac{1}{n-4} \sum_{i=1}^{4} \sum_{j=1}^{7} (b_0 + b_1 \cdot H_i + b_2 \cdot \Theta_j + b_3 \cdot H_i \cdot \Theta_j - C_{i,j})^2 = 0.227. \tag{25}$$

Then according to (8), (9), (10), (11) and (12) the approximated standard uncertainties of the estimated coefficients usually are given by [11, 22, 23]:

$$u_A(b_k') = \sqrt{\frac{S^2}{n \cdot [\boldsymbol{M}^{-1}]_{k,k}}}, \tag{26}$$

$$u_A(b_0') = \sqrt{\frac{S^2}{n \cdot 1}} = 0.0901; \qquad u_A(b_1') = \sqrt{\frac{S^2}{n \cdot \overline{h^2}}} = 0.0190; \tag{27}$$

$$u_A(b_2') = \sqrt{\frac{S^2}{n \cdot \overline{\theta^2}}} = 9.013 \cdot 10^{-3}; \quad u_A(b_3') = \sqrt{\frac{S^2}{n \cdot \overline{h^2} \cdot \overline{\theta^2}}} = 1.904 \cdot 10^{-3}. \tag{28}$$

Note. Standard uncertainties (26) are only approximated and they are related to the estimated values of coefficients. Because $n = 28$ (measurement results) and $m = 4$ (number of function coefficients) based on [26] the standard uncertainties of the coefficients $\beta' = \left(\beta'_0, \beta'_1, \beta'_2, \beta'_3\right)^T$ are larger than the standard uncertainties (26) of estimator $\mathbf{b}' = \left(b'_0, b'_1, b'_2, b'_3\right)^T$ of $\sqrt{24/22} = 1.044$ times, i.e. less then 5%.

Therefore, approximations of standard uncertainties (25) can be used in this task.

Using influence coefficients (15) and standard uncertainties of influences factor the combined standard uncertainty dependently temperature at different humidity is:

$$u_c(H;\Theta) = \sqrt{\begin{array}{l}(C_{Hm}\cdot u_B(H_m))^2 + (C_{\Theta m}\cdot u_B(\Theta_m))^2 + (C_{\Theta}\cdot u_B(\Theta))^2 + (C_{Cmeas}\cdot u_B(C_{meas}))^2 \\ +(C_{b0}\cdot u_A(b_0))^2 + (C_{b1}\cdot u_A(b_1))^2 + (C_{b2}\cdot u_A(b_2))^2 + (C_{b3}\cdot u_A(b_3))^2\end{array}}. \quad (29)$$

At different humidity dependencies of combined standard uncertainty of correction humidity on temperature is shown in Fig. 5a. Effectiveness of correction is determined as a ratio of systematic error of uncorrected result to the standard uncertainty of correction:

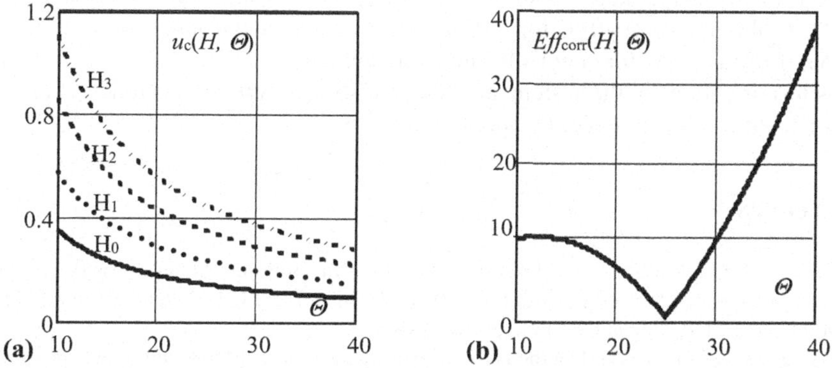

Fig. 5 Combined standard uncertainty of correction dependent from temperature at different humidity(a) and (b) the effectiveness of correction of the systematic temperature effect

$$Eff_{corr}(H,\Theta) = \frac{\Delta H(H,\Theta)_{syst}}{u_c(H;\Theta)}, \quad (30)$$

The dependence of correction efficiency is shown in Fig. 5b. Practically independent on humidity value effectiveness depend only on temperature changing. At mean temperature (25 °C) the measuring result is correct without correction. We can see that if the temperature deviates from the mean value, the effectiveness of correction is about from 10 to 40, i.e. by correcting of the systematic effects of temperature, the accuracy of moisture measurement can be increased by about of 10 to 40.

4 Conclusions

When measuring the moisture content of bulk materials using the capacitance method, changes in the temperature of the test samples cause a significant systematic effect on the results, even within the range of ±4% (for moisture content of about 11%) to ±12% (for moisture content of about 24%).

In order to realize the correction of the temperature influence, experimental tests of 4 reference samples were carried out, the values of the sensor output capacitance were measured at 7 values of temperature. From the experimental results using LSM, the coefficients of bilinear capacity dependence on moisture and temperature were determined. Based on the obtained dependence, the correction functions for the effect of temperature on the moisture measurement result were determined.

A detailed analysis of the uncertainty of the correction as a function of 8 influencing factors: the measurement of the sensor capacitance, the average moisture content of the 4 standards, the temperature of the test material and the average temperature of the standards and the 4 coefficients of the approximation function was performed. The estimated composite standard uncertainty of correction depends on the moisture content value and on the temperature deviation from the mean value and ranges from about 1.1% at 10 °C to about 0.3% at 40 °C for 24% moisture content and from about 0.4% at 10 °C to about 0.1% at 40 °C for about 11% moisture content.

Using corrections of the systematic effects of temperature, the accuracy of moisture measurement can be increased by about of 10 to 40.

References

1. Nilsson, L.O.: Methods of Measuring Moisture in Building Materials and Structures. Springer, Cham (2018). https://doi.org/10.1007/978-3-319-74231-1State-of-the-Art Report of the RILEM Technical Committee 248-MMB
2. Shapiro, M.V.: Devices and methods for measuring the moisture content of grain and products of its processing: author's ref. dis. for science. degree of Cand. tech. Science: special. 05.11.13 "Devices and methods of control and determination of the composition of substances", p. 18, M.V. Shapiro. Kharkiv (2002)
3. Ivakh, R.: Systematization of methods for measuring the humidity of bulk materials. In: Ivakh, R., Dorozhovets, M., Pytel, I. (eds.) Measuring Equipment and Metrology, no. 62, pp. 97–101 (2003)
4. Binda, L., Squarcina, T., Van Hees, R.: Determination of moisture content in masonry materials: calibration of some direct methods. In: Proceedings of the international congress on deterioration and conservation of stone, Berlin, 30 September–4 October 1996, pp. 587–599 (1996)
5. CIB W40 (2012) Heat, air and moisture transfer terminology. Parameters and concepts. de Freitas, V.P., Barreira, E. (eds.) CIB – W040 heat and moisture transfer in buildings, CIB report 369
6. Wang, N., Daun, J.K., Wallis, R.: Collaborative study on a method for determination of moisture content in pulses (AACC method 44–17). Cereal Food World **50**(1), 23–26 (2005)
7. Lin, H.Y., Chiueh, L.-C., Shih, D.Y.-C.: Detection of genetically modified soybeans and maize by the polymerase chain reaction method. J. Food Drug Anal. **8**(3), 200–207 (2000)

8. Fuchs, A., Moser, M.J., Zangl, H., Bretterklieber, T.: Using capacitive sensing to determine the moisture content of wood pellets – investigations and application. Int. J. Smart Sens. Intell. Syst. **2**(2), 293–308 (2009)
9. Berliner, M.: Measurements of Humidity, 2nd edn., p. 400. Energia (1973). (in Russian)
10. Ivakh, R.M.: Capacitive primary converters of dielectric constant of loose materials: the dissertation on competition of a scientific degree of the candidate of technical sciences: 05.11.05 – devices and methods of measurement of electric and magnetic sizes. National University: Lviv Politechnic. (in Ukraine)
11. Evaluation of measurement data – Guide to the expression of uncertainty in measurement Joint Committee for Guides in Metrology, JCGM 100: 2008
12. Pavese, F.: On the difference of meaning of 'zero correction': zero value versus no correction, and of the associated uncertainties. In: Advanced Mathematical and Computational Tools in Metrology IX, pp. 297–309. World Scientific (2012)
13. Fuchs, A., Zangl, H., Holler, G.: Capacitance-based sensing of material moisture in bulk solids: applications and restrictions. In: Mukhopadhyay, S.C., Gupta, G.S. (eds.) Smart Sensors and Sensing Technology, vol. 20, pp. 235–248. Springer, Cham (2008). https://doi.org/10.1007/978-3-540-79590-2_16
14. Wu, S., Zhang, B., Tian, Y., Zhou, S., Ma, H.: A grain moisture model based on capacitive sensor. J. Phys. Conf. Ser. **1074**, 012120 (2018)
15. Lataste, J.F., et al.: Spatial distributions. In: Nilsson, L.O. (ed.) Methods of Measuring Moisture in Building Materials and Structures, vol. 26, pp. 173–190. Springer, Cham (2018). https://doi.org/10.1007/978-3-319-74231-1_19
16. DSTU (GOST) 13586.5-93 (Ukrainian standard) Grain. Method of moisture content determination (method of drying and weighing)
17. ISO 712:1998. Cereals and cereal products – Determination of moisture content
18. ISO 6540: 2021. Determination of moisture content (on milled grains and on whole grains)
19. ISO 24557: 2009 Determination of moisture content – Air-oven method
20. http://izmeritelnyepribory.ru/view/klimaticheskie_kamery/ilka_tv-1000.html
21. Digital meter LCR E7-12 Russia: Operating instructions.
22. Draper, N., Smith, H.: Applied Regression Analysis, 2nd edn. Wiley, New York (1981)
23. Rawlings, J.O., Pantula, S.G., Dickey, D.A.: Applied Regression Analysis: A Research Tool, 2nd edn. Springer, New York (1998)
24. Dorozhovets, M.: Evaluation of the influence of systematic distortions on the uncertainty parameters using approximation by the least squared method. Meas. Autom. Monit. **12**, 21–25 (2006). (in Polish)
25. Warsza, Z.L., Puchalski, J.: Uncertainty bands of the regression line for data with type a and type b uncertainties of dependent variable *Y*. In: Szewczyk, R., Zieliński, C., Kaliczyńska, M. (eds.) Automation 2021: Recent Achievements in Automation, Robotics and Measurement Techniques, vol. 1390, pp. 342–363. Springer, Cham (2021). https://doi.org/10.1007/978-3-030-74893-7_32
26. Dorozhovets, M.: Forward and inverse problems of type A uncertainty evaluation. Measurement **165**, 108072 (2020)

Instrumental Covariance and Its Impact on the Uncertainty of Tested Parameters of Industrial Objects

Yevhen Volodarskyi[1], Zygmunt L. Warsza[2(✉)], Larysa Kosheva[3], and Alex Sautin[1]

[1] Department of the Experimental Studies Automation, National Technical University of Ukraine «KPI», Kyiv, Ukraine
vet-1@ukr.net

[2] Łukasiewicz Research Network -Industrial Research Institute for Automation and Measurements PIAP, Al. Jerozolimskie 202, 02-486 Warszawa, Poland
zlw1936@gmail.com

[3] Department of Biocybernetics and Aerospace Medicine, National Aviation University of Ukraine, Kyiv, Ukraine
l.kosh@ukr.net

Abstract. The analysis shows that many parameters of industrial facilities are determined by computational operations on the results of direct measurements. Such measurements can be arranged according to parallel and parallel-serial structure of the measurement system. Considering that there are interference in the experimental study of industrial facilities, preference is given to a parallel-serial measurement structure, which provides high noise immunity, and also uses only one measuring channel (MC). The bias of the real characteristic of the measuring conversion during the direct measurement of the input quantities, which can take one of the possible values in the given dynamic range, remains constant. In this case, a stochastic relationship arises between the input quantities; hereinafter, this relationship will be called instrumental covariance. The influence of instrumental covariance on the estimated uncertainty of tested parameter depends on arithmetic operations performed on functionally related results of direct measurements. Many such parameters are determined using multiplication and division operations. A simple example is determining the power and resistances from direct current and voltage measurements. The influence of the ratio between the input values on the additional component of the estimated uncertainty of the object parameter due to instrumental covariance is analyzed. Recommendations on how to reduce the impact of instrumental covariance are also given.

Keywords: Instrumental covariance · Measuring channel · The real characteristic of transformation · Measurement uncertainty · Direct measurements

R. Szewczyk et al. (Eds.): AUTOMATION 2022, AISC 1427, pp. 370–379, 2022.
https://doi.org/10.1007/978-3-031-03502-9_36

1 Introduction

Industrial facilities are characterized by many parameters that are determined by calculation, based on results of direct measurements of their informative parameters [3].

A characteristic feature of measurements for industrial facilities is the presence of all kinds of interference and the variability of the conditions for conducting experimental operations. This, first, leads to a change in the characteristics of the transformation of measuring instruments, which are used in the control/monitoring of the correct functioning of objects. To characterize the accuracy of measurements, the uncertainty of the results is used [2]. For industrial measurements, Type B uncertainty prevails. Therefore, if no special requirements are put forward, preference is given to the serial-parallel structure of the organization of the measurement procedure, when the informative parameters of the signals of an industrial facility are converted into a certain unified value, which is sequentially fed into the measuring channel.

With serial-parallel structural organization, the results of direct measurements of informative parameters of quantities that are used to determine the value the desired parameter of the object, are subjected to functional transformation in accordance with the model equation. In this case, the same measuring channel is used to obtain the measured values. According to [1, 2], the input quantities are considered, which are correlated when the same standard, sensor, measuring instrument, standard data and even the measurement method are used for their assessment.

The input quantities themselves and the results of their direct measurement are independent. However, carrying out mathematical operations on the measurement results in accordance with the model equation in order to determine the desired parameter leads to the appearance of a correlation, which we will call instrumental.

In some sources, for example [4], this type of stochastic connection is called logical correlation. But this term does not correspond to the physics of this phenomenon.

The case of estimating the uncertainty of the result obtained from the regression dependence of type B is considered in [5].

The presence of instrumental correlation in determining the desired parameter due to the fact that the results of direct measurement depend not only on the input value, which takes one of the possible values, but also on the measurement error. Since the measurement uses the same measuring channel, the values of the accuracy characteristics when obtaining the results of direct measurements, will be the same.

Thus, the mathematical operation, as a result of which the required parameter is determined, is performed over random quantities, which have a common influence value. This is the error of direct measurement. In industrial conditions, the predominant will be the additive error, the influence of which on the result of determining the desired parameter and we will further analyze.

It is necessary to assess the influence of the type of the real characteristic of the measuring channel on the accuracy of determining the desired parameter, which is considered in [6], to analyze the measurement equation, to estimate the total standard uncertainty in determining the desired parameter of the object, taking into account the instrumental covariance. To assess the contribution of this component to the uncertainty of the obtained result, to consider the possibilities to reduce the influence of instrumental covariance.

2 Calculation of Uncertainties of Correlated Measurements

According the internationally accepted Guide to the Expression of Uncertainty in Measurement of acronym GUM [1, 2] with few Supplements, the accuracy of the measurements is described now by uncertainties defined in the International Vocabulary of Metrology VIM [3]. The final result of the measurements of the tested quantity named measurand, is the estimator of its value, most often the arithmetic mean of repeated observations under the permissible random conditions. The assessment of its accuracy is the expanded uncertainty as the range within which the measured value lies with a given probability. Before calculations the final values of these parameters, effects of systematic errors are analyzed using formulas for small increments resulting from the total differential of the measurement equations. of the values of these errors are known, their influence on the measurement result is eliminated by making corrections.

The expanded uncertainty is the interval determined from the complex uncertainty $u = \sqrt{u_A^2 + u_B^2}$ as a standard deviation of the distribution of the difference of possible values of the measurement result from the estimated value of the measurand. Geometrical components of u are named as uncertainties of types A and B. The uncertainty of type A describes the scatter of repeated measurement results in the actually made experiment. The type B uncertainty estimated the mean measure of the influence of systematic errors of unknown values, constant or slowly varying, which may occur in a given experiment. The common influence of these errors are randomized and heuristically estimated with using the existing knowledge on possible their mean and limited values.

The type B uncertainty of measuring instruments is usually assessed on the basis of their maximum permissible error (MPE) $E_{\%}$ given in their technical data. As example, for digital meters the uncertainty u_B of the multiplicative errors is taken as the standard deviation $E/\sqrt{3}$ of a uniform distribution with a span, equal to $2E$ Then the uncertainty u_B of these instruments increases linearly with the value of the measured quantity. There is also small additive error ± LSB.

In multivariable measurements occurring in a common tested object, or processed by a common system, and when several examined objects are subject to common external influences, the uncertainties of the measured quantities depend on the autocorrelation of their observations and on the cross-correlations between the sets of their errors described by type A and type B uncertainties. If the results of the measurements of these quantities are or will be subsequently used together, their correlation coefficients should also be determined. For example, in indirect multi-parameter measurements, they have an influence on the uncertainties of these t measurement results in output. The variance propagation in multi-parameter measurements is described by the covariance matrix. Information on estimating the uncertainties of multivariable measurements is provided in Supplement 2 of the GUM and upgrading in few authors ‘ works, e.g. in [7–9, 12].

Also important for the accuracy of the final results are deviations resulting from the inadequacy of the mathematical model and interferences existing in the test object. They can be also treated uniformly as uncertainties.

So, if several values of the same quantity or of different quantities are measured in the tested object and in the same time, a correlation may occur between the random sets of their observations, which affects the uncertainty of the results calculated from such measurements. We will briefly explain the influence of the correlation coefficient ρ on the example of the uncertainty of the sum and the difference of two correlated quantities x_1, x_2 measured with uncertainties u_1, u_2 and the correlation coefficient ρ, i.e.:

$$u_{\sum} = \sqrt{u_1^2 + 2\rho u_1 u_2 + u_2^2}, u_D = \sqrt{u_1^2 - 2\rho u_1 u_2 + u_2^2}$$

The distributions of errors described by uncertainties of type A and B are related to a different time scale. It should be assumed that are not correlated with each other.

The correlation coefficient ρ for the combined uncertainty u of a sum $u_{\sum}$ of correlated components is [6–8]: $\rho = \frac{\rho_A u_{1A} u_{2A} + \rho_B u_{1B} u_{2B}}{u_1 u_2}$.

Similar relations are for the relative uncertainties $\delta_{x1 \cdot x2}$, $\delta_{x1/x2}$ of the product $y = x_1 \cdot x_2$ and of the quotient $y = x_1/x_2$ of two values:

$$\delta_{x1 \cdot x2} = \sqrt{\delta_1^2 + 2\rho\delta_1\delta_2 + \delta_2^2}, \delta_{x1/x2} = \sqrt{\delta_1^2 - 2\rho\delta_1\delta_2 + \delta_2^2},$$

$$\rho = \frac{\rho_A \delta_{1A} \delta_{2A} + \rho_B \delta_{1B} \delta_{2B}}{\delta_1 \delta_2}$$

Value of the correlation coefficient ρ_A between two sets of errors described with uncertainties u_A, is calculated by the classic formula [1, 2, 6–8] from values of observations synchronously performed, and it can take values in the entire range (–1, 1).

The coefficient ρ_B is determined similarly as u_B, to a large extent heuristically on the knowledge obtained. As MPE error $E_{\%}$ of the meter is the same for any measured value, then two values measured by the same meter should have the same maximal multiplicative errors. Some influences give also additive errors, then the exact estimation of the correlation coefficient ρ_B is difficult, depends on each external and internal interactions, measured values and requires special long-term statistical studies. Then for measurements by the same meter and in the same conditions can be taken $\rho_B \approx 1$, by different meters and in different conditions $\rho_B \approx 0$.

If two different values are measured with the same or a similar device, and in the same permissible conditions, then for the uncertainty $u_{B\sum}$ of their sum it can be assumed that the correlation coefficient $\rho_B = 1$, and for the difference $\rho_B = -1$. Then uncertainties type B of the sum and of the difference are: $u_{B\sum} = u_{B1} + u_{B2}$ and $u_{BD} = |u_{B1} - u_{B2}|$.

For measurements of two values with different instruments or measurement systems and/or under different conditions, the sets of their observations should be treated as uncorrelated, i.e. $\rho_B = 0$, and uncertainties type B of sum and difference will be equal $u_{B\sum} = u_{BD} = \sqrt{u_{B1}^2 + u_{B2}^2}$. More about uncertainties of correlated measurands is in [6–10].

3 Uncertainty of the Result of Multiplication Operations

Many parameters of industrial facilities are determined using operations multiplication and division [11]. A simple example is determining power and resistance from direct current and voltage measurements.

Let us consider the estimation of the uncertainty of a parameter calculated using the simplest arithmetic operations on the results of measurements of the quantities X_1, X_2, in particular, multiplication. The model equation in the considered case has the form:

$$y = ax_1 * bx_2, \quad (1)$$

where x_1, x_2 – values of quantities, which determine the required parameters of the research object, *a*, *b* – are coefficients that can take any values other than zero.

The block diagram of the implementation of the measuring and computational procedure for determining the desired parameter is shown in Fig. 1.

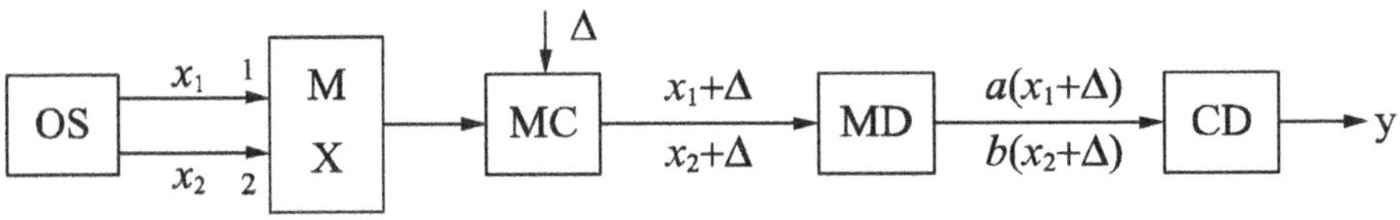

Fig. 1. Block diagram of the implementation of the measuring and computational procedure for determining the desired parameter; OS – object of study, MX – multiplexer, MC –measuring channel, MD –multiplier device, CD – computing device Δ – bias characteristic measuring transformation

To analyze the components of the uncertainty of the calculated result, we will proceed from a measurement equation of the form:

$$(y + \Delta y) = a(x_1 + \Delta) * b(x_2 + \Delta), \quad (2)$$

where Δy – deviation of the result; Δ – bias characteristic measuring transformation, which we do not know, but with a probability of 0.99 is in normalized limits.

We transform Eq. (2) and get:

$$(y + \Delta y) = (x_1 + \Delta)ab(x_2 + \Delta) = abx_1x_2 + abx_1\Delta + abx_2\Delta + ab\Delta^2. \quad (3)$$

After simple mathematical transformations and neglecting small quantities, we obtain equation to reject the result of determining the desired parameter due to the presence of displacement specifications transformations measuring channel:

$$\Delta y = ab\Delta(x_1 + x_2). \quad (4)$$

Combined standard uncertainty $u_c(y)$ by type B of determining the desired parameter of the object based on the results of measuring the parameters of the input quantities from their subsequent multiplication based on expression (4) will be written in the form:

$$u_c(y) = a^2b^2u^2(\Delta)(x_1 + x_2)^2, \quad (5)$$

where $u(\Delta)$ – standard uncertainty of bias characteristics of the measuring transformation.

The presence of a bias in the characteristic and functional transformation of the results of direct measurements of the informative parameters of the object leads to an increase in the uncertainty of the result. Additional component combined standard uncertainty is due to the presence of instrumental covariance. It displays the stochastic relationship between the input quantities X_1 and X_2, which arises when determining the parameter.

Let's consider the expression separately:

$$(x_1 + x_2)^2 = x_1^2 + x_2^2 + 2x_1x_2, \tag{6}$$

where term 2 x_1x_2 due to the presence instrumental covariance.

We denote ratio between measured values:

$$m = x_2/x_1, \tag{7}$$

then expression (6) will be rewritten as.

$$(x_1 + x_2)^2 = x_1^2(1 + m^2 + 2m). \tag{8}$$

When substituting it into expression (5), we can conclude that the increase in the uncertainty of the result due to instrumental covariance depends on the ratio between the parameters of the input quantities x_1 and x_2. The first two terms in brackets in expression (8) characterize the uncertainty of the result in the absence of instrumental correlation.

Such situation arises, when each informative parameter of the input quantity is measured separately, that is, individual measuring channels are used that have homogeneous, but not correlated measurement conversion characteristics.

Let us estimate the fraction of the result uncertainty introduced by the instrumental covariance in the case of serial-parallel organization of the measurement of the informative parameters of the quantities by which the desired parameter of the object is determined. For this, we represent expression (5) in the form:

$$u_c^2(y) = B(1 + \eta), \tag{9}$$

where

$$B = a^2b^2u^2(\Delta)x_1^2\left(1 + m^2\right), \tag{10}$$

those, the measurement is carried out using the same type of parallel measuring channels for each input quantity.

The fraction of standard uncertainty due to instrumental covariance is towards uncertainty in parallel measurement of informative parameters of input quantities will be:

$$\eta = \frac{2m}{1 + m^2}. \tag{11}$$

We investigate how influences ratio m between input quantities on the value of instrumental covariance. The research results are presented in Table 1.

Table 1 Influence of the m value on the value of instrumental covariance

M	0.1	0.2	0.3	0.5	0.8	**1.0**	3.0	5.0	7.0	10.0	15.0
η	0.2	0.38	0.55	0.8	0.98	**1.0**	0.6	0.4	0.3	0.2	0.1

As follows from Table 1, a 100% increase in the combined standard uncertainty due to instrumental covariance will be at $m = 1$. With increasing m, the contribution of instrumental covariance decreases, and at $m > 15$ it will increase the combined uncertainty by no more than 10% for the case with independent measurement of each of the informative quantities. At $m < 1$, only from a value of 0.3, a decrease in the ratio between x_1 and x_2, the contribution of instrumental covariance begins to decrease significantly.

From the analysis it follows that the influence of instrumental covariance on the accuracy of determining the desired parameter of the object decreases with a change in m. Therefore, after preliminary estimation of the actual ratio, the scaling (division) of the parameter of one of the input quantities is introduced.

It is desirable to choose the division factor k, based on Table 1, so that the values supplied to the input of the measuring channel differ by an order of magnitude. The larger the difference, the less the influence of the instrumental covariance. Then the result of the calculation is multiplied by $1/k$.

4 Uncertainty of the Result Obtained Using the Division Operation

Let us analyze the components of the uncertainty of the desired parameter, determined from the results of direct measurements of the informative parameters of the input quantities using the division operation. In this case, the model equation has the form:

$$y = \frac{ax_1}{bx_2}. \tag{12}$$

To analyze the uncertainty of the calculated result in this case, we will proceed from the measurement equation:

$$(y + \Delta y) = \frac{a(x_1 + \Delta)}{b(x_2 + \Delta)} = \frac{ax_1 + \Delta a}{bx_2 + \Delta b} \tag{13}$$

or:

$$(y + \Delta y) = \frac{ax_1}{bx_2}\left(1 + \frac{\Delta}{x_1}\right)\left(1 + \frac{\Delta}{x_2}\right) = \frac{ax_1}{bx_2} + \frac{ax_1}{bx_2}\left(\frac{\Delta}{x_2} - \frac{\Delta}{x_1}\right), \tag{14}$$

whence, the expression for the increment Δy due to the presence of a bias of the characteristic measuring transformation will be:

$$\Delta y = \frac{ax_1}{bx_2}\Delta\left(\frac{1}{x_1} - \frac{1}{x_2}\right). \tag{15}$$

Thus, the uncertainty of the results obtained depends from the ratio of the bias of the characteristic of the measuring transformation Δ and the input quantity.

The squared combined standard uncertainty of the desired parameter is defined as:

$$u^2(y) = \frac{a^2 x_1^2}{b^2 x_2^2} u^2(\Delta) \left(\frac{1}{x_1^2} + \frac{1}{x_2^2} - 2\frac{1}{x_1 x_2} \right). \tag{16}$$

The presence of the term $2\frac{1}{x_1 x_2}$ in expression (16) is due to instrumental covariance. In contrast to the previously considered case, this component does not increase, but decreases the combined standard uncertainty.

Considering the relation (7) between the values of the measured quantities m, the expression in brackets in the formula (14) will take the form:

$$\left(\frac{x_1^2 + x_2^2}{x_1^2 x_2^2} - 2\frac{1}{x_1 x_2} \right) = \left(\frac{x_1^2 + m^2 x_1^2}{x_1^2 m} - 2\frac{1}{x_1^2 m} \right). \tag{17}$$

Using expression (15), we analyze the influence of instrumental covariance on the accuracy of the parameter estimate.

$$u_c^2(y) = B_1(1 - \eta), \tag{18}$$

where.

$$B_1 = \frac{a^2(1 + m^2)}{b^2 m^2 x_1^2} u^2(\Delta) \tag{19}$$

– standard uncertainty squared of the result calculated by the model Eq. (10), provided that there is no instrumental covariance.

It should be noted that in this case, the presence of instrumental covariance leads to a decrease in the uncertainty in determining the desired parameter of the object.

5 Numerical Example

Let's take a look at the simplest example. Suppose that when determining the resistance *R* characterizing the performance of the object, we have, respectively, the voltage $x_1 = 5$ V and the current $x_2 = 1\,A$. For this, a measuring channel with the limited values of the bias of the transformation characteristic $\Delta = \pm 0.2$ is used, the standard measurement uncertainty of which is $u(\Delta) = 0.115$. Then the standard uncertainty of the desired parameter in the absence of instrumental covariance will be equal to the square root of expression (19) and will be 0.58. Thus, the value of the required parameter *R* will be in the range 3.88–6.16 Ω with a probability of 0.95.

In the presence of instrumental covariance, based on (18), we have that the required resistance R will be in the range 4.29–5.71 Ω. To check, we will calculate the required resistance R = 5.2/1.2 = 4.33.

As follows from Table 1, the effect of the bias characteristic of measuring transformation Δ to the result there will be no for $\eta = 1$, that is, for $m = 1$.

In these connections, it is necessary to construct an algorithm for determining the desired parameter in order to ensure this condition. This can be achieved in the following way. Initially, the ratio between the input quantities x_1 and x_2 is determined. According to the obtained result a scaling factor of $1/m$ of the larger of the input values is introduced. To determine the desired parameter, the result of division is multiplied by m.

The "residual" uncertainty in determining the desired parameter will correspond to the uncertainty in establishing the relationship m between the parameters of the input quantities.

6 Conclusions

Based on the specifics of measurements in industrial conditions, the parameters of objects are determined on basis of the results obtained from direct measurements of the informative parameters of the input quantities. For this, a sequentially parallel measurement circuit is used, and the desired parameter is obtained as a result of subsequent computational operations. The presence of measuring and computational operations in determining the desired parameter leads to a stochastic connection – instrumental covariance.

The analysis of the uncertainty components of the desired parameter, which is calculated using multiplication/division operations, showed that the relative change in the uncertainty of the result due to instrumental covariance depends on the ratio between the values of the parameters of the input measured quantities.

Installed conditions for reducing this influence during multiplication. By doing division operations the presence of instrumental covariance always reduces the combined standard uncertainty. An approach is considered that makes it possible to minimize the uncertainty of the result when dividing.

References

1. JCGM 100:2008, GUM 1995 with minor corrections Evaluation of measurement data—Guide to the Expression of Uncertainty in Measurement. Bureau International des Poids et Mesures, France
2. ISO/IEC Guide 98-3:2008 Uncertainty of measurement—Part 3: Guide to the expression of uncertainty in measurement. International Organization for Standardization, Switzerland
3. International vocabulary of metrology. Basic and general concepts and associated terms (VIM). – ICGM, 2012, p. 88
4. Zakharov, I.P.: Estimating measurement uncertainty on the basis of observed and logical correlation. Meas. Tech. **50**(8), 808–816 (2007). https://doi.org/10.1007/s11018-007-0154-8
5. Volodarsky, E., Warsza, Z., Kosheva, L.A., Idźkowski, A.: Transforming the conversion characteristic of a measuring system used for technical control. In: Szewczyk, R., Kaliczyńska, M. (eds.) Recent Advances in Systems, Control and Information Technology, vol. 543, pp. 524–534. Springer, Cham (2017). https://doi.org/10.1007/978-3-319-48923-0_56
6. Warsza, Z.L., Puchalski, J.: Estimation of vector uncertainties of multivariable indirect instrumental measurement systems on the star circuit example. J. Phys.: Conf. Ser. 1065 (2018). https://doi.org/10.1088/1742-6596/1065/5/052026, XXII World Congress IMEKO 2018 Belfast. IOP Conf. Series

7. Warsza, Z.L., Puchalski, J.: Method of the estimation of uncertainties in multiparameter measurements of correlated quantities. In: Proceedings of V Workshop: Applied Methods of Statistical Analysis, Statistical Computation and Simulation, AMSA 2019, 18–20 September Novosibirsk NSTU, pp.47–59. ISSN 2313-870X (2019). https://amsa.conf.nstu.ru/amsa2019/
8. Warsza, Z.L., Puchalski, J.: Estimation of uncertainties in indirect parameter measurements of correlated quantities. In: Proceedings of 12th International Conference on Measurement 2019, pp. 51–57. Slovak Academy of Sciences, Slovakia (2019). ISBN 978-80-972629-2-1
9. Warsza, Z.L., Puchalski, J.: Estimation of uncertainties in indirect multivariable measurements part 1. Case of correlated quantities. In: Szewczyk, R., Zieliński, C., Kaliczyńska, M. (eds.) Automation 2020: Towards Industry of the Future, vol. 1140, pp. 309–325. Springer, Cham (2020). https://doi.org/10.1007/978-3-030-40971-5_29
10. Warsza, Z.L., Puchalski, J.: Uncertainty bands of the regression line for autocorrelated data of dependent variable Y. In: Szewczyk, R., Zieliński, C., Kaliczyńska, M. (eds.) Automation 2021: Recent Achievements in Automation, Robotics and Measurement Techniques, vol. 1390, pp. 364–386. Springer, Cham (2021). https://doi.org/10.1007/978-3-030-74893-7_33
11. Volodars'kiy, Є.T., Kosheva, L.O., Dobrolyubova, M.V.: Informatsiyno-vimiryuval'na sistema ta neviznachenist' (Information-measuring system and uncertainty). Ukrains'kiy metrologichniy zhurnal (Ukrainian Metrological Journal) **3A**, 30–34 (2020). WoS 000604400000005
12. Warsza, Z.L., Puchalski, J.: Method of estimation uncertainties of indirect multivariable measurement including accuracy of processing function as extension of GUM-S2. In: Pavese, F., Forbes A.B., Zhang N.F., Chunovkina A.G. (eds.) Advanced Mathematical and Computational Tools in Metrology and Testing AMCTM XII, Series on Advances in Mathematics for Applied Sciences, vol. 90, pp. 451–463. World Scientific Publishing Company, London 2021

Application of the Polynomial Maximization Method for Estimation Parameters of Autoregressive Models with Asymmetric Innovations

Serhii Zabolotnii[1], Oleksandr Tkachenko[2], and Zygmunt L. Warsza[3](✉)

[1] Cherkasy State Business College, Cherkasy, Ukraine
[2] Ivan Franko National University of Lviv, Lviv, Ukraine
[3] Łukasiewicz Research Network – Industrial Research Institute for Automation and Measurements PIAP, Al. Jerozolimskie 202, 02-486 Warszawa, Poland
zlw1936@gmail.com

Abstract. This paper considers the application of the Polynomial Maximization Method to find estimates of the parameters of autoregressive model with non-Gaussian innovation. This approach is adaptive and is based on the analysis of higher-order statistics. Analytical expressions that allow finding estimates and analyzing their uncertainty are obtained. Case of asymmetry of the distribution of autoregressive innovations is considered. It is shown that the variance of estimates of the Polynomial Maximization Method can be significantly less than the variance of the estimates of the linear approach (based on Yule-Walker equation or Ordinary Least Squares). The increase in accuracy depends on the values of the cumulant coefficients of higher orders of innovation residuals. The results of statistical modeling by the Monte Carlo method confirm the effectiveness of the proposed approach.

Keywords: Polynomial maximization method · Autoregressive models · Non-Gaussian distributions · Cumulant coefficients · Variance · Higher order statistics

1 Introduction

Autoregressive (AR) models are a way of describing time series, widely used to predict data of geophysical, economic, biomedical, sociological, and other origins. Traditionally, such models are synthesized and analyzed based on the assumption of the normality (Gaussian distribution) of the random component of the AR model, which is often called «innovation». However, a large number of researchers note the idealization of such an assumption, which greatly simplifies the solution of the problem of estimating informative parameters, but often does not correspond to the realities of practical tasks [1–6]. In such situations, to improve the estimation accuracy, a transition is made from simple nonparametric estimates based on the method of moments (Yule-Walker equations) or

R. Szewczyk et al. (Eds.): AUTOMATION 2022, AISC 1427, pp. 380–390, 2022.
https://doi.org/10.1007/978-3-031-03502-9_37

the least squares method (OLS) to a more implementation-complex parametric approach based on the maximum likelihood method (MLM). It is known that one of the key elements of this approach is the need to solve an additional type identification problem and estimate the parameters of the probability distribution to describe the random component of the AR model.

Currently, a large number of different types of distributions are used: exponential and gamma distribution [1], Poisson [1, 2], a family of exponential power distributions, including asymmetric modification [3], inverse Gaussian and Student's t distribution [4], mixes based on Gaussian [2] or Laplace distributions [5], etc. Such a large variety can make a difficult to choose of the right type of distributions, and in addition, the problem of analyzing the performance of resulting estimates is an important one [6].

An alternative approach to probabilistic description is the use of higher-order statistics (moment or cumulant functions) or their poly-spectra. A practical example of using this approach for constructing and estimate parameters of various predictive models (including AR) is the library of functions (HOSA toolbox) for the MATLAB package [7]. Based on the apparatus of cumulative description, a generalized higher ranks AR model for linear prediction of non-Gaussian processes [8] has been theoretically substantiated, and original methods for estimating parameters are being developed [9].

In this paper, for the first time, it is proposed to use the polynomial maximization method (PMM) [10] to finding the parameter estimates of AR-models. This relatively new estimation method was originally developed for solving problems of statistical radio-physics when processing signals with continuous time but was later adapted for the discrete case. It is based on the representation of the logarithm of the probability distribution density in the form of functional series. Conceptually, PMM is similar to MLM, as it also uses the principle of maximizing the functional from sample data in the vicinity of the true value of the estimated parameter. However, to form such a functional, not the probability distribution density is used, but the description of random variables in the form of a finite number of averaged statistics (moments, cumulants, etc.), which are much easier to obtain. Let us note the functionality of this method, which can be applied not only to find estimates of the scalar parameter [11], but also to find the moments of change point in the properties of a random sequence in a posteriori formulation of the problem [12], as well as to estimate the vector parameter linear [13] and polynomial [14] regression.

The general idea of using PMM for estimating the parameters of AR models is based on the well-known mathematical analogy between an autoregressive model and linear (in terms of parameters) one-factor regression. Since the errors are statistically independent of observations, the latter can be used as predictors of linear regression to find estimates of the parameters of which linear estimates (Yula-Walker or OLS) can be used. At the same time, the inadequacy of the Gaussian hypothesis is not critical from the point of view that the linear estimates remain unbiased and consistent (although they are no longer effective), and the probabilistic properties of the AR of the residuals do not differ significantly from the properties of non-Gaussian errors. This fact opens up the possibility of overcoming the condition of having the necessary information about the value of higher-order statistics to describe the random component of AR models, because instead of the a priori values of these parameters one can use a posteriori estimates of such residuals. Thus, this approach can be interpreted as a two-stage adaptive procedure for estimating the parameters of AR models.

Note that the idea of such an adaptive approach based on the use of higher-order statistics in the process of solving the problem of estimating the parameters of time series models is not new. In particular, in [15], the use of a posteriori values of the asymmetry coefficients and kurtosis of the AR residuals after the use of the Yule-Walker estimates is considered. However, the fundamental difference is that the obtained a posteriori information is used for a more complex procedure for approximating the random component of the AR model based on an asymmetric exponential power distribution followed by adaptive refinement of estimates of informative parameters using the MLM.

This study is a continuation of works [11, 13, 14], which considered the use of PMM to solve the problem of estimating the parameters, respectively, of the constant component (shift of the distribution center) and linear and polynomial regressions under conditions of asymmetrically distributed statistical data. Its main purpose is to generalize the results of these works for an autoregressive model, t compare the efficiency of the estimates obtained by Monte Carlo modelling, and also to test the proposed approach to real statistical data of economic origin.

2 Mathematical Formulation of the Problem

Let there be a vector $\boldsymbol{X} = \{x_1, x_2, ..., x_N\}$ containing the values of the time series described by the autoregressive observation model

$$x_v = a_0 + \sum_{p=1}^{P} a_p x_{v-p} + \xi_v, \ v = \overrightarrow{1, N}, \tag{1}$$

To generalize the entry, we use the vector $\boldsymbol{\theta} = \{a_0, a_1, ...a_P\}$ including a constant component a_0 and P parameters (autoregressive coefficients). Countdowns ξ_v the random component of the AR model (1) are a sequence of centered ($E\{\xi\} = 0$), independent and equally distributed random variables. In this case, the distribution of the random component ξ differs from the Gaussian (normal) law and has significant asymmetry. An additional restriction constraint is that the random variable ξ has finite central moments μ_r up to the 4th order.

The general task is to find estimates of the vector parameter $\boldsymbol{\theta}$ based on statistical analysis of a set of samples $\boldsymbol{X}$. In this case, the values of the probability characteristics of the random component of the AR model may be a priori unknown.

3 Theoretical Results

3.1 Theoretical Foundations of PPM

To solve this problem, you can use a modification of the PMM designed to find estimates of the vector parameter under unequally distributed statistical data. It is based on the property of maximizing a functional as a stochastic polynomial of a general form [10]:

$$L_{SN} = \sum_{v=1}^{N} \sum_{i=1}^{S} \phi_i(x_v) \int^{a} k_{iv}(a) dz - \sum_{i=1}^{S} \sum_{v=1}^{N} \int^{a} \Psi_{iv} k_{iv}(a) dz, \tag{2}$$

It is assumed that random variables x_v are described by a sequence of mathematical expectations [10]:

$$\Psi_{iv} = E\{\phi_i(x_v)\},\ i = \overrightarrow{1, S},\ v = \overrightarrow{1, N} \tag{2}$$

which are double differentiated by the estimated parameter a.

By the well-known analogy with the MLM, the parameter estimate can be found from the solution of an equation of the form [10]:

$$\left.\frac{d}{da}L_{SN}\right|_{a=\hat{a}} = \left.\sum_{i=1}^{S}\sum_{v=0}^{N} k_{iv}[\phi_i(x_v) - \Psi_{iv}]\right|_{a=\hat{a}} = 0, \tag{4}$$

where the optimal coefficients k_{iv} that maximize functional (2) are found from solving a systems of linear algebraic equations [10]:

$$\sum_{j=1}^{S} k_{jv}F_{(i,j)v} = \frac{d}{da}\Psi_{iv},\ i = \overrightarrow{1, S},\ v = \overrightarrow{1, N}, \tag{5}$$

where $F_{(i,j)v} = \Psi_{(i,j)v} - \Psi_{iv}\Psi_{jv},\ \Psi_{(i,j)v} = E\{\phi_i(x_v)\phi_j(x_v)\},\ i, j = \overrightarrow{1, S}$.

This approach can be extended to the case of finding estimates of a vector parameter $\boldsymbol{\theta} = \{a_0, a_1, ...a_P\}$. In such a situation, it is necessary to use $P + 1$ polynomials general form (2) for each component of the vector parameter. Thus, the desired estimates can be found as solution to a systems of equations of the form [10]:

$$\left.\sum_{i=1}^{S}\sum_{v=1}^{N} k_{iv}^{(p)}[\phi_i(x_v) - \Psi_{iv}]\right|_{a_p=\hat{a}_p} = 0,\ p = \overrightarrow{0, P}\ldots \tag{6}$$

To solve the problem in the formation of the stochastic polynomial (2) as basic functions, we can use power transformations of sample values, i.e. $\phi_i(x_v) = x_v^i$. In this case, the sequence of mathematical expectations (3) is a set of initial moments of the corresponding order.

3.2 PPM-Estimates of AR Parameters with Polynomial Degree S = 1

By using a stochastic polynomial of order $S = 1$ PMM-estimates of the vector parameter $\boldsymbol{\theta}$ can be found from the solution a system of equations of the form:

$$\sum_{v=1}^{N}\left\{k_{1,v}^{(p)}\left[x_v - \left(a_0 + \sum_{p=1}^{P} a_p x_{v-p}\right)\right]\right\} = 0,\ p = \overrightarrow{0, P}. \tag{7}$$

where are the optimal coefficients, which are found as solution of system (5)

$$k_{1,v}^{(p)} = \frac{1}{\mu_2}\frac{\partial}{\partial a_p}\left[\sum_{p=0}^{P} a_p x_{v-p}\right] = \frac{x_{v-p}}{\mu_2},\ p = \overrightarrow{0, P}, \tag{8}$$

where μ_2 – central moment of the 2nd order of the random component ξ.

Note that after the substitution of (8) into (7), the resulting system for finding the PMM-estimates parameters at $S = 1$ becomes equivalent to the linear system of estimates of the least squares method. This means that the accuracy of both methods is the same [13, 14]. It is known that the efficiency of such OLS estimates is significantly reduced when the distribution of the random component of the AR model differs from the Gaussian distribution. Therefore, a new approach to non-linear estimation of AR parameters based on the use of quadratic power stochastic polynomials is discussed.

3.3 PPM-Estimates of AR Parameters with Polynomial Degree S = 2

Using stochastic polynomials of order $S = 2$, the PMM-estimates for the AR model (1) can be found by solving a system of equations of the form:

$$\sum_{v=1}^{N}\left\{k_{1,v}^{(p)}\left[x_v-\left(a_0+\sum_{p=1}^{P}a_p x_{v-p}\right)\right]+k_{2,v}^{(p)}\left[[x_v]^2-\left(a_0+\sum_{p=1}^{P}a_p x_{v-p}\right)^2-\mu_2\right]\right\}=0,\ \ p=\overrightarrow{0,P}, \tag{9}$$

where: the optimal coefficients $k_{i,v}^{(p)}$, $i = \overrightarrow{1,2}$ provide the minimization of the variance of the estimates parameter when using the degree of the polynomial $S = 2$.

These coefficients are found as a solution of an appropriate system of the form (5) and can be represented as functions depending on the central moments of the random component ξ AR models:

$$k_{1,v}^{(p)}=\frac{\mu_4-\mu_2^2+2\mu_3\left(a_0+\sum\limits_{p=1}^{P}a_p x_{v-p}\right)}{\mu_2\left(\mu_4-\mu_2^2\right)-\mu_3^2}x_{v-p},\ k_{2,v}^{(p)}=-\frac{\mu_3^2}{\mu_2\left(\mu_4-\mu_2^2\right)-\mu_3^2}x_{v-p}, \tag{10}$$

By substituting the coefficients (10) into (9), after certain transformations, the system of equations for finding estimates can be written in the form:

$$\sum_{v=1}^{N}\left\{x_{v-p}\left[A\left(a_0+\sum_{p=1}^{P}a_p x_{v-p}\right)^2+B_v\left(a_0+\sum_{p=1}^{P}a_p x_{v-p}\right)+C_v\right]\right\}=0,\ \ p=\overrightarrow{0,Q-1}, \tag{11}$$

where

$$A=\mu_3,\ B_v=\mu_4-\mu_2^2-x_v\mu_3,\ C_v=x_v^2\mu_3-x_v\left(\mu_4-\mu_2^2\right)-\mu_2\mu_3, \tag{12}$$

depends on sampled values x_v and focal points $\mu_2 - \mu_4$ random component ξ.

Obviously, for the degree of the polynomial $S = 2$ PMM-estimates can be found only numerically, for example, using the Newton-Raphson iterative procedure [14].

It is proved that non-linear PMM parameters estimates are consistent and asymptotically unbiased [10]. In order to obtain analytical expressions describing the variances of PMM estimates, a matrix of the amount of extracted information about the parameter components is used when applying stochastic polynomials (1) of order S. Such a matrix consists of the elements [10]:

$$J_{SN}^{(p,q)} = \sum_{v=1}^{N} \sum_{i=1}^{S} k_{i,v}^{(p)} \frac{\partial}{\partial a_q} \Psi_{iv}, \quad p, q = \overrightarrow{0, Q-1}. \tag{13}$$

In a statistical sense, the amount of information obtained is conceptually a concept close to the amount of Fischer information. It is shown that the variances $\sigma^2_{(a_p)S}$ of PMM-estimates of vector parameter $\boldsymbol{\theta}$ in the asymptotic case (for $N \to \infty$) can be obtained as elements of the main diagonal of the variation matrix, which is inverse to the matrix composed of elements (13).

Using expressions for the optimal coefficients (10), it can be shown that the elements of the variation matrix of quadratic PMM-estimates for $S = 2$ differ from the elements of the matrix of linear estimates for $S = 1$ by a certain coefficient. It is the same for all components of the estimated vector parameter $\boldsymbol{\theta}$ and can be represented as [13]:

$$g_{2(Var)} = \frac{\sigma^2_{(a_p)S=2}}{\sigma^2_{(a_p)S=1}} = 1 - \frac{\gamma_3^2}{2 + \gamma_4}. \tag{14}$$

The transition from the moments to cumulant description in expression (14) is due to the fact that the deviations of the values of the cumulant coefficients of higher orders $\gamma_r = \kappa_r / \kappa_2^{r/2}$ from zero indicates the degree of difference from the Gaussian distribution of the random component of the AR model. Taking into account the well-known inequality $\gamma_4 + 2 \geq \gamma_3^2$ we can conclude that the coefficient of variance reduction $g_{2(Var)}$ is a dimensionless quantity that belongs to the range (0; 1]. Thus, with an increase in the asymmetry of the distribution of the random component of the AR model, the magnitude of the relative variance reduction can be quite significant.

4 Statistical Modeling

To verify the theoretical results, a set of functions has been developed in the R-Studio environment. Based on the method of multiple Monte Carlo tests, it implements a comparative analysis of the accuracy of different methods for estimating the parameters of AR models, the random component of which has an asymmetric distribution. It should be noted that the expression for the variance ratio (14), used as a theoretical criterion for the efficiency of PMM-estimates, is derived for two important practical constraints:

1) if there is a priori information about the values of the parameters (moments or cumulants up to the 4th order) determining the probabilistic properties of the random component of AR models;
2) with a statistical sample size asymptotically tending to infinity ($N \to \infty$).

To overcome the first limitation, an adaptive approach is used, which is based on the above stated fact that the statistical properties of OLS residuals practically do not differ from the properties of the initial random component of the AR model. Therefore, when forming the system of PMM equations, instead of the a priori values of the central moments $\mu_2 - \mu_4$ you can use their posterior estimates.

To minimize the influence on the reliability of the factor associated with the finiteness of real samples, in addition to the experimental values of the variance ratios $\hat{g}_{2(Var)}$ additionally, two other statistical performance criteria are considered: the ratio of the mean square error $\hat{g}_{2(MSE)}$ and the ratio of absolute shifts $\hat{g}_{2(Bias)}$ estimates (relative to the true value) of informative parameters obtained by different methods.

When implementing statistical modeling, the simplest first-order AR(1) model (with a constant $a_0 = 1$ and information parameter $a_1 = 0.8$) was used as the object of research. Parameter values were estimated by different approaches: the least-squares method, the method based on the solution of Yule-Walker equations, the maximum likelihood method (optimized for a Gaussian model), and a quadratic modification of the polynomial maximization method. As an asymmetric random component of the AR model, we used a sequence of independent and equally distributed random variables having a gamma distribution with different values of the shape parameter, which were determined by the degree of asymmetry. In this case, the parameters (central moments up to the 4th order) of the random component, which are necessary to find the adaptive PMM-estimates, were considered a priori unknown, and in accordance with the adaptive algorithm, their posterior estimates based on the LSM residuals were used.

The set of experimental values of efficiency coefficients (presented in Tables 1, 2 and 3) were obtained for a series in $M = 10^4$ repeated experiments. An analysis of these results shows that the experimental values of the values of all the relative performance criteria under consideration are in the range from 0.35 to 0.8. Accordingly, this indicates a significant (20–65%) increase in the accuracy of the resulting estimates of the informative parameter of the studied AR model. Obviously, the value of the gain depends on the degree to which the distribution of the random component differs from the Gaussian distribution, which is numerically expressed primarily by the asymmetry coefficient. In this case, the analysis of the presented empirical values of the ratios of variances of estimates $\hat{g}_{2(Var)}$ indicates the adequacy of analytical calculations, since the maximum relative discrepancy between theoretical and experimental values is insignificant (at $N = 200$ does not exceed 3%).

Table 1. Relative efficiency of PMM-estimates at $S = 2$ compared to OLS-estimates (PMM estimates at $S = 1$)

Gamma distribution parameter (*shape*)	Theoretical values			Monte Carlo statistical simulation results								
	γ_3	γ_4	$g_{2(Var)}$	$\hat{g}_{2(Var)}$			$\hat{g}_{2(MSE)}$			$\hat{g}_{2(Bias)}$		
				N			*N*			*N*		
				50	100	200	50	100	200	50	100	200
1	2	6	0.5	0.48	0.46	0.47	0.37	0.41	0.45	0.43	0.41	0.44
2	1.4	3	0.6	0.65	0.59	0.59	0.55	0.55	0.57	0.6	0.55	0.56
4	1	1.5	0.71	0.78	0.73	0.7	0.69	0.7	0.68	0.74	0.71	0.69

Table 2. Relative efficiency of PMM-estimates at $S = 2$ comparison with estimates based on the Yule-Walker equations

Gamma distribution parameter (*shape*)	Theoretical values			Monte Carlo statistical simulation results								
	γ_3	γ_4	$g_{2(Var)}$	$\hat{g}_{2(Var)}$			$\hat{g}_{2(MSE)}$			$\hat{g}_{2(Bias)}$		
				N			*N*			*N*		
				50	100	200	50	100	200	50	100	200
1	2	6	0.5	0.49	0.45	0.47	0.35	0.38	0.42	0.34	0.34	0.36
2	1.4	3	0.6	0.67	0.62	0.58	0.55	0.54	0.54	0.48	0.48	0.45
4	1	1.5	0.71	0.8	0.74	0.7	0.59	0.65	0.66	0.6	0.57	0.57

Table 3. Relative efficiency of PMM-estimates at $S = 2$ compared to maximum likelihood estimates (Gaussian model).

Gamma distribution parameter (*shape*)	Theoretical values			Monte Carlo statistical simulation results								
	γ_3	γ_4	$g_{2(Var)}$	$\hat{g}_{2(Var)}$			$\hat{g}_{2(MSE)}$			$\hat{g}_{2(Bias)}$		
				N			*N*			*N*		
				50	100	200	50	100	200	50	100	200
1	2	6	0.5	0.48	0.45	0.47	0.37	0.41	0.44	0.42	0.41	0.44
2	1.4	3	0.6	0.67	0.59	0.58	0.55	0.55	0.56	0.58	0.55	0.56
4	1	1.5	0.71	0.79	0.73	0.7	0.69	0.7	0.68	0.72	0.71	0.69

5 Experiment with Real Data

As an example of approbation of the developed approach to finding the PMM-estimates of parameters of AR models we used a set of time series data in the form of daily changes in the Dow Jones index (127 values from July 01, 2002 to December 31, 2002, Fig. 1a). Similar statistical data have been analyzed in [16], according to which for many rather long periods of time there is a slowly decaying positive selective autocorrelation function. And to ensure stationarity, it is advisable to consider the behavior of differential changes in this index (see Fig. 1b). As shown in Fig. 1c the resulting time series is adequately described by a first order AR model. However, the distribution of the random component differs from the Gaussian law and has a significant asymmetry of the distribution (Fig. 1d).

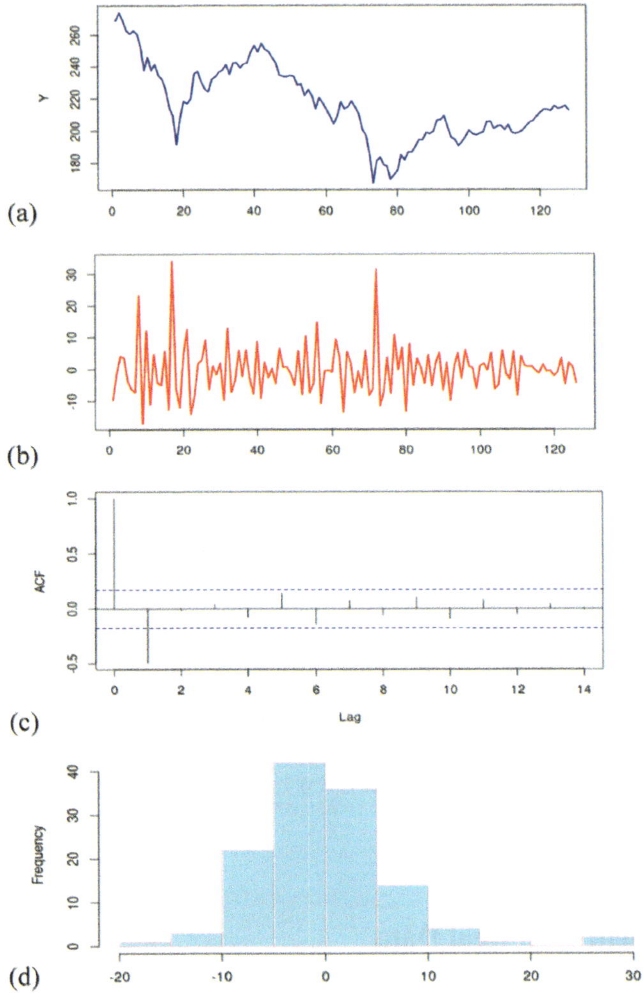

Fig. 1. Dow jones index: (a) raw data; (b) differential sequence; (c) autocorrelation function; (d) residuals histogram

The linear estimates of the informative parameter obtained by the classical least squares methods, the method based on the Yule-Walker equations, the maximum likelihood method (optimized for the Gaussian model) do not practically differ and are $\hat{a}_1^{(1)} = -0.49$. The estimate of this parameter by the quadratic modification by the polynomial maximization method is $\hat{a}_1^{(2)} = -0.43$. In this case, the values of the numerical characteristics of AR residues (skewness $\hat{\gamma}_3 = 1.1$ and kurtosis $\hat{\gamma}_4 = 3.7$) indicate a significant difference from the Gaussian distribution and permit to estimate potential asymptotic gain of the proposed approach at the level $\hat{g}_{2(Var)} = 0.77$.

Thereby, the practical importance of considering the non Gaussianity of real statistical data (in the form of asymmetry and kurtosis of residuals) consists in obtaining autoregressive models with less uncertainty of its parameters. This improves the predictive accuracy of the resulting models accordingly.

6 Conclusions

The totality of the results obtained confirms the possibility of effective application of the polynomial maximization method for solving the problems of finding estimates of the informative parameters of autoregressive models for those situations when the nature of innovations (random component) has an asymmetric distribution.

Overall, the proposed approach to finding parameter estimates can be interpreted as adaptive and compromise in terms of practical implementation, as the proposed computational procedures potentially have less analytical complexity compared to the approach based on the maximum likelihood method and provide an improved accuracy (reduced variance of estimates) compared to linear estimates (least squares, Yula-Walker, etc.).

The next stage of the research should be to compare the effectiveness of the adaptive estimates of the polynomial maximization method and maximum likelihood, optimized for the corresponding non-Gaussian distributions.

References

1. Grunwald, G.K., Hyndman, R.J., Tedesco, L., Tweedie, R.L.: Theory and methods: non-Gaussian conditional linear AR (1) models. Aust. N. Z. J. Stat. **42**(4), 479–495 (2000)
2. Ozaki, T., Iino, M.: An innovation approach to non-Gaussian time series analysis. J. Appl. Probab. **38**(A), 78–92 (2001)
3. Bondon, P.: Estimation of autoregressive models with epsilon-skew-normal innovations. J. Multivar. Anal. **100**, 1761–1776 (2009)
4. Hürlimann, W.: On non-Gaussian AR(1) inflation modelling. J. Stat. Econom. Methods. **1**(1), 93–109 (2012)
5. Nguyen, H.D., McLachlan, G.J., Ullmann, J.F.P., Janke, A.L.: Laplace mixture autoregressive models. Stat. Probab. Lett. **110**, 18–24 (2016)
6. Akkaya, A.D., Tiku, M.L.: Time series AR(1) model for short-tailed distributions. Statistics **39**(2), 117–132 (2005)
7. Tikhonov, V.: Generalized autoregressive model of non-Gaussian processes. Radiotekhnika **132**, 78–82 (2003). (in Russian)
8. Swami, A., Mendel, J.M., Nikias, C.: Higher-order spectral analysis toolbox. In: MATLAB User Guide. The Math Works Inc. (2001)
9. Al-Smadi, A.: A new coefficient estimation method for autoregressive systems using cumulants. Int. J. Circuit Theory Appl. **29**(5), 511–516 (2001)
10. Kunchenko, Y.: Polynomial Parameter Estimations of Close to Gaussian Random variables. Shaker, Aachen (2002)
11. Warsza, Z.L., Zabolotnii, S.: Estimation of measurand parameters for data from asymmetric distributions by polynomial maximization method. In: Szewczyk, R., Zieliński, C., Kaliczyńska, M. (eds.) Automation 2018. AISC, vol. 743, pp. 746–757. Springer, Cham (2018). https://doi.org/10.1007/978-3-319-77179-3_74

12. Zabolotnii, S.W., Warsza, Z.L.: Semi-parametric estimation of the change-point of parameters of non-Gaussian sequences by polynomial maximization method. In: Szewczyk, R., Zieliński, C., Kaliczyńska, M. (eds.) Challenges in Automation, Robotics and Measurement Techniques. AISC, vol. 440, pp. 903–919. Springer, Cham (2016). https://doi.org/10.1007/978-3-319-29357-8_80
13. Zabolotnii, S.W., Warsza, Z.L., Tkachenko, O.: Estimation of linear regression parameters of symmetric non-Gaussian errors by polynomial maximization method. In: Szewczyk, R., Zieliński, C., Kaliczyńska, M. (eds.) Automation 2019, vol. 920, pp. 636–649. Springer, Cham (2020). https://doi.org/10.1007/978-3-030-13273-6_59
14. Zabolotnii, S., Tkachenko, O., Warsza, Z.L.: Application of the polynomial maximization method for estimation parameters in the polynomial regression with non-Gaussian residuals. In: Szewczyk, R., Zieliński, C., Kaliczyńska, M. (eds.) Automation 2021: Recent Achievements in Automation, Robotics and Measurement Techniques. AISC, vol. 1390, pp. 402–415. Springer, Cham (2021). https://doi.org/10.1007/978-3-030-74893-7_36
15. Bondon, P., Song, L.: AR processes with non-Gaussian asymmetric innovations. In: European Signal Processing Conference, pp. 1–5 (2013)
16. Brockwell, P.J., Davis, R.A.: Introduction to Time Series and Forecasting. Springer, New York (1996). https://doi.org/10.1007/b97391

Measurement Uncertainty Evaluation of Parameters Describing the Calibrated Curves

Igor Zakharov[1,2], Pavel Neyezhmakov[2], Valery Semenikhin[1,2], and Zygmunt L. Warsza[3(✉)]

[1] Kharkov National University of Radio-Electronics, Kharkov, Ukraine
[2] National Scientific Center, Institute of Metrology, Kharkov, Ukraine
[3] Lukasiewicz Research Network – Industrial Research Institute of Automation and Measurement PIAP, Warsaw, Poland
zlw1936@gmail.com

Abstract. Relationships characterizing measuring instruments are determined in a measuring experiment called calibration. The values of the parameters describing calibrated curves are measured with associated uncertainties. Identification of a given type of dependency is made based on the obtained values of the measured parameters. The most popular method of this identification is the least squares method, well implemented for polynomial relationships. In metrological practice, when identifying a non-linear calibrated relationship, it very often happens that increasing the polynomial degree within reasonable limits does not lead to a significant reduction of the approximation error. In that case, the primary dependence is transformed into a linear one by changing the variables. Then parameters of such a linearized relationship are determined by the least squares' method. The article discusses the solution of estimating the uncertainty of parameters identified for the non-polynomial dependence, considering the instrumental uncertainties for the measured values.

Keywords: Calibration dependence · Change of variables · Least squares method · Instrumental uncertainties · Measured parameters of calibrated curve

1 Introduction.

The problem of plotting calibration dependences of measuring instruments is solved using a measurement experiment, during which values are measured with some uncertainties. The identification of a given type of dependence is carried out based on the obtained values of the measured quantities for the dependence. The most common method for solving the identification problem is the least squares method (LSM), which is well implemented for polynomial dependencies [1–11]. In the metrological practice, one often must deal with the case when an increase in the degree of the polynomial within reasonable limits does not lead to a significant decrease in the approximation error when identifying a nonlinear calibration dependence. In this case, the transformation of the original dependence into a linear dependence is applied by the change of variables. The

R. Szewczyk et al. (Eds.): AUTOMATION 2022, AISC 1427, pp. 391–398, 2022.
https://doi.org/10.1007/978-3-031-03502-9_38

parameters of the linearized dependence are subsequently found using the LSM method. The paper discusses the solution to the problem of uncertainty evaluation of the non-polynomial dependence identification accounting for the instrumental uncertainties of the measured values.

2 Transformation of Non-Polynomial Calibration Dependence: Basic Relations

The measuring instrument (MI) response is measured at N calibration points of the considered range X_1, X_2, ..., X_N to plot the calibration dependence of the MI. At each X_j calibration point measurements of the MI readings Y_{ji} ($i = 1, 2,\ldots, n$) are carried out and repeated n times. Based on the measurements, a calibration dependence is plotted and its approximation is carried out by a mathematical dependence. Finding the parameters (identification) of the dependence is most often performed by the LSM method, which provides the best results when the approximation dependence is a polynomial.

In metrological practice, one often must deal with the case when an increase in the degree of the polynomial within reasonable limits does not lead to a significant decrease in the approximation error when identifying a nonlinear calibration dependence. In this case, the transformation of the original dependency

$$Y = f(X, A, B) \tag{1}$$

within the linear dependance

$$y = a + bx \tag{2}$$

by the change of variables [2]:

$$x = \Phi(X); \tag{3}$$

$$y = \Psi(Y). \tag{4}$$

is applied. Expressions (1) for nonlinear functions (1) reduced to linear ones are given in the first column of Table 1, the second column of which contains expressions (3) and (4) for the functions.

The estimates $\hat{a}$, $\hat{b}$ of the parameters of the linearized dependence (4) are subsequently found using the LSM [3]. When evaluating the uncertainty of the calibration dependence (1), one should consider the uncertainty of the values of the output signal u_Y of the measurement instrument MI. These values of uncertainty are evaluated both by type A (due to random errors in measurements with the MI) and by type B (due to instrumental measurement errors during calibration, as well as the uncertainty in determining values u_X of calibration points, evaluated by type B). It should be considered that since the values of Y and X are directly measured with the help of the MI, the instrumental uncertainties u_Y and u_X must be recalculated into instrumental uncertainties u_y and u_y when identifying the linearized dependence (2) just as:

$$u_x = \frac{\partial \Phi(X)}{\partial X} u_X = c_X u_X; \tag{5}$$

$$u_y = \frac{\partial\Psi(Y)}{\partial Y} u_Y = c_Y u_Y \quad (6)$$

The expressions for the sensitivity coefficients c_X, and c_Y for different initial dependencies (1) are given in the fifth column of Table 1.

After identifying the linear dependence, it is necessary to perform its inverse transformation into the original one, according to which it will be necessary to calculate the uncertainty of finding the value X for any value Y. The values of the coefficients $\hat{A}$ and $\hat{B}$ recalculated from the obtained coefficients are given in Table 1. Expressions for sensitivity coefficients $c_x = \frac{\partial X}{\partial x}$, linking uncertainties u_X and u_x for different initial dependencies (1), are given in the fifth column of Table 1. The expressions for sensitivity coefficients, linking uncertainties, i.e.:

$$u_X = c_x u_x \quad (7)$$

are given in the last column of Table 1.

3 Algorithm for Identifying the Required Dependence Using a Linear Dependence of a General Form

The general linear dependence has the form [3]:

$$y = a_0 + b(x - \overline{x}). \quad (8)$$

Using dependence (8) instead of (2) eliminates the correlation between estimates $\hat{a}_0$ and $\hat{b}$ [2]. The algorithm for estimating the measurement uncertainty for the dependence will consist of the following points.

According to the experimental points $\overline{Y}_j = \frac{1}{n}\sum_{i=1}^{n} Y_{ji}$, the dependence

$$\overline{Y}_j = f(X_j) \quad (9)$$

is plotted, which is approximated by one of the functions $Y = f(X, A, B)$ from the first column of Table 1. The initial dependence (9) is linearized using the expressions $x = \Phi(X)$ and $y = \Psi(Y)$ from the second column of Table 1. The estimates of the linearized dependence (2) coefficients a and b are calculated using the expressions [3]:

$$\hat{b} = \sum_{j=1}^{N} \overline{y}_j (x_j - \overline{x}) / \sum_{j=1}^{N} (x_j - \overline{x})^2; \quad (10)$$

$$\hat{a} = \frac{1}{N}\sum_{j=1}^{N} \overline{y}_j - \frac{\hat{b}}{N}\sum_{j=1}^{N} x_j = \frac{1}{Nn}\sum_{j=1}^{N}\sum_{i=1}^{n} y_{ji} - \hat{b}\overline{x}, \quad (11)$$

where

$$\overline{x} = \frac{1}{N}\sum_{j=1}^{N} x_j; \qquad \overline{y}_j = \frac{1}{n}\sum_{i=1}^{n} y_{ij} \quad (12)$$

According to the obtained values $\hat{a}$ and $\hat{b}$, the values of the coefficients $\hat{A}$ and $\hat{B}$ are calculated, using the formulas from the third column of Table 1, obtaining a model of the original dependence $Y = f(X, \hat{A}, \hat{B})$.

Table 1. Basic expressions for implementing the variable replacement method

$Y = f(X, A, B)$	$x = \Phi(X)$ $y = \Psi(Y)$	$\hat{A}$, $\hat{B}$	$X = f^{-1}(Y, \hat{A}, \hat{B})$	c_X ; c_Y	c_x
$Y = A\exp(BX)$	$x = X$ $y = \ln Y$	$\hat{A} = e^{\hat{a}}$ $\hat{B} = \hat{b}$	$X = \dfrac{\ln Y - \ln \hat{A}}{\hat{B}}$	$c_X = 1$ $c_Y = 1/Y$	$c_x = 1$
$Y = AX^B$	$x = \ln(X)$ $y = \ln(Y)$	$\hat{A} = e^{\hat{a}}$ $\hat{B} = \hat{b}$	$X = \exp\left(\dfrac{\ln Y - \ln \hat{A}}{\hat{B}}\right)$	$c_X = 1/X$ $c_Y = 1/Y$	$c_x = e^x$
$Y = A + B\ln(X)$	$x = \ln(X)$ $y = Y$	$\hat{A} = \hat{a}$ $\hat{B} = \hat{b}$	$X = \exp\left(\dfrac{Y - \hat{A}}{\hat{B}}\right)$	$c_X = 1/X$ $c_Y = 1$	$c_x = e^x$
$Y = A + \dfrac{B}{X}$	$x = 1/X$ $y = Y$	$\hat{A} = \hat{a}$ $\hat{B} = \hat{b}$	$X = \dfrac{\hat{B}}{Y - \hat{A}}$	$c_X = -1/X^2$ $c_Y = 1$	$c_x = -\dfrac{1}{x^2}$
$Y = \dfrac{1}{A + BX}$	$x = X$ $y = 1/Y$	$\hat{A} = \hat{a}$ $\hat{B} = \hat{b}$	$X = \dfrac{1/Y - \hat{A}}{\hat{B}}$	$c_X = 1$ $c_Y = -1/Y^2$	$c_x = 1$
$Y = \dfrac{X}{A + BX}$	$x = 1/X$ $y = 1/Y$	$\hat{A} = \hat{a}$ $\hat{B} = \hat{b}$	$X = \dfrac{\hat{A}Y}{1 - \hat{B}Y}$	$c_X = -1/X^2$ $c_Y = -1/Y^2$	$c_x = -\dfrac{1}{x^2}$

4 Measurement Uncertainty Evaluation Using the Identified Dependence

For a given value Y, the value X is calculated according to the corresponding formula $X = f^{-1}(Y, \hat{A}, \hat{B})$ from the fourth column of Table 1.

The uncertainty of finding the value x is calculated by formulas [3]:

$$u_{LSM}(x) = \sqrt{\left[\frac{1}{N} + (x - \overline{x})^2 / \sum_{j=1}^{N} (x_j - \overline{x})^2\right] \cdot \left[u^2(\overline{y}) + b^2 u_B^2(x)\right]}, \qquad (13)$$

if $x_j, j = 1, \ldots, N$ are not correlated ($\operatorname{cov}(x_j, x_k) = 0$ for any $j, k = 1, \ldots, N$ if $j \neq k$).

If values $x_j, j = 1, \ldots, N$ are correlated (aggravated by the same constant error), the standard uncertainty $u_{LSM}(x)$ is calculated just as:

$$u_{LSM}(x) = \sqrt{\left[\frac{1}{N} + (x - \overline{x})^2 / \sum_{j=1}^{N} (x_j - \overline{x})^2\right] \cdot u^2(\overline{y}) + b^2 u_B^2(x)}. \qquad (14)$$

In this case, the values $u(\overline{y})$ are found by the formula:

$$u(\overline{y}) = \sqrt{u_A^2(\overline{y}) + u_B^2(\overline{y})}, \tag{15}$$

where:

$$u_B(\overline{y}) = c_Y u_B(Y); \qquad u_B(x) = c_X u_B(X) \tag{16}$$

$$u_A(\overline{y}) = \sqrt{\frac{1}{(Nn-2)n}\sum_{j=1}^{N}\sum_{i=1}^{n}(y_{ji} - \hat{a} - \hat{b}x_j)^2}, \tag{17}$$

Moreover, the sensitivity coefficients c_X and c_Y are calculated by the formulas from the fifth column of Table 1. Uncertainty $u_{LSM}(x)$ is recalculated into uncertainty $u_{LSM}(X)$ according to the formula (7), using values of sensitivity coefficients c_x from the last column of Table 1.

5 Example of the Calibration of Measuring Transducers

To confirm the solutions presented in the paper, an experiment was carried out with a diode converter at the input of which a square-wave voltage (X value) was applied using the G5-60 generator, and the converted (Y value) was measured at the output using a V7-23 voltmeter (Fig. 1).

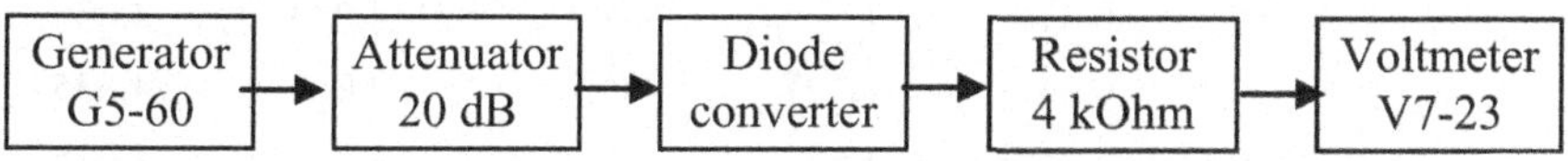

Fig. 1. Determining the transfer function of a diode transducer

The initial measured and calculated values of the input and output quantities of the diode converter are given in Table 2.

The function $Y = AX^B$ was taken as an approximating dependence. The linearized dependence and calculations are given in Table 3.

Measurement uncertainty evaluation using the identified dependence are in Table 4.

Table 2. Initial measured and calculated values

X_j, V	1	2	3	4	5
Y_{ji}, mV	0.65	5.10	20.24	56.46	103.04
	0.66	5.08	20.25	56.42	103.02
	0.65	5.09	20.26	56.41	103.04
	0.66	5.09	20.26	56.41	103.04
	0.65	5.09	20.25	56.42	103.03
$\overline{Y}_j$, mV	0.654	5.090	20.252	56.424	103.03
$u_B(Y_j)$, mV	0.029	0.034	0.069	0.032	0.023
$\overline{X}$, V	3.000				
$u_B(X_j)$, V	0.018	0.036	0.053	0.070	0.088

Table 3. Linearized dependence and calculations

$x_j = \ln(X_j)$	0	0.693	1.099	1.386	1.609
$y_{ji} = \ln(Y_{ji})$	−0.431	1.629	3.008	4.034	4.635
	−0.415	1.625	3.008	4.033	4.635
	−0.431	1.627	3.009	4.033	4.635
	−0.416	1.627	3.009	4.033	4.635
	−0.431	1.627	3.008	4.033	4.635
$\overline{y}_j$	−0.425	1.627	3.008	4.033	4.635
$u_A(\overline{y}_j)$	0.0037	0.0006	0.0002	0.00016	3.88E−05
$\overline{x}$	0.958				
$(x_j - \overline{x})^2$	0.917	0.070	0.020	0.184	0.425
$\sum_{j=1}^{5} x_j = 4.797$			$\sum_{j=1}^{5} (x_j - \overline{x})^2 = 1.62$		
$\overline{y}_j(x_j - \overline{x})^2$	0.407	−0.430	0.425	1.729	3.022
Estimates of coefficients	$\hat{b} = 3.189$	$\hat{a}_0 = 2.576$	$\hat{a} = -0.478$	$\hat{A} = 0.620$	$\hat{B} = 3.189$
$\sum_{j=1}^{5} \overline{y}_j(x_j - \overline{x})^2$	5.152				
$\hat{X}_j$, V	1.017	1.935	2.984	4.114	4.969
$\hat{x}_j$	0.017	0.660	1.093	1.414	1.603

Table 4. Measurement uncertainty evaluation using the identified dependence

X_j, V	1	2	3	4	5
$(y_{ji} - \hat{a} + \hat{b}x_j)^2$	0.0022	0.0107	0.000329	0.00815	0.00039
	0.0039	0.0115	0.000311	0.00802	0.00040
	0.0022	0.0111	0.000294	0.00799	0.00039
	0.0039	0.0111	0.000294	0.00799	0.00039
	0.0022	0.0111	0.000311	0.00802	0.00039
$\sum_{j=1}^{5} (\overline{y}_j - \hat{a} - \hat{b}x_j)^2 = 0.1137$			$u_A(\overline{y}) = 0.0314$		
$u_B(\overline{y}_j)$	0.045	0.0067	0.0034	0.00057	0.00022
$u_B(x_j)$	0.019	0.018	0.018	0.018	0.018
$u(\overline{y}_j)$	0.055	0.032	0.032	0.031	0.031
$u_{LSM}(x_j)$	0.071	0.032	0.030	0.036	0.044
$u_{LSM}(X_j)$	0.072	0.063	0.089	0.1483	0.217

6 Conclusions

An algorithm for identification of non-polynomial calibration dependences, which can be reduced to linear dependences by change of variables has been developed. The parameters of linear dependences are found by the Least Squares Method. The advantage of the proposed approach compared to existing ones, including the Monte Carlo method, is the simplicity of implementation without special software, e.g. [5].

The procedure for measurement uncertainty evaluation using the identified dependence is given. The instrumental measurement uncertainties of the responses of the calibrated measuring instrument and the values of the calibration points are considered.

An example of identification of the transfer function of a diode converter by the developed method, which well illustrates all stages of its application is considered.

References

1. EA-4/02 M: Evaluation of the Uncertainty of Measurement in Calibration (2013)
2. Laaneots, R., Mathiesen, O.: An Introduction to Metrology, p. 271. TUT Press, Tallinn (2006)
3. Rabinovich, S.G.: Evaluating Measurement Accuracy: A Practical Approach, 3rd edn., p. 313. Springer, Cham (2017). https://doi.org/10.1007/978-3-319-60125-0
4. COOMET R/GM/32:2017. Calibration of measuring instruments. Algorithms of processing measurement results and uncertainty evaluation
5. Vodyka, S., Zaharov, I.: Using PUMA for coefficients estimation of calibration dependences. In: Proceedings of XI International Ph. D. Workshop OWD 17–19 October 2009, Silesian TU, pp. 97–99 (2009)
6. Siraja, T.: Certification of algorithms for constructing calibration curves of measuring instruments. In: Advanced Mathematical and Computational Tools in Metrology and Testing X. Advanced in Mathematics for Applied Sciences, vol. 86, pp. 368–376. World Scientific Publications, London (2015). ISBN: 978-981-4678-61-2

7. Zakharrov, I., Neyezhmakov, P.: Peculiarity of measurement instruments verification by results of their calibration. In: Proceedings of 11th International Conference of Measurements, Smolenice, Slovakia, 29–31 May 2017, pp. 19–22 (2017). ISBN: 978-80-972629-0-7
8. Vitkovsky, V., Wimmer, G.: Inverse and direct prediction and its effect on measurement uncertainty in polynomial comparative calibration. In: Proceedings of 12th International Conference of Measurements, Smolenice, Slovakia, 27–29 May 2019, pp. 62–65 (2019). ISBN: 978-80-972629-2-1
9. Wimmer, G., Vitkovsky, V.: Two-dimensional linear comparative calibration and measurement uncertainty. In: Proceedings of 12th International Conference of Measurements, Smolenice, Slovakia, 27–29 May 2019, pp. 66–69 (2019). ISBN: 978-80-972629-2-1
10. Chunovkina, A., Stepanov, A.: Estimation of linear regression confidence bands in case of correlated noise. In: Proceedings of 12th International Conference of Measurements, Smolenice, Slovakia, 27–29 May 2019, pp. 58–61 (2019). ISBN: 978-80-972629-2-1
11. Vitkovsky, V., Wimmer, G.: Generalized polynomial comparative calibration: parameter estimation and application. In: Yurish, S.Y. (ed.) Advances in Measurements and Instrumentation: Reviews, chap. 1, vol. 1, pp. 15–52. IFSA International Frequency Sensor Association Publishing (2018). ISBN: 978-84-09-07321-4, e-ISBN: 978-84-09-07322-1

Author Index

R. Szewczyk et al. (Eds.): AUTOMATION 2022, AISC 1427, pp. 399–400, 2022.
https://doi.org/10.1007/978-3-031-03502-9

The manufacturer's authorised representative in the EU is Springer Nature Customer Service Centre GmbH, Europaplatz 3, 69115 Heidelberg, Germany. If you have any concerns regarding our products, please contact ProductSafety@springernature.com

Printed and bound by CPI Group (UK) Ltd, Croydon, CR0 4YY
30/09/2026
02225320-0015